W9-CNV-246

A C S   S Y M P O S I U M   S E R I E S **663**

Technology for Waterborne Coatings

J. Edward Glass, EDITOR

North Dakota State University

Developed from a symposium sponsored by the
Division of Polymeric Materials: Science and Engineering, Inc.

American Chemical Society, Washington, DC

Library of Congress Cataloging-in-Publication Data

Technology for waterborne coatings / J. Edward Glass, editor.

    p.   cm.—(ACS symposium series, ISSN 0097–6156; 663)

    "Developed from a symposium sponsored by the Division of Polymeric Materials: Science and Engineering, Inc., at the 210th National Meeting of the American Chemical Society, Chicago, Illinois, August 20–25, 1995."

    Includes bibliographical references and indexes.

    ISBN 0–8412–3501–5

    1. Emulsion paint—Congresses.

    I. Glass, J. E. (J. Edward), 1937– . II. American Chemical Society. Division of Polymeric Materials: Science and Engineering. III. American Chemical Society. Meeting (210th: 1995: Chicago, Ill.) IV. Series.

TP934.T43   1997
667'.9—dc20   97–5741
                                                        CIP

This book is printed on acid-free, recycled paper.

Foreword

THE ACS SYMPOSIUM SERIES was first published in 1974 to provide a mechanism for publishing symposia quickly in book form. The purpose of this series is to publish comprehensive books developed from symposia, which are usually "snapshots in time" of the current research being done on a topic, plus some review material on the topic. For this reason, it is necessary that the papers be published as quickly as possible.

Before a symposium-based book is put under contract, the proposed table of contents is reviewed for appropriateness to the topic and for comprehensiveness of the collection. Some papers are excluded at this point, and others are added to round out the scope of the volume. In addition, a draft of each paper is peer-reviewed prior to final acceptance or rejection. This anonymous review process is supervised by the organizer(s) of the symposium, who become the editor(s) of the book. The authors then revise their papers according to the recommendations of both the reviewers and the editors, prepare camera-ready copy, and submit the final papers to the editors, who check that all necessary revisions have been made.

As a rule, only original research papers and original review papers are included in the volumes. Verbatim reproductions of previously published papers are not accepted.

ACS BOOKS DEPARTMENT

Contents

Preface

MANY INDUSTRIAL SUPPLIERS in the mid-1960s, even those supplying solvent to the coating industry, were actively engaged in research in zero volatile organic component (VOC) UV and powder coatings. For most suppliers, this effort declined significantly or was terminated by the early 1970s because growth in these areas was slow and demanded new equipment. However, even with growth, the number of surfaces to which UV and powder coatings can be applied is limited. Much of the research in the 1970s and 1980s was devoted to the high-solids area. Some of this research suggested that elimination of the solvent in high-solids formulations was achievable by replacing the solvent with nonpolluting supercritical fluids such as carbon dioxide. After a short time, it was concluded that nearly half of the solvent used in high solids had to be added back in the critical fluid approach. In the 1990s, the emphasis in coatings research has turned strongly to the waterborne area.

In accordance with the rapid changes in technology, the Polymeric Materials: Science and Engineering, Inc. (PMSE), division of the American Chemical Society (ACS) abandoned its general practice of holding broad symposia that included all areas of coatings technology. In the 1990s, symposia have focused on specific areas of coatings science. The first symposium on waterborne coatings was held at the 203rd National Meeting of the ACS in San Francisco, California, April 5–10, 1992. This book originates from the second symposium in this important field, presented at the 210th National Meeting of the ACS in Chicago, Illinois, August 20–24, 1995.

The 15 chapters in this book discuss conventional latices of the type used in architectural, photographic, and paper coatings, and aqueous epoxy and polyurethane dispersions used in original equipment manufacturer coatings. Topics include their synthesis from the view of their chain-growth or step-growth mechanism, the type of stabilizer employed in the production of the disperse phase, and the compositional influences of the resin particle on their interfacial energies and morphology. Chapters are also included on the radiation curing of applied aqueous dispersions, on the curing of films through carbodiimide chemistry, and on polyurethane films formed from nonisocyanate precursors. The dispersions and film properties of the different resin types are discussed, as well as surfactant orientations at the film substrate and air interfaces.

The prior art of particle coalescence and film formation of latex particles is reviewed, and recent studies on the particle coalescence of step-growth oligomer dispersions (polyurethanes and epoxies) are discussed. The phenomenon of film formation is presented from uniquely different perspectives in the chapters on waterborne alkyd dispersions and high-clay-content paper coatings.

Drying is an important part in the film formation process, and a chapter devoted to this subject is included in this text. Chapters devoted to the interactions of dispersions and dispersion rheology and to the spray application of waterborne coatings are also included.

Acknowledgments

Funding for support of the symposium speakers was provided by the PMSE division of the ACS; the S. C. Johnson, Aqualon division of Hercules, Inc.; Union Carbide Corporation; and Rohm and Haas Company. Their support is gratefully acknowledged. Appreciation is also expressed to the reviewers for their comments and contributions to this book.

J. EDWARD GLASS
Department of Polymers and Coatings
North Dakota State University
Fargo, ND 58105

January 16, 1997

Chapter 1

Waterborne Coatings with an Emphasis on Synthetic Aspects: An Overview

Melissa A. Grunlan, Lin-Lin Xing, and J. Edward Glass[1]

Department of Polymers and Coatings, North Dakota State University, Fargo, ND 58105

This chapter provides an overview of water-borne coatings to complement the chapters that follow. It provides the basic chemical reactions that many outside the field will not be familiar with, and general principles that those practicing in the field may have forgotten. The most active areas in water-borne coatings are in the synthesis of latices and water-reducible acrylics by a chain-growth polymerization mechanism and water-borne polyurethane, polyester, and epoxy resins synthesized by step-growth polymerizations. Water-borne epoxy resins are discussed in detail in chapter 5. The other resin families are discussed in this and in the chapter to follow. The emphasis in this chapter will address the synthetic aspects as imposed by the polymerization process on the stability of the disperse phases and their film properties.

Synthetic polymer coatings originally were applied from organic solvents. Solution viscosities, at a given concentration, are proportional to the polymer's molecular weight. Film properties, such as tensile strength and toughness also are proportional to the polymer's molecular weight. Therefore, to obtain the low viscosities needed for the application of a coating the concentration of the polymeric resin was necessarily low, and the organic solvent emitted to the atmosphere (VOC, Volatile Organic Component) was high. As society changed and environmental concerns were emphasized, coating technology has advanced in several directions. Two areas considered VOC free are UV cure(1) and powder(2) coatings. UV curing of solvent-free formulations is restricted, with only a few exceptions, to flat surfaces, with film thickness and pigmentation limitations. Powder coatings are subject to Faraday effects and can not be applied to many surfaces.

The effort to produce synthetic rubber during World War II led to the development of an emulsion polymerization process, and with it, water-borne coatings

---

[1]Corresponding author

as they are designed in the architectural and paper coatings area. Such an emulsion polymerization recipe employs a surfactant, a monomer with low solubility in water, and a redox free radical initiator in the aqueous phase (recipe given in Table I in chapter 2). When a monomer with low solubility in the aqueous phase (e.g., styrene and butadiene) is employed, the emulsion polymerization occurs by a micellar process. After the initiator has reacted with the monomer in the aqueous phase, the monomer propagates to an oligomeric form and then enters a micelle, formed by the aggregation of excess of surfactant used in the polymerization recipe. The chain growth polymerization (chapter 3 in reference 3) of the olefinic monomer then occurs in the micelle. When the chain growth polymerization of olefinic monomers occurs in bulk or solvent, the free-radical concentration of the propagating species should be kept below 10^{-8} molar to minimize termination reactions. Polymerization in a micellar environment isolates the propagating species and permits higher concentrations of propagating radicals. Thus, higher rates of polymerization (R_p) with simultaneous higher molecular weights (X_n, degree of polymerization) are realized, as described by the relationships (chapter 4 in reference 3) in equations 1 and 2.

$$R_p = 10^3 N n k_p [M]/N_A \tag{1}$$

where: n = average number of radicals per micelle plus particle
          k_p = propagation rate constant
          $[M]$ = monomer concentration
          N_A = Avogadro number

$$X_n = r_p/r_i = N k_p [M]/R_i \tag{2}$$

where: X_n = degree of polymerization
          r_p = rate of growth of a polymer chain
          r_i = rate at which primary radicals enter the polymer particle
          N = steady state concentration of micelles plus particles
          k_p = propagation rate constant
          $[M]$ = monomer concentration
          R_i = rate of initiation

and

$$N = k(R_i/\mu)^{2/5}(a_s S)^{3/5} \tag{3}$$

where: k = constant
          μ = rate of volume increase of a polymer particle
          a_s = interfacial surface area occupied by a surfactant molecule
          S = total concentration of surfactant in the system

A more developed description of the process is given in chapter 2 by a major contributor to this technology, John Gardon. The high rates of polymerization, with ease of removal of the heats of polymerization in an aqueous environment and the ease in handling of a low glass transition temperature particle, are major assets of the emulsion polymerization process.

As the market was developed in the sixties, the compositions of the latices varied from styrene/butadiene, to methylmethacrylate/acrylic esters (for their exterior durability), to low cost vinyl esters. With the transition in latex compositions to vinyl acetate, the oligomers became more hydrophilic and participated less in the micellar polymerization process. They continued to polymerize in the aqueous phase and then aggregate into a polymer colloid. Consequently, the rate for an emulsion polymerization of vinyl acetate is not dependent on the surfactant concentrations as cited for styrene in equations 1 and 3. For a more detailed treatment of the nonmicellar emulsion polymerization where there is essentially no polymerization in micelles, (i.e., the Hansen, Ugelstad, Fitch, Tsai theory) the reader is referred to references 4 and 5. For a discussion of the influence of monomer and surfactant hydrophilicity on the median particle size and particle size distribution of the latex, the reader is referred to references 6 and 7, respectively.

The use of seeding techniques to produce latices for more specialty applications is discussed in chapter 2. Compositional influences on interfacial tensions and how this affects the latex's morphology are discussed in chapter 3. What follows below in this chapter relates more to commodity latex production for use in architectural and paper coating markets.

Surface Stabilizers In Latex Chain-Growth Synthesis

The stabilities of latices in their early development were related to the electrostatic repulsions of particles due to catalyst fragments and adsorbed surfactants on their surfaces. Commercial materials demand more stability to electrolyte concentration, mechanical deformation and temperature variations than these entities can provide. Therefore, early in their development, the stabilities of latices were increased by oligomeric surface stabilizers, or by chemically grafted water-soluble polymer segments. These surface segments illustrated in Figure 1 provided electrosteric and steric stabilization. The most universal approach was the inclusion of methacrylic or acrylic acid in the latex synthesis recipe, generally in the second stage of a starved, semi-continuous process. The acids polymerize in sequence runs of acrylic or methacrylic acid, not possible in the step-growth synthesis of polyester and polyurethane disperse phases, discussed in the later section of this chapter. The repeating acid segments position near or on the surface of the latex and expand to provide electrosteric stabilization of the particle when neutralized. The parameters that influence the positioning of the acid surface stabilizers near the water/latex interface are the hydrophilicity of the copolymer and the latex's glass transition temperature(8). Thus, for a methyl methacrylate latex having equivalent glass transition temperatures, the surface acid content and its extent of swelling (reflected by sedimentation coefficients in Figures 2 and 3) will be greater in a copolymer composition having a more hydrophilic acrylate comonomer (where more is required of ethyl acetate than of butyl acrylate to achieve a given glass transition temperature, Figure 2). With the more hydrophobic acrylate comonomer, more of the 2 wt.% methacrylic acid charged is retained within the latex particle. The importance of the hydrophilicity of the latex composition also is evident in styrene and methyl methacrylate latices (not illustrated), both containing 50 % ethyl acrylate. The methacrylic acid segments are buried within the more hydrophobic styrene latex synthesized by a conventional batch process. In other compositions having a constant hydrophobicity but varying glass transition temperature, the importance of the glass transition temperature is evident (Figure 3). Such stabilizers have been important in small particle (100 nm) latices used in formulations containing a variety of pigments, because of their high pigment binding power. The importance of the oligomeric surface stabilizers in photographic processes are discussed in chapter 12. The theoretical treatment of the parameters influencing the stability of this type of latex particle and for those discussed in the next paragraph containing only nonionic grafted segments are treated in chapter 6 by Professor Goodwin.

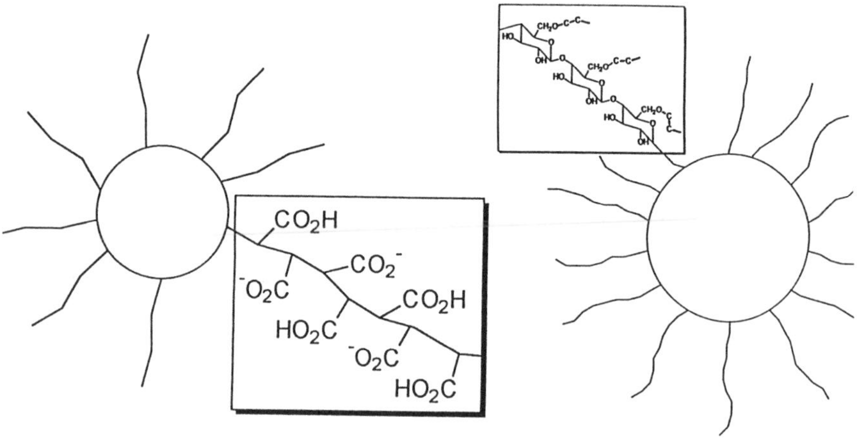

Figure 1. Electrosteric and Steric Stabilized Latices

While oligomeric surface acid stabilizers are used in vinyl acetate latices in the paper coatings area, where water sensitivity is not a factor in formulations containing high clay contents, and in methacrylic and styrene based latices in architectural latex coatings, vinyl acetate latices in low pigment volume concentration architectural coatings are water sensitive and surface acids are not the method of

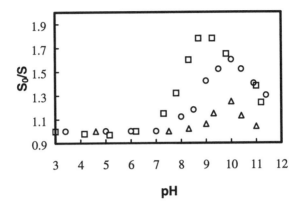

Figure 2. Expansion of acrylic copolymer latexes bearing 2% acid. S_0 / S (relative rate sedimentation of particle) verses dispersion pH. MMA/MA /BA: methyl methacrylate / ethyl acrylate / butyl acrylate. ☐   MMA/EA/BA, 40/52/6, Tg 15°C;   O MMA/EA/BA, 40/58/0, Tg 18°C; △ MMA/EA/BA, 49/0/49, Tg 15°C (Adapted from ref. 8.).

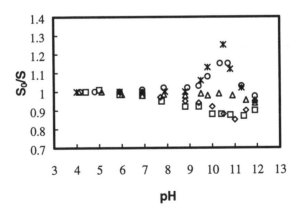

Figure 3. Expansion of Butyl Acrylate/methyl Methacrylate Copolymer Acid Bearing 2% Latexes. S_0/S (relative rate of Sedimentation of Particle) verses Dispersion pH. ✽ 50/50, Tg 15°C; O 37.5/62.5, Tg 33°C; △ 31/69, Tg 44°C; ☐ 25/75, Tg 54°C; ◊ 0/100, Tg 105°C (Adapted from ref. 8.).

stabilization. An alternative approach with the more hydrophilic vinyl ester monomers is to chemically graft segments of water-soluble polymers to the growing polymer chain. Generally, cellulose ethers are used in the synthesis of vinyl acetate latices for architectural coatings; vinyl acetate/vinyl alcohol copolymers are grafted to latices used in adhesive applications. Grafting of the nonionic stabilizers to the latex surface also provided a means of increasing viscosities at high shear rates (10^4 s^{-1}) for vinyl acetate latices, and there was a desire to impart this to large particle size acrylic latices. The general technical understanding of the sixties impeded this development on a wide commercial scale. It was known that the vinyl acetate radical was high in energy and therefore responsible for the grafting of poly(vinyl acetate) to the cellulose ether. Efforts by most to graft acrylics or styrene to cellulose ethers were unsuccessful and this was generally considered to be related to the lower energy of these propagating radicals. In fact, the lowest energy radicals in these systems were those generated at the carbon acetal linkage of the cellulose derivative or next to the carbonyl group of the acetate segments of the vinyl alcohol copolymer (Figure 4). The primary deterrent to the commercial production of methacrylic latices was the relatively uninterrupted propagation of the methacrylate radical without transfer reactions. The high energy vinyl acetate radical transferred with everything in the system except water, and there was therefore, sufficient water-soluble polymer chemically grafted to ensure adequate stability of the final vinyl acetate latex. Once the error in understanding of the mechanism was understood (*9*), water-soluble chain transfer agents, such as triethanolamine, were added to the polymerization recipe and methyl methacrylate latices, with sufficient cellulose ether grafts for stability, were successfully prepared.

During this period, polymer dispersions also were prepared in organic solvents using a different steric stabilizer technique. These Non-Aqueous Dispersions (*10*) of course contained VOC's of now unacceptable levels. Both these and the water-borne

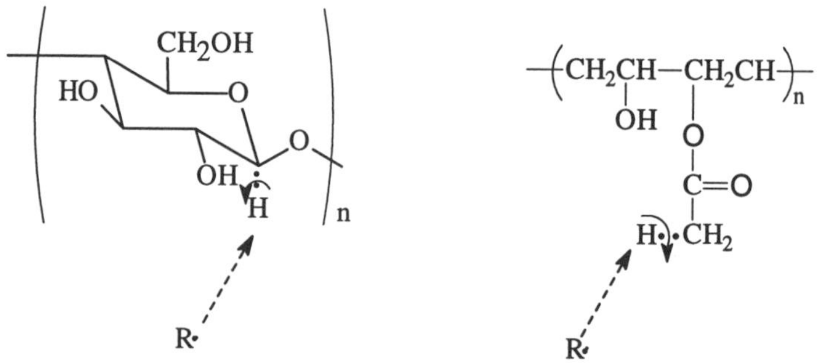

Figure 4. Low Energy Radicals Generated on Cellulose Derivative and Vinyl Alcohol Copolymer.

latices discussed above must coalesce to form a continuous film, and to do so with sufficiently high glass transition temperature latices to provide reasonable film durability after application, requires a coalescing aid. This contributes to the VOC of aqueous disperse phase coatings and will be addressed in chapter 4.

Disperse Phases in Original Equipment Manufacturing Coatings

Latex coatings used in the paper and architectural coating areas are thermoplastic. Coatings for industrial markets, as in automotive and appliances, require more durability, generally in a more hostile environment. The harder surface of a crosslinked thermoset film is required. The approach to achieving a low VOC coatings in this area has been the "high solids" route. Oligomers in the 3,000 to 5,000 molecular weight range are prepared. The synthesis of such resins is considered first in this chapter, to understand how the synthesis influences their conversion to aqueous dispersions and the stability of the particles generated. Examples of their preparation by step-growth synthesis are given in Scheme 1 for the preparation of polyesters (1A), noted for their low costs and flexibility; for epoxies(1B), noted for their solvent resistance and adhesion; and for urethanes(1C), noted for their abrasion resistance and flexibility. The chain growth synthesis of methyl methacrylate oligomers, noted for their durability, containing 2-hydroxyethyl methacrylate units randomly placed within the chain are illustrated in Scheme 1D.

The reactive groups from these synthesis are generally hydroxyl units that are crosslinked with multifunctional compounds such as melamine or the isocyanate groups of an isocyanurate compound (Scheme 2). Examples of such reactions are illustrated in Scheme 2 between a polyester with terminal hydroxyls and a melamine resin (2A), between a methyl methacrylate chain containing multiple hydroxyl units and the isocyanate units of an isocyanurate (2B), and between an epoxy and an acid terminated polyester (2C) or with amine functionalities (not illustrated).

Group Placement in Chain-Growth Oligomers. Methyl methacrylate oligomers (i.e., o-MMA, "water-reducible (W-R) acrylics") are used in OEM applications for their proven durability. They differ from an architectural latex formulation in that these oligomers contain 2-hydroxyethyl methacrylate (HEMA) or acrylate (HEA) units, 7 wt.% or less, that provide the crosslinking sites through their hydroxyl units (Scheme 1D). They also contain acrylic acid units (not illustrated) to provide electrosteric stability to the particles produced on the introduction of water. W-R acrylics also differ in that they generally do not contain traditional surfactants. Methacrylic acid is generally used in the synthesis of acrylic latices because a significant amount of water-soluble poly(acrylic acid) would be formed and reside in the aqueous phase. W-R acrylics are synthesized in an organic media and the lower cost acrylic acid is used. There has been considerable discussion about the desirability of uniform crosslink density networks in OEM films (Figure 5). This is difficult to obtain in a free-radical,

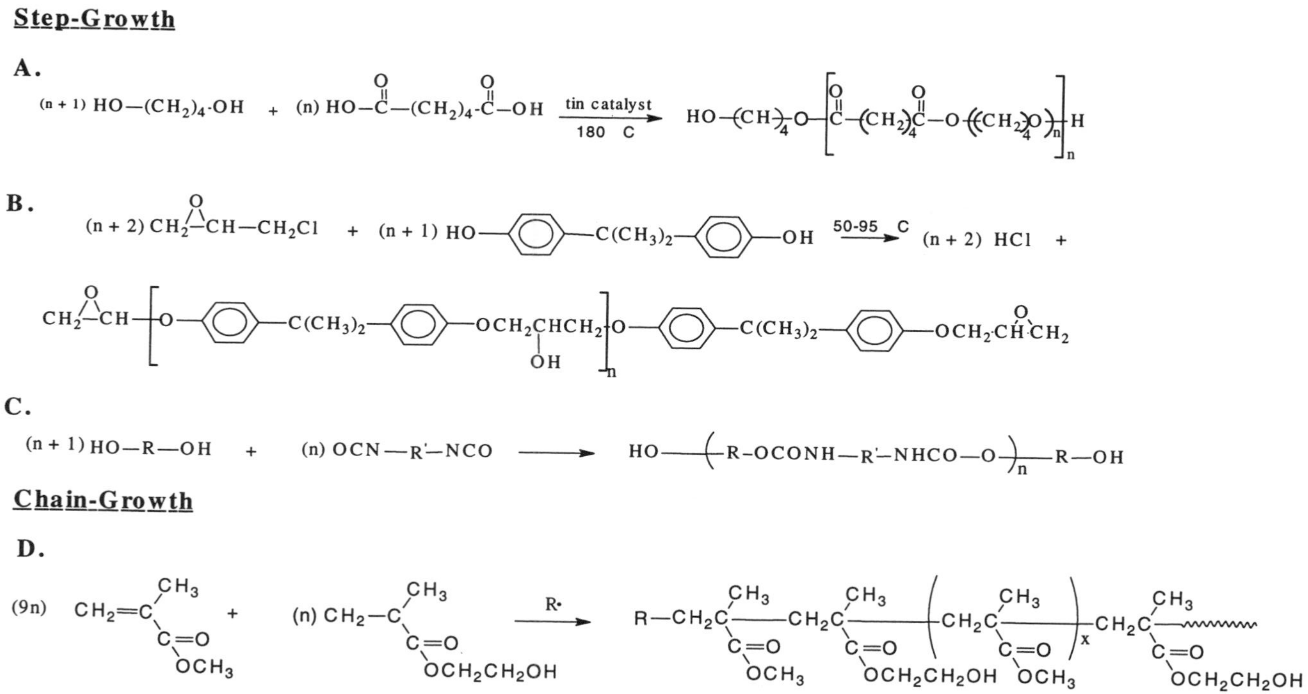

Step-Growth

A.

$(n+1)$ HO—$(CH_2)_4$—OH + (n) HO—$\overset{\displaystyle O}{C}$—$(CH_2)_4$—$\overset{\displaystyle O}{C}$—OH $\xrightarrow[\text{180 C}]{\text{tin catalyst}}$ HO—$(CH_2)_4$—O$\left[\overset{\displaystyle O}{C}$—$(CH_2)_4\overset{\displaystyle O}{C}$—O$(CH_2)_4$—O$\right]_n$H

B.

$(n+2)$ $CH_2\overset{O}{\triangle}CH$—$CH_2Cl$ + $(n+1)$ HO—⬡—C(CH_3)_2—⬡—OH $\xrightarrow{50\text{-}95\ C}$ $(n+2)$ HCl +

$CH_2\overset{O}{\triangle}CH$—⬡—O—C(CH_3)_2—⬡—$\left[OCH_2CHCH_2\overset{}{\underset{OH}{}}$O—⬡—C(CH_3)_2—⬡—$\right]_n$OCH_2CH—CH_2$\overset{O}{\triangle}$

C.

$(n+1)$ HO—R—OH + (n) OCN—R'—NCO $\longrightarrow$ HO—$\left($R—OCONH—R'—NHCO—O$\right)_n$R—OH

Chain-Growth

D.

$(9n)$ $CH_2=\overset{CH_3}{\underset{\underset{OCH_3}{\overset{O}{C}}}{C}}$ + (n) $CH_2=\overset{CH_3}{\underset{\underset{OCH_2CH_2OH}{\overset{O}{C}}}{C}}$ $\xrightarrow{R\cdot}$

R—$CH_2\overset{CH_3}{\underset{\overset{O}{C}}{C}}$...

Scheme 1. Preparation of Oligomers in the 3,000 to 5,000 Molecular Weight Range (A) Polyester (B) Epoxy (C) Urethanes (D) Methylmethacrylate with 2-Hydroxyethyl Methacrylate Units.

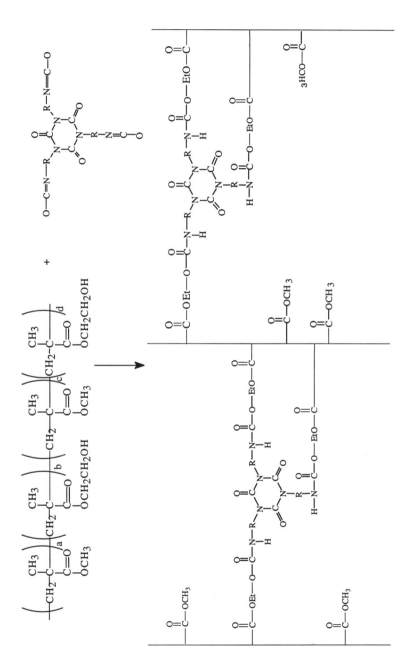

Scheme 2A. Crosslinking between a Methyl Methacrylate (with hydroxyl units) and an Isocyanurate.

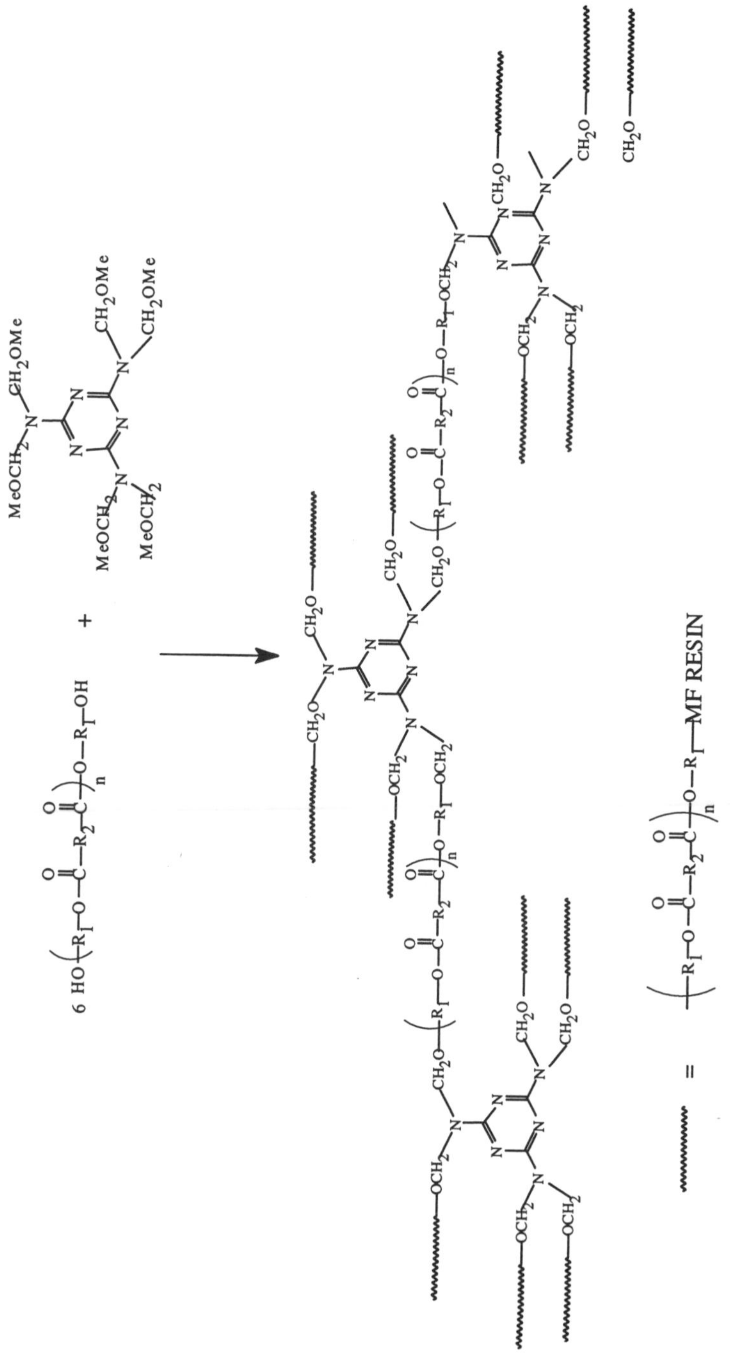

Scheme 2B. Crosslinking between a Hydroxy-terminated Polyester and a Melamine Resin.

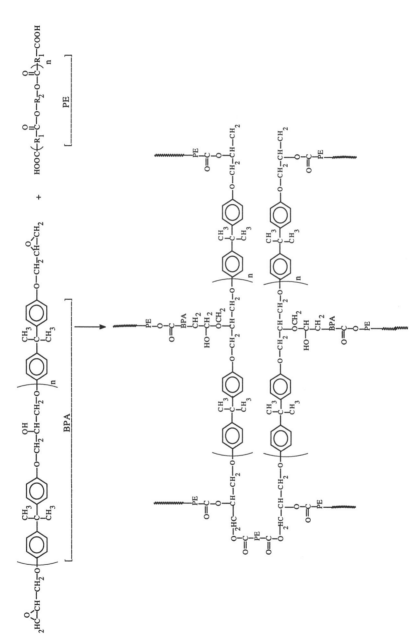

Scheme 2C. Crosslinking between an epoxy and an acid terminated polyester.

chain-growth oligomerization. The inclusion of the hydroxy containing monomer in a given oligomeric chain is dependent on the monomer's relative reactivity ratio and concentration (chapter 5 in reference 3). Thus, in a low molecular weight chain, a HEMA or HEA unit may not be incorporated in many methacrylate oligomers and many chains may have more than two hydroxy containing HEMA or HEA units. In addition, there is a very low probability that hydroxyl units would be positioned at the terminal positions of the methacrylate chains, considered important to the formation of a uniform network structure on curing with multifunctional crosslinkers. To this latter goal, researchers developed an anionic initiator (*11*), a ketene silyl acetal, that affected the synthesis of a monodisperse molecular weight o-MMA resin, typical of an anionic initiator, but also with the ability to place hydroxyl group at both ends of the chain. Cost is the primary parameter in coating formulations, and this innovative approach has not achieved a dominant commercial position.

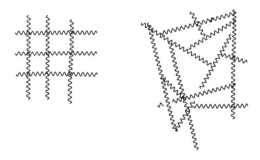

Figure 5. Uniform and Non-uniform Crosslink Density Networks.

Water-reducible acrylics, used in thermosetting applications, are prepared in an organic solvent (such as t-butanol) and "reduced" to an aqueous disperse phase by dilution with water. This type of resin was at one time the primary resin used in automotive coatings, but they were humidity sensitive and this caused major problems during production. The data in Figure 6A illustrates the influence of dilution of such a resin with t-butanol, the solvent used in synthesizing the low molecular weight W-R acrylic, and with water on the solution or dispersion viscosity. The dilution of a traditional, high molecular weight acrylic latex with water also is illustrated. In two of the curves, the viscosity decreases with dilution, as would be expected; however, with water addition of the t-butanol acrylic resin solution abnormal viscosity behavior is observed. Initially the viscosity decreases due to dilution; however, with continued dilution the viscosity increases. The addition of water to the t-butanol W-R acrylic solution causes a phase inversion and it is in this area that a viscosity abnormality occurs (Figure 6A). The magnitude of the viscosity abnormality at an intermediate dilution level is dependent on the molecular weight of the W-R acrylic, on the monomer acid level in the resin (*12, 13*) (Figure 6B), and on the extent of neutralization. Prior to dilution of the ca. 70 wt.% resin in butanol the acrylic acid segments   are   partially   neutralized   with   an   amine   such   as   2-

(dimethylamino)ethanol(DMAE). The abnormality arises in part from the decreasing ratio of solvent to water, causing the nonpolar segments of the resin to associate. The solvent is soluble in the resin and some of it swells the aggregates formed on dilution of the t-butanol with water. The swollen aggregates increase in number with continued dilution and the viscosity increases with increasing volume fraction as would be expected from the principles defined by the Mooney (*14*) or Kreiger-Dougherty (*15*) equations (discussed in chapter 6 of this text). With increasing water dilution the occluded solvent distributes more to the continuous water phase and the aggregates begin to shrink in size. This complements the dilution effect and the viscosity begins to decrease again. A similar aggregation phenomenon has been reported in aqueous polyester dispersions (*16*).

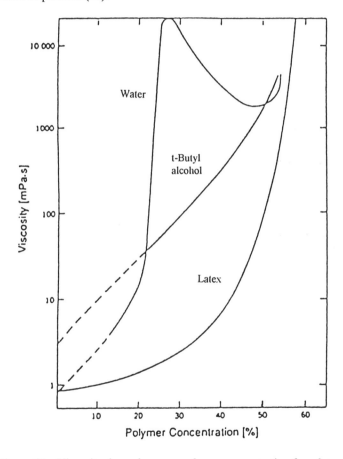

Figure 6A. Viscosity dependence on polymer concentration for a latex and for water and t-butyl alcohol solutions of a 10/90 acrylic acid/butyl methacrylate copolymer with 75% effective neutralization with dimethylaminoethanol (Reproduced with permission from ref. 12).

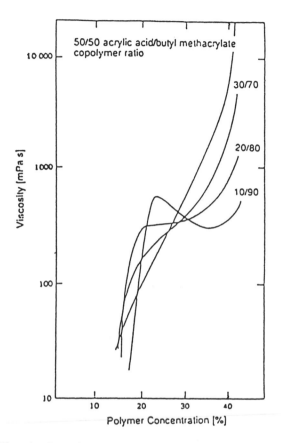

Figure 6B. Viscosity dependence on polymer concentration for dilutions in water of acrylic acid butyl methacrylate with different copolymer ratios and 75% effective neutralization with dimeylaminoethanol (Reproduced with permission from ref. 12).

The viscosity abnormality disappears as the acid content of the resin is increased (Figure 6B). As the acid percentages in the resin is increased, fewer of the acid monomer units are isolated and trapped in the aggregated resin's interior, where with their solvation by water and reaction with the solubilized base, they help swell the particle. Interestingly, it has been observed in studies at NDSU that approximately 25 % of the acid monomer is trapped within the acrylic disperse phase in both methacrylate latices and in water-reducible acrylics. The surface acid units in W-R acrylics are not desirable crosslinking sites with multifunctional isocyanates and melamine resins. The reaction of isocyanates with carboxylic acids leads to carbon dioxide liberation and possible film defects. A reaction with melamine lead to ester linkages and possible hydrolysis during the life of the film. Reaction of the latter with the hydroxyl of the HEMA units leads to stable ether crosslinks. The goal is to minimize the number of carboxylate groups as far as particle stability in the aqueous phase will permit and to maximize the number of hydroxyl groups of the HEMA or HEA units for crosslinking sites.

There is a second problem to be addressed in these resins. The melamine crosslinking reactions are inhibited by the amine used to neutralize the carboxylate groups. This can lead to poor surface gloss when the monomers in the surface, where an amine of low volatility may evaporate, can polymerize without polymerization occurring throughout the film where amine is still present. The resulting immobilized surface film wrinkles forming an irregular surface when the subsurface, later free of amine by evaporation, polymerizes, and shrinks.

Group Placement in Step-Growth Synthesis. The majority of polymers used in OEM applications arise from step-growth polymerizations (chapter 2 in reference *3*). There are disadvantages and advantages to a step-growth polymerization. The most notable disadvantages are the broad molecular weight distributions (illustrated for isocyanate reaction products (*17*) in Figure 7), and high molecular weights are not obtained until high conversions are reached. Lower molecular weights can be volatile and are toxic, especially isocyanates. The broad distribution in molecular weights is a concern if the viscosity of the resins is important. The distribution is not a primary concern if the polymers are taken to high conversion and molecular weight, but the latter is generally not the case for the resins used in thermosetting applications. An advantage in a step-growth polymerization, to some extent countering the broad molecular weight distribution, is that by controlling the stoichiometry, the functional groups (i.e., hydroxyl, isocyanate, epoxy, or carboxylate groups) can be placed in the terminal position (Scheme 1), prior to crosslinking to the thermoset coating.

In the production of aqueous polyester dispersions, high reaction temperatures are required for polyesterification (>180 °C). This restricts the wide use of "flexible components" such as poly(tetramethylene oxide) in polyesters. The sulfonate anion (Figure 8B) promotes charge stabilization across a broad pH range (*18*) and the structure given in 8B is the anionic stabilizer used in commercial polyesters. The hydrolytic stability of aqueous polyester dispersions are addressed in this book. Epoxy resins are covered in chapter 5. The systems most discussed in the open literature are water-borne urethanes.

Polyurethanes are formed by the reaction of diisocyanates with diols (*19*, Scheme 1C), but in processing and application, many more isocyanate reactions can be involved. They are given in Scheme 3. The reaction of an isocyanate with an alcohol unit provides the desired urethane linkage (3A), reaction with a carboxylate group (3B) leads to a carbamic acid intermediate that decomposes to carbon dioxide and to an amine, that can react with other isocyanate in the system to form urea linkages(3C). The liberation of CO_2 in a coating system can lead to surface defects and is undesirable. Recent studies indicate that the rate of CO_2 production during the urethane network formation can be controlled by crosslinking conditions (*20*). The reaction of an isocyanate with water is illustrated in 3D. It also should be mentioned that because both urethane and urea linkages still have an active hydrogen, they can further react

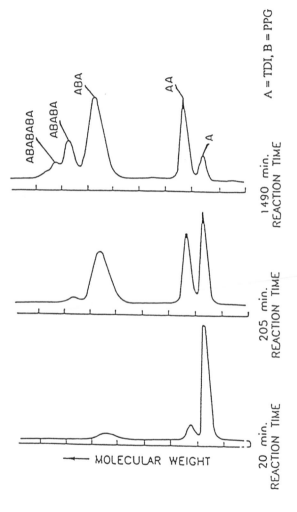

Figure 7. Size exclusion chromatograms of Urethane elastomers.

A. Reaction with Alcohols: most widely used in coatings

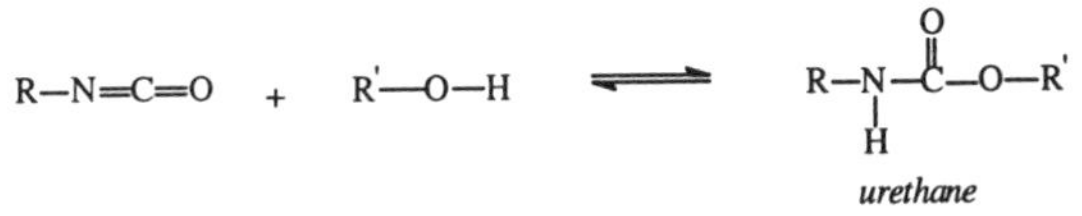

$$R-N=C=O \quad + \quad R'-O-H \quad \rightleftharpoons \quad R-\underset{H}{\overset{O}{N}}-\overset{\|}{C}-O-R'$$

urethane

B. Reaction with Carboxylic Acids : Polyols (especially polyesters) must have low acid content

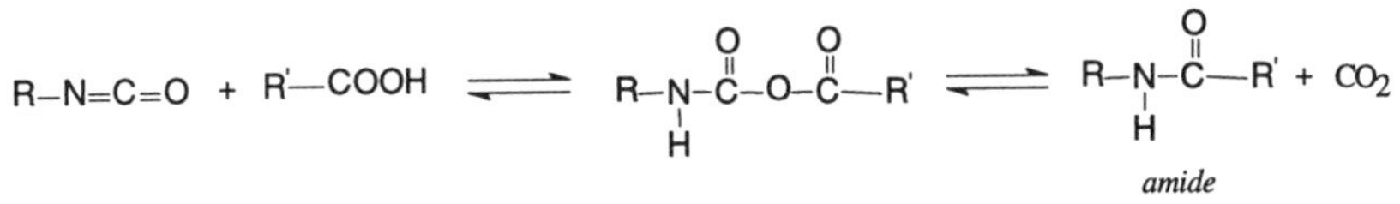

$$R-N=C=O \; + \; R'-COOH \; \rightleftharpoons \; R-\underset{H}{\overset{O}{N}}-\overset{\|}{C}-O-\overset{O}{\overset{\|}{C}}-R' \; \rightleftharpoons \; R-\underset{H}{\overset{O}{N}}-\overset{\|}{C}-R' + CO_2$$

amide

C. Reaction with Amines : fast reaction limits use

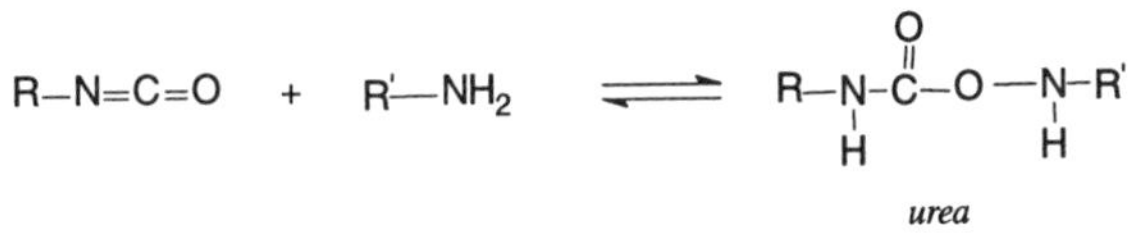

$$R-N=C=O \quad + \quad R'-NH_2 \quad \rightleftharpoons \quad R-\underset{H}{\overset{O}{N}}-\overset{\|}{C}-O-\underset{H}{N}-R'$$

urea

D. Reaction with Water : water should be removed from polyols

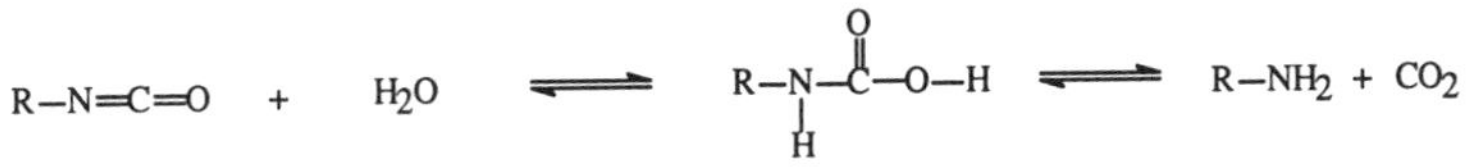

$$R-N=C=O \quad + \quad H_2O \quad \rightleftharpoons \quad R-\underset{H}{\overset{O}{N}}-\overset{\|}{C}-O-H \quad \rightleftharpoons \quad R-NH_2 + CO_2$$

E. Reaction with urethanes

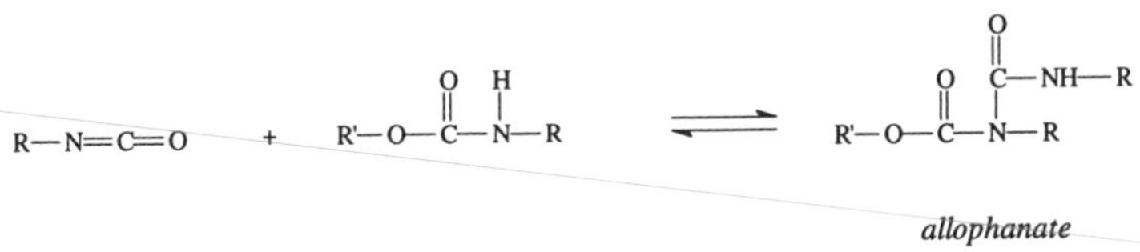

$$R-N=C=O \quad + \quad R'-O-\overset{O}{\overset{\|}{C}}-\underset{}{\overset{H}{N}}-R \quad \rightleftharpoons \quad R'-O-\overset{O}{\overset{\|}{C}}-\underset{}{N}-R \;\; (C-NH-R)$$

allophanate

F. Reaction with Urea

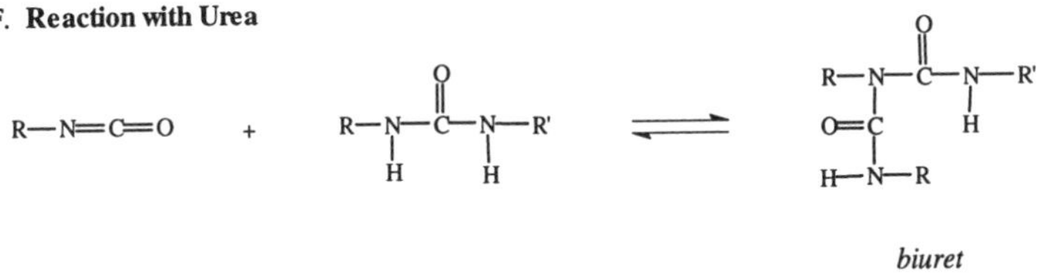

$$R-N=C=O \quad + \quad R-\underset{H}{\overset{O}{N}}-\overset{\|}{C}-\underset{H}{N}-R' \quad \rightleftharpoons \quad$$

biuret

Scheme 3. Reactions of isocyanate with (A) alcohol (B) carboxylate (C) amine (D) water (E) urethane (F) urea with an isocyanate to form an allophonate (3E) and a biuret (3F), respectively. However, these products typically are only formed in baked, uncatalyzed systems.

The methodology of producing an aqueous disperse polyurethane phase (PUDs)(*21*), which can generically represent the step-growth synthesis of OEM type resins, is given in Schemes 4 - 6. First, a medium molecular weight isocyanate terminated prepolymer is formed by reacting a diol with a stoichiometric excess of diisocyanate (*21*, Scheme 4). The second step is to chain extend this prepolymer, where problems in viscosity build up and the fast rate of reaction of the isocyanate units with the amine groups are encountered, and methodologies then vary with the process used. They include a solvent process using acetone (Scheme 5), a ketimine/ ketazine process (Scheme 6), a prepolymer mixing process, and a hot melt process. The latter processes employ technologies that reduce or eliminate the use of organic solvents.

In the acetone solvent process (*21*) the prepolymer is chain extended with a sulfonate based diamine, to give the desired high molecular weight (Scheme 5). Acetone prevents excessive viscosity build up. Water is then added to the polymer solution until water becomes the continuous phase. Viscosity abnormalities such as those reported with water-reducible acrylics have not been detailed. The solvent is removed by distillation, adding to production costs, to obtain dispersions low in VOC. The ionic sulfonate group acts as a charge-repulsion stabilizer. Although this process yields reproducible PU dispersions, it is limited to those that are soluble in acetone, and consequently, the resulting urethane films are not very solvent resistant.

One of the alternatives to the solvent process is the Ketimine/Ketazine process. It does not require distilling off a solvent or high shear equipment; the molecular weight builds in water so a viscosity build up is avoided and the solvent resistance of the final film can approach that cast from solvent. The reactions are given in Scheme 6. This process, however, is sensitive, and subject to product variations.

Surface Stabilizers in Step-Growth Polymerizations

Like traditional latices, "water-reducible, step-growth polymers" require anionic and/or nonionic surface stabilizers to maintain dispersion stability at reasonable volume solids in a coating formulation. This is accomplished in step-growth polymers by the inclusion of a carboxylate (Figure 8A), sulfonate (8B), or nonionic oxyethylene (8C) group, placed within the chain by a diol or diamine structure (illustrated in Schemes 4 and 5). These disperse phases are not as stable as the acrylic disperse phases, for the surface stabilizera are isolated as one charge and there is no steric contribution with these isolated anionic groups. They do not occur in sequence runs as observed in chain-growth acrylic polymers, and therefore, the particles are more susceptible to flocculation by mechanical and thermal forces and by electrolyte concentrations. There is also a particle size distribution difference among these different disperse phases, with the W-R acrylic being much smaller than those obtained from an emulsion polymerization.

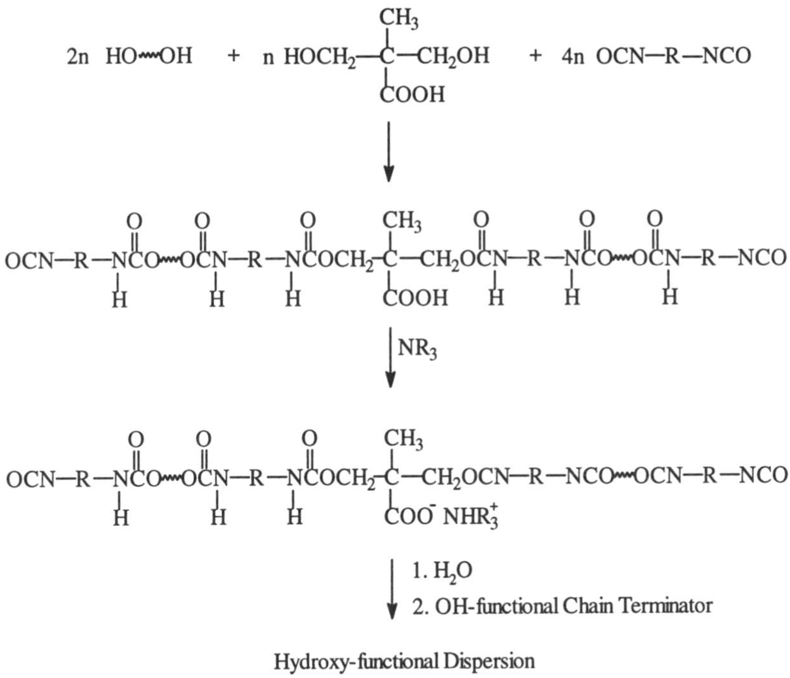

$$2n \ HO\text{\textasciitilde}OH \quad + \quad n \ HOCH_2\overset{\overset{\displaystyle CH_3}{|}}{\underset{\underset{\displaystyle COOH}{|}}{C}}CH_2OH \quad + \quad 4n \ OCN-R-NCO$$

$$\downarrow$$

$$OCN-R-N\overset{O}{\overset{||}{C}}O\text{\textasciitilde}O\overset{O}{\overset{||}{C}}N-R-N\overset{O}{\overset{||}{C}}OCH_2\overset{\overset{\displaystyle CH_3}{|}}{\underset{\underset{\displaystyle COOH}{|}}{C}}CH_2O\overset{O}{\overset{||}{C}}N-R-N\overset{O}{\overset{||}{C}}O\text{\textasciitilde}O\overset{O}{\overset{||}{C}}N-R-NCO$$

(with H beneath each N)

$$\downarrow NR_3$$

$$OCN-R-N\overset{O}{\overset{||}{C}}O\text{\textasciitilde}O\overset{O}{\overset{||}{C}}N-R-N\overset{O}{\overset{||}{C}}OCH_2\overset{\overset{\displaystyle CH_3}{|}}{\underset{\underset{\displaystyle COO^- \ NHR_3^+}{|}}{C}}CH_2OCN-R-NCO\text{\textasciitilde}OCN-R-NCO$$

(with H beneath each N on left side)

$$\downarrow \begin{array}{l} \text{1. } H_2O \\ \text{2. OH-functional Chain Terminator} \end{array}$$

Hydroxy-functional Dispersion

Scheme 4. Prepolymer Process: Preparation of hydroxy-functional dispersion.

When an acrylic latex is thickened with a nonadsorbing polymer such as hydroxyethyl cellulose, the latex will flocculate by a depletion flocculation mechanism (discussed in detail in chapter 6). This is usually not observed visually. With a step-growth disperse phase, separation is observed within hours(22). This does not happen when the thickener is a Hydrophobically-modified, Ethoxylated URethane (HEUR) polymer.

HEUR polymers have been discussed in detail with respect to their synthesis, solution behavior and structural influences on architectural coatings rheology (23, chapters 10, 17, and 24, respectively). The adsorption of an HEUR thickener on a latex is determined by its chemical composition (that defines the energy of the surface), the concentration of surface acids and the amount of surfactant present (24). The molecular weight, concentration and effective hydrophobe terminal size on the HEUR are also important. The relative concentration and size of the surfactant and of the effective hydrophobe terminal size of the HEUR are the key variables in the adsorption process. There is no free surfactant used in the synthesis of a step-growth disperse phase, and therefore no impedance to adsorption and stabilization.

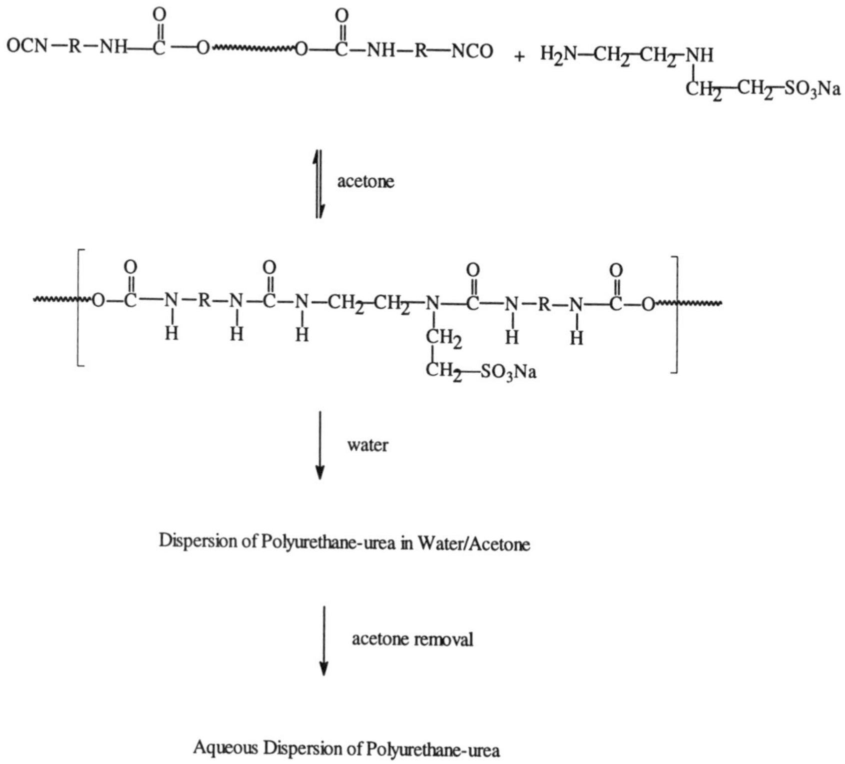

$$\text{OCN−R−NH−}\overset{\overset{\textstyle O}{\|}}{C}\text{−O}\wwww\text{O−}\overset{\overset{\textstyle O}{\|}}{C}\text{−NH−R−NCO} + H_2N\text{−CH}_2\text{−CH}_2\text{−NH}$$
$$\underset{\quad\quad CH_2\text{−CH}_2\text{−SO}_3Na}{}$$

acetone

$$\left[\wwww O\text{−}\overset{\overset{\textstyle O}{\|}}{C}\text{−}\underset{\underset{\textstyle H}{|}}{N}\text{−R−}\underset{\underset{\textstyle H}{|}}{N}\text{−}\overset{\overset{\textstyle O}{\|}}{C}\text{−}\underset{\underset{\textstyle H}{|}}{N}\text{−CH}_2\text{−CH}_2\text{−}\underset{\underset{\textstyle CH_2}{|}}{N}\text{−}\overset{\overset{\textstyle O}{\|}}{C}\text{−}\underset{\underset{\textstyle H}{|}}{N}\text{−R−}\underset{\underset{\textstyle H}{|}}{N}\text{−}\overset{\overset{\textstyle O}{\|}}{C}\text{−O}\wwww\right]$$

water

Dispersion of Polyurethane-urea in Water/Acetone

acetone removal

Aqueous Dispersion of Polyurethane-urea

Scheme 5. Acetone Process: Preparation of Aqueous Dipsersion of Polyurethane-Urea.

As noted earlier, the broad molecular weight distribution in step-growth oligomers is off-set by the ability to place the reactive groups in the terminal position. As noted in the preceding paragraph, the greater instability of step-growth dispersions can be improved by steric stabilization through adsorption of HEUR associative thickeners. By changing R and R' in Scheme 1, it is possible to vary the crosslinking, chain flexibility, and intermolecular forces among the chains. It has been well documented (25) that increasing the stiffness of the mainchain will increase the modulus or glass transition temperature of a polymer and suppress secondary transitions responsible for impact strength and flexibility in a polymer. The most unique aspect of step-growth polymers is the latitude in designing hardness and flexibility into the resin, discussed below, in an academic example. More practical approaches are discussed in chapter 9.

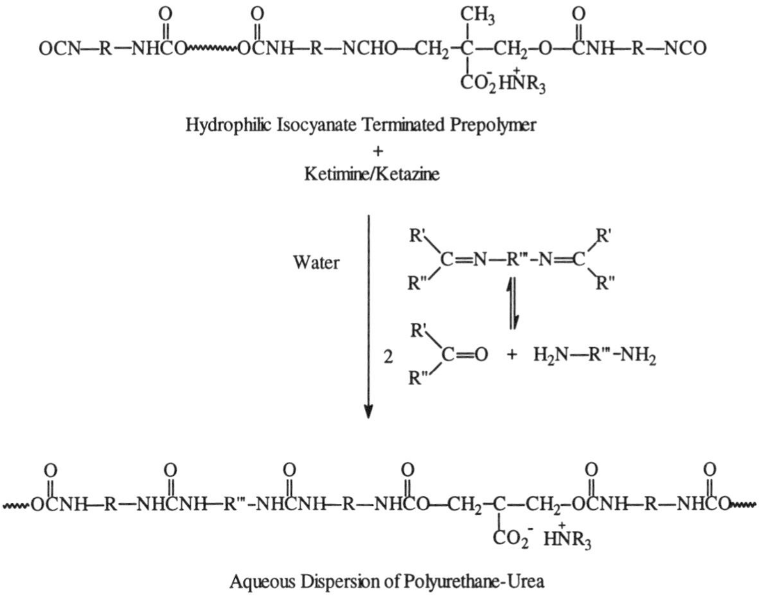

$$OCN-R-NH\overset{O}{\overset{\|}{C}}O\text{\small wwww}O\overset{O}{\overset{\|}{C}}NH-R-NCHO-CH_2-\underset{\underset{CO_2^-H\overset{+}{N}R_3}{|}}{\overset{CH_3}{\overset{|}{C}}}-CH_2-O-\overset{O}{\overset{\|}{C}}NH-R-NCO$$

Hydrophilic Isocyanate Terminated Prepolymer
+
Ketimine/Ketazine

Water

$$\underset{R''}{\overset{R'}{>}}C=N-R'''-N=C\underset{R''}{\overset{R'}{<}}$$

$$2\ \underset{R''}{\overset{R'}{>}}C=O\ +\ H_2N-R'''-NH_2$$

$$\text{\small wwwO}\overset{O}{\overset{\|}{C}}NH-R-NH\overset{O}{\overset{\|}{C}}NH-R'''-NH\overset{O}{\overset{\|}{C}}NH-R-NH\overset{O}{\overset{\|}{C}}O-CH_2-\underset{\underset{CO_2^-\ H\overset{+}{N}R_3}{|}}{\overset{}{C}}-CH_2-O\overset{O}{\overset{\|}{C}}NH-R-NH\overset{O}{\overset{\|}{C}}O\text{\small wwww}$$

Aqueous Dispersion of Polyurethane-Urea

Scheme 6. Ketimine/KetazineProcess: Preparation of aqueous dispersion of polyurethane-urea.

Influence of Phase Transitions on Film Properties

Designing hardness and flexibility in an applied film is accomplished by varying the carbon structure between the difunctional units, and it is best demonstrated in the open literature with polyurethanes(PUs). The oligomeric diol carrying the low glass transition temperature segments can be represented by polycaprolactonediol, polyesterdiols or poly(tetramethylene oxide)diols (PTMO). Urethane linkages to aromatic segments are unstable in sunlight and in alkaline aqueous media; therefore, the diisocyanates used in coatings are aliphatic. The diisocyanates represent the hard segments with high glass or melting transition temperatures. Due to the difference in polarity of the segments the PUs exhibit a phase separated morphology with distinct hard and soft segment domains. The good mechanical properties (e.g., stiffness with flexibility) of PUs have been attributed to this phase separation at temperatures between the transition temperatures of the soft and hard segments. In addition, PUs aqueous dispersions (PUDs) have the ionic stabilizers needed for disperse phase stability, that in the final film can result in coulombic forces of attraction (as they do in ionomers (26)). These columbic forces between ionic centers are claimed (27) to enhance the mechanical properties of PUD films over the conventional thermoplastic PUs. The ionic group attractions complement hydrogen bonding among the urethane and urea groups (attributed to result in the excellent abrasion resistance of PUs) to achieve unique film properties. These types of systems in real coatings will be discussed in chapter 9. This chapter will close with an example that delineates the

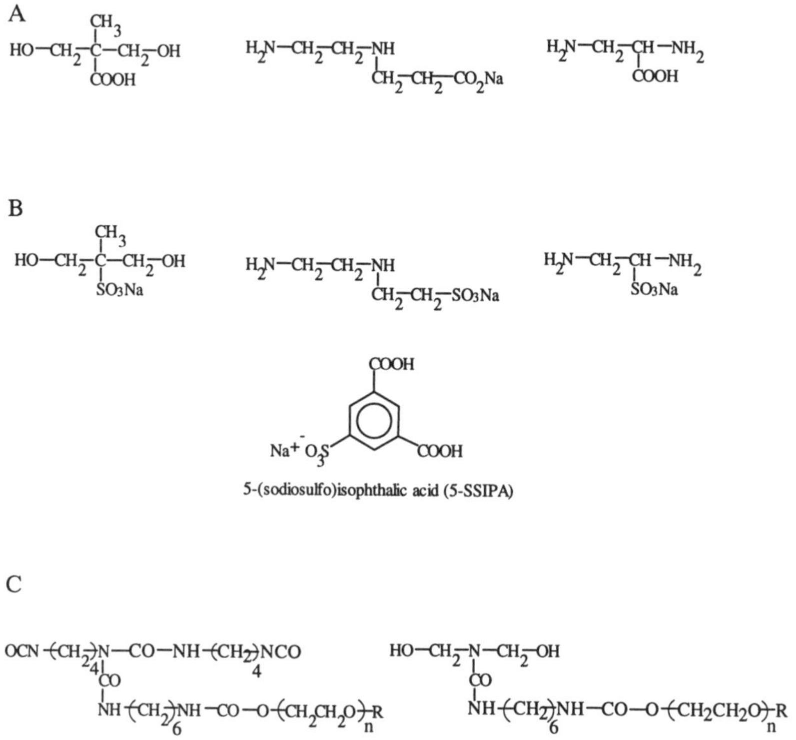

A

$$HO-CH_2-\underset{\underset{COOH}{|}}{\overset{\overset{CH_3}{|}}{C}}-CH_2-OH \qquad H_2N-CH_2-CH_2-\underset{\underset{CH_2-CH_2-CO_2Na}{|}}{NH} \qquad H_2N-CH_2-\underset{\underset{COOH}{|}}{CH}-NH_2$$

B

$$HO-CH_2-\underset{\underset{SO_3Na}{|}}{\overset{\overset{CH_3}{|}}{C}}-CH_2-OH \qquad H_2N-CH_2-CH_2-\underset{\underset{CH_2-CH_2-SO_3Na}{|}}{NH} \qquad H_2N-CH_2-\underset{\underset{SO_3Na}{|}}{CH}-NH_2$$

5-(sodiosulfo)isophthalic acid (5-SSIPA)

C

$$OCN-(CH_2)_4-\underset{\underset{NH-(CH_2)_6-NH-CO-O-(CH_2CH_2O)_n-R}{|}}{\overset{\overset{}{|}}{N}}-CO-NH-(CH_2)_4-NCO$$

$$HO-CH_2-\underset{\underset{NH-(CH_2)_6-NH-CO-O-(CH_2CH_2O)_n-R}{|}}{\overset{\overset{}{|}}{N}}-CH_2-OH$$

Figure 8. Surface stabilizers. (A) carboxylate (B) sulfonate (C) nonionic oxyethylene groups.

latitude in properties that can be obtained with changes in the synthesis component ratios.

The following materials (28) highlight the latitude in film properties that can be achieved in a step-growth polymer. The polymers synthesized are illustrated in (Figure 9A). The simplest way to achieve phase separation of the hard segments is to increase the amount of these units; however, too high a concentration of hard units can embrittle the film. In the materials illustrated in Figure 9A, the hydrophilicity and ability to crosslink are accomplished through pyridine (BIN) units in the hard segment that are reacted with iodopropane (IP, to increase the hydrophobicity of the hard segment) or diiodopropane (DIP, to increase crosslinking among the hard segments). The hard segment concentration was varied: 41, 52 and 62 wt.%, and their tensile properties are illustrated in Figure 9B. The materials containing the broadest spectrum of properties obtainable are best illustrated in the 54 wt.% hard segment materials. PTMO in these studies is poly(tetramethylene oxide). Their dynamic mechanical properties are illustrated in Figure 10. The glass transition temperature of the soft segment decreases, as the alkyl group is attached to the pyridine ring, suggesting that the alkyl group improves the phase separation and thereby provides a purer soft segment. The -130° C peak is associated with the crankshaft transition of the four methylene units (25), contributing to impact strength; the transition near -50°C is the glass transition temperature of the polyether (PTMO). The dihalide crosslinks two pyridine groups giving a broad rubbery plateau transition in the plus 100° C temperature range. These types of transitions are not easily inserted in chain-growth polymers, and provide an architectural scenario to fine-tune film properties in step-growth polymers.

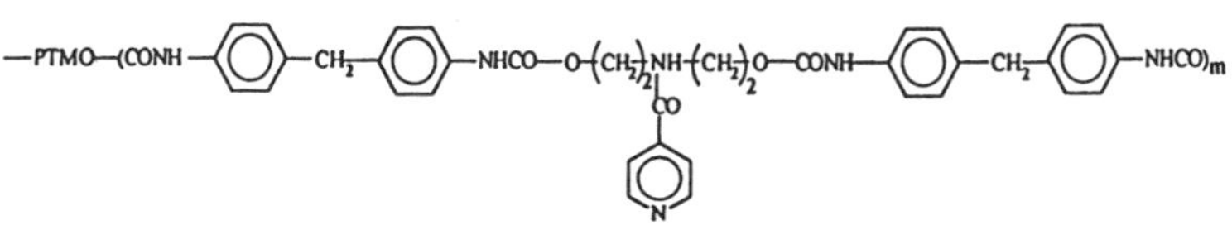

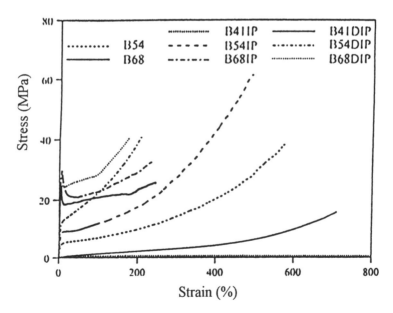

Figure 9A. PTMO-MDI-BIN.

Figure 9B. Tensile properties of materials illustrated in Figure 9A.

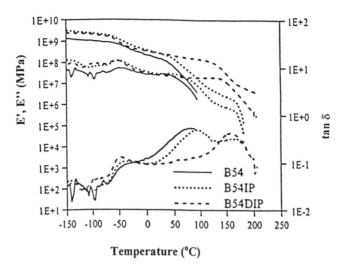

Figure 10. Dynamic mechanical properties of materials illustrated in Figure 9A.

Conclusions

The general principles associated with the synthesis of organic binders and the stability of their aqueous disperse phases are reviewed in this chapter.

For application in commodity architectural and paper coatings areas, oligomeric surface structures are required. This is achieved by copolymerizing sequence runs of oligomeric acid or by grafting water-soluble polymers to the binder during the synthesis of the emulsion polymer. The procedures utilized to achieve this goal for different monomers are described in this chapter. Modification and development of different latex surface morphologies are discussed in the two chapters that follow.

More durable coatings are required in industrial applications and this is accomplished after application by crosslinking the applied oligomers to thermoset resins, and the approach to resin synthesis and dispersion stability, prior to application, is more complex. The majority of polymers used as thermosets arise from step-growth polymerizations, that have disadvantages and advantages relative to chain growth polymers. The most notable disadvantages are the broad molecular weight distributions and the volatility and toxicity of lower molecular weights, especially isocyanates. The broad distribution in molecular weights is a concern if the viscosity of the resins is important. The isolation of the acid stabilizing moieties in a step-growth synthesis also produces a more unstable dispersion, but this can be compensated for by using hydrophobe-modified, ethoxylated urethane thickeners that provide steric stabilization through adsorption on the surfaces of step-growth polymer disperse phases.

An advantage of a step-growth polymer, to a large extent countering their disadvantages, is that by controlling the stoichiometry the functional groups (e.g., hydroxyl, isocyanate, epoxy, or carboxylate groups) can be placed in the terminal position, prior to crosslinking the oligomers to a thermoset coating. Another advantage is that chain flexibility and intermolecular forces can be designed into the resin to provide high and low temperature transitions. This is the most unique aspect of step-growth polymers. These types of transitions are not easily inserted into chain-growth polymers, and they provide an synthesis scenario to fine-tune film properties, such as hardness and flexibility, into step-growth polymers.

Literature Cited

1. Hoyle, C.E.; Kinstle, J.F., Eds., *Radiation Curing of Polymeric Materials;* ACS Symposium Series 417, American Chemical Society: Washington, DC, 1990.
2. Jilek, J.H., *Powder Coatings;* Federation of Societies for Coatings Technology: Blue Bell, PA, 1991.
3. Odian, G., Eds., *Principles of Polymerization;* 3rd Edition, John Wiley & Sons, Inc.: New York, 1991, Chapter 4. .
4. Fitch, R.M.; Tsai, C.H. in *Polymer Colloids*; Fitch, R.M., Ed.; Plenum press; 1971; pp.73 and pp.103.
5. Hansen, F.K., in *Polymer Latexes: Preparation, Characterization, and Application*, ACS Symposium Series 492, Daniels, E.S.; Sudol, E.D.; El-Aasser, M.S.; Eds., American Chemical Society, Washington, DC, 1992; Chapter 2.
6. Yeliseyeva, V.I., in *Emulsion Polymerization*, Pirma, I., Ed., Academic Press, New York, 1982; Chapter 7.
7. Dunn, A.S., in *Emulsion Polymerization*, Pirma, I., Ed., Academic Press, New York, 1982; Chapter 6, and in *Polymer Latexes: Preparation, Characterization, and Application,* ACS Symposium Series 492, Daniels, E.S.; Sudol, E.D.; El-Aasser, M.S.; Eds., American Chemical Society, Washington, DC, 1992; Chapter 4.
8. Hoy, K.L., *Journal of Coatings Technology,* **1979**, *51 (651)*, 27.
9. Craig, D.H., in *Water-Soluble Polymers: Beauty with Performance*, Advances in Chemistry Series 213, Glass, J.E., Ed., American Chemical Society, Washington, DC, 1986; Chapter 18. Several additional studies were reported in ACS-PMSE national meeting reprints after this publication and the study was later republished in *J. Coatings Techn.*
10. Barrett, K. E. J., Ed., *Dispersion Polymerization in Organic Media*, Wiley, London, 1975.
11. Webster, O.W., Sogah, D.Y., in *Comprehensive Polymer Science*, Eastman, G.C.; Ledwith, A.; Russo, S.; Sigwalt, P., Eds., Pergamon Press, NY, 1989; Chapter 10.
12. Richards, B.M.; Masters Thesis, North Dakota State University, Fargo, ND, 1977.
13. Brandenburger, L.B.; Ph.D. Thesis, North Dakota State University, 1977.
14. Mooney, M.; *J. Polym.Sci.*, **1959**, *34*, 599.

15. Krieger, I.M., *Adv. Colloid and Interface Sci.*, **1972**, *3*, 111.

16. Raynolds, P.W., Proceedings of Water-borne and Higher Solids Coatings Symposium, **1990**, New Orleans, LA, pp. 121-129.

17. Thompson, C.M.; Taylor, S.G.; McGee, W.M.; *J.Polym. Sci. Part A; Polym. Chem.* **1990**, *28*, 333.

18. Kuo, T.; Blount, W.W.; Moody, K.M.; *Proc. ACS Div. Polymer Materials: Science and Engineering,* **1992**, 66, 168-169.

19. Dieterich, D., *Progress in Organic Coatings*, **1981**, *9*, 281.

20. Potter, T.A.; Rosthauser, J.W.; Schmelzer, H.G., Proceedings of Water-borne and Higher Solids Coatings Symposium, **1986**, New Orleans, LA, pp. 162-165.

21. Rosthauser, J.W.; Nachtkamp, K., "Water-borne Polyurethanes," K.C. Frisch; Klempner, D.; Eds., *Advances in Urethane Technology*, **1987**, *10*, 121.

22. Kaczmarski, J.P.; Fernando, R.H.; Glass, J.E.; *Journal of Coatings Technology,* **1993**, *65 (818)*, 39.

23. Wetzel, W.H.; Chen, Mao; Tarng, M.R.; Kaczmarski, J.P.; Lundberg, D.J.; Ma. Zeying; Alahapperuna, K. and Glass, J.E.; in *Hydrophillic Polymers: Performance with Environmental Acceptability*, Advances in Chemistry Series 248, Glass, J.E., Ed., American Chemical Society, Washington, DC, 1996. Chapters 10, 17, 24.

24. Ma, Z.; Chen, M. and Glass, J.E.; *Colloids and Surfaces*, **1996**, *112*, 163.

25. Heijboer, J.; *Brit. Poly. J.*, **1969**, *1*, 3.

26. Lundberg, R.D.; Encyclopedia of Chemical Technology, Supplement Vol., John Wiley & Sons, Inc. 1984. p. 546.

27. Yang, W.P., *Proc. ACS Div. Polymer Materials: Science and Engin.* **1992**, *66*, 216.

28. Velankar, S; Yang, C.Z.; Couper, S.L.; *Polymer Preprints,* **1996**, *37*(1), 400.

Chapter 2

A Perspective on Resins
for Aqueous Coatings

John L. Gardon

Research and Development, Akzo Nobel Coatings, Inc., Troy, MI 48007

The binders in traditional aqueous coatings are **latexes made by free radical polymerization** in water. The modern seeded semibatch synthesis allows precise particle size control and versatile modification of the morphology of latex particles. Core-shell structures provide unique film formation characteristics, gloss, rheology and crosslink acceptance. These advances in the classical latex technology have a sophisticated scientific foundation. In contrast, various other rapidly evolving classes of aqueous resins are based on largely empirical developments. **Polyurethane latexes** are the products of the reaction between isocyanate terminated prepolymers and aqueous diamines and, just as classical latexes, have very high molecular weights. A great variety of other **aqueous dispersion polymers** are synthesized in nonaqueous media, have low molecular weights and are dispersed into water by mechanical means. This class of dispersion polymers encompass alkyds, polyesters, acrylics and epoxies and forms films only when crosslinked. **Novel crosslinking chemistry** includes two component aqueous urethanes and epoxies, and systems for crosslinking carboxyl, carbonyl and acryloyl functional polymers.

The need for solvent reduction has led to an exciting technology competition between aqueous, high solids, powder, radiation cure and coil coatings (1-4). For example, modern high solids automotive topcoats are being partially replaced by aqueous and even by powder coatings. Discussion of the competition between the five low VOC technologies is beyond the scope of this paper. It should suffice to describe below issues associated with the rapidly advancing technologies of aqueous binder resins.

This subject matter excludes additives, thickeners, and reactive diluents. Resins which dissolve in water are ignored because all modern aqueous binders belong to one of three major classes of polymeric colloids: classical latexes, urethane latexes and dispersion polymers.

Classical latexes are made by aqueous free radical polymerization. Modern synthesis methods now allow precise particle size control and well defined modification of the morphology of the particles.

Urethane latexes are similar to classical latexes in that they have high molecular weights (>100,000) achieved by reactions in aqueous media and their particle sizes are defined primarily by the chemistry of the aqueous reactions. However, they are products of polycondensation between isocyanate terminated prepolymers and aqueous diamines and are in this respect fundamentally different from classical latexes.

Dispersion polymers differ greatly from latexes. They have low (<20,000) molecular weights and are synthesized in nonaqueous media. Their particle sizes are strongly dependent on the shearing forces employed as they are mechanically dispersed into water.

Semantics is often a problem in the coatings industry. The three classes of polymeric colloids are known by various synonyms and it is difficult to maintain consistent terminology. This paper identifies as **latexes** a class of polymers which are also known as "emulsion polymers", "primary emulsions", "PADs" (polyacrylic dispersions), "PUDs" (polyurethane dispersions), and simply "dispersions". The term **dispersion polymers** is used in this paper as a synonym for "water-reducibles", "secondary emulsions", plain "emulsions", "colloidal solutions" and "hydrogels".

End Uses of Latexes

In the United States, classical latex-based architectural coatings are sold at an annual volume of about $ 4-5 billion; this is about 75 % of all architectural coatings (5). The dominance of latexes in low-to-medium gloss architectural coatings dates back over three decades, and is the consequence not of environmental regulations, but of superior performance relative to solvent-borne alkyds. In Europe, unlike in the United States, high gloss architectural paints are very popular, resulting in a higher utilization of solvent-borne alkyds rather than aqueous latexes. Extensive use of classical latexes also predates environmental regulation in textile, paper and leather finishes and industrial maintenance coatings.

New paint applications for latexes have been driven by environmental regulations and, coincidentally, by increased requirements for paint quality (6,7). In almost all instances the new latex paints exceed the quality of the low solids solvent-borne paints they replace. The now numerous industrial and specialty latex coatings tend to be based on resins far more sophisticated than the older architectural latexes. Phase separated acrylics, rheology controlled acrylics and urethane latexes provide paint properties heretofore beyond the range of classical latexes. These new applications include business machine coatings, coil coatings, industrially applied wood coatings (furniture, paneling, kitchen cabinets, joinery), general metal primers, automotive basecoats, coil coatings, floor coatings, traffic paints, adhesives and inks.

The Scientific Basis of Modern Industrial Production Methods for Classical Latexes

Emulsion polymerization is recognized as a well defined scientific discipline (8-12) originating with the epochal paper of Smith and Ewart (13). Their set of mechanistic assumptions and mathematical approximations accurately apply only to certain simple recipes involving single shot isothermal reactions as illustrated in Table I. The theory accurately predicts all measurable properties: particle size, molecular weight and time dependence of conversion. The understanding of such simple systems gives useful insights for the more complex industrial processes which are quite different from the model recipe.

Table I. Typical Classical Latex Recipe

Methyl Methacrylate	40
Water	60
$K_2S_2O_8$	0.1
Sodium Lauryl Sulfate	0.17

Isothermal Polymerization at 55° C

| Final Particle Diameter | 106 nm (electron microscopy) |

| Final Molecular Weight | 4.11 million |

Conversion-Time Relationship

Minutes	10	20	40	50	70	100
Conversion (%)	7	14	42	58	90	98

The part of the theory dealing with particle nucleation is discussed below. This applies only to recipes with anionic surfactants present at levels above the critical micelle formation concentration. Additionally, the monomer must be a poor chain transfer agent (not vinyl acetate and vinyl chloride) and a good solvent for its polymer (not acrylonitrile and vinylidene chloride). No comonomer should yield a water soluble homopolymer. The monomer/water ratio must be at least above 10/90. The initiator must be water soluble and have a half life of a few hours.

The theory assumes that during the interval of particle nucleation the organic phase is comprised of monomer droplets, monomer swollen micelles and newly nucleated latex particles. The surface area of the droplets is negligibly small so that surfactant adsorbed on their surfaces and radicals entering them can be ignored. The droplets act as an inert monomer reservoir. Radicals are being formed in the aqueous phase at a constant rate **R**. Radicals either enter monomer swollen micelles nucleating new particles or enter existing particles without causing *soap*

particle nucleation. As long as micelles are present, the particle surfaces remain saturated with surfactant molecules. Consequently the interfacial area between the organic phase, i.e. particles and micelles, and water remains constant and equals the size S of the monolayer that can be formed by the surfactant molecules. Initially, when there are no particles present, the rate of particle nucleation, dN_t/dt equals **R**. After particles are formed and grow, their surface area, A_t becomes significant and the fraction of **R** radicals which do not nucleate new particles becomes A_t/S. The corresponding differential equation describing particle nucleation is shown below:

$$dN_t/dt = R\,[\,1 - (A_t/S)\,]$$

The growing surface of particles, A_t, is calculable from the growing particle number, N_t and a parameter **K** describing the volume growth of a particle containing one radical. This volume growth is the result of polymerization in particles which remain saturated by monomers.

Gardon's extension (14) of the Smith Ewart theory contains three exactly defined parameters. The initiator parameter **R** is the product of initiator concentration, decomposition rate constant and Avogadro's number. The surfactant parameter **S** is the product of surfactant concentration, area per molecule and Avogadro's number. The monomer parameter **K** is calculable from the propagation constant and the monomer volume fraction in the particles, ϕ_m.

Particle nucleation stops when all surfactant becomes adsorbed on latex particles; that is, when A_t becomes big enough to equal S. The theory provides the final number of particles **N,** the diameter of the particles **D** and the critical volume of converted polymer P_{cr} when particle nucleation stops.

$$N = 0.208\ S^{0.6}\ (R/K)^{0.4}$$
$$D \sim (1/N)^{1/3}$$

The proportionality in the equation above is exactly defined by the geometry of the system and is not spelled out here.

$$P_{cr} = 0.209\ S^{1.2}\ (K/R)^{0.2}\ (1-\phi_m)$$

The aesthetic appeal of the theory is that it predicts colloidal properties resulting from chemical reactions. The postulated mechanism and the mathematical approximations are very simple. The three parameters in the equations are not adjustable and are determined from independent experiments not involving emulsion polymerization. Under these circumstances it is satisfying that the theoretical predictions of the experimental **N** or **D** values are of the right order of magnitude for a large body of data (12). However, the theory goes beyond predicting only the right order of magnitude of **N** or **D**. There is another large body of data obtained with styrene, methyl methacrylate and chloroprene (12) showing quantitative agreement between theory and experiment as illustrated in Table II,

Table II. Examples for Particle Size Results

Monomer/Water Ratio: 40 / 60 (or 20 / 80*)
Initiator: $K_2S_2O_8$ (KPS)
Surfactant: SLS

	Temp.	% on Water		Diameter (nm)	
	^{o} C.	SLS	KPS	Theory	El. Micr.
Methyl	40	3.44	0.905	56	72
Methacrylate	40	2.00	0.035	72	86
	40	0.514	0.206	104	102
	40	0.128	0.081	157	190
	40	0.10	0.018	198	191
	55	0.244	0.165	124	115
Styrene	40	0.66	0.23	109	96
	60	0.23	4.3	84	86
	*80	0.124	0.125	66	74
	*80	0.5	0.125	50	50

taken from Gardon (15). It is not known why there are discrepancies in some sets of data and none in others.

An important prediction of the theory is that the value of P_{cr} is very small, so that in most practical recipes particle nucleation must stop at a few percent conversion (14,15). There is a large body of experimental data exhibiting constancy of particle numbers after a very short initial reaction interval (12). This is illustrated in Figure 1 based on the data taken from Klevens (15,16). Here there is no variation in the ratio between particle diameter and the cube root of conversion, i.e. in the particle number.

Numerous authors (14, 17-20) have theoretically and experimentally proven that there are nucleation mechanisms other than micellar. Homogeneous nucleation is by precipitation of radicals initially growing dissolved in water or by formation of transient surfactant molecules. According to theoretical considerations (14) this nucleation mechanism should lead to larger **D** and smaller P_{cr} values than micellar nucleation. Particles can also be nucleated by radicals entering monomer droplets. Finally, there is "negative nucleation" when particles partially coagulate.

Production experience indicates that micellar particle nucleation is best for achieving reproducibility of particle sizes. It is easy and desirable to avoid non-micellar particle nucleation by dosing an adequate amount of anionic surfactant in the precharged seed, as explained below.

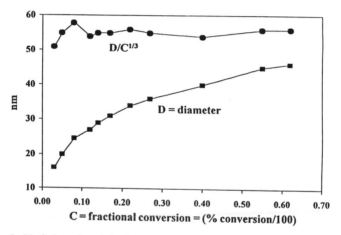

Figure 1. Variation of particle diameter with conversion in the copolymerization of styrene and isoprene 36/54 at 50° C, at an initial monomer/water ratio of 36/54 and with 2.15% potassium myristate in water. The ratio between the diameter and the cube root of fractional conversion indicates constancy of particle number.

Seeded Semibatch Latex Synthesis

Reproducible production of batches in several ton size at 40-55 % solids content requires good control of particle size, of colloidal stability and of reaction temperature, among many other variables.

Accurate control of the reaction temperature is quite difficult in a large production reactor if the recipe calls for a single charge of all reactants as in Tables I and II and in Figure 1. Such a single shot batch process is now obsolete. In modern industrial practice the rate of exothermic heat generation is adjusted by feeding the monomer to the reactor in a slow continuous stream. Uniform distribution of the monomer can be enhanced by feeding it in the form of a water-in-oil emulsion which inverts in the reactor. This methodology has the additional advantage of maximizing colloidal stability by continuous introduction of submicellar levels of surfactants combined with the gradual addition of monomer.

Gradual addition is superior to the single charge process for comonomers with greatly differing reactivity ratios because it yields more uniform monomer composition within the particles.

Early versions of the gradual addition recipes suffered from poor particle size control. It is now recognized that it is best to place into the reactor a relatively dilute and fully converted seed latex before starting the gradual addition of the rest of the monomer.

The seed can be "in situ" or "external". The "in situ" version is prepared in the same reactor where the latex product is finalized by gradual addition. The "external" seed is separately prepared in a large quantity, and aliquots of it are used for multiple batches of latex products.

It is often advantageous to prepare the seed to contain 10-20% solids by recipes analogous to those of Tables I and II, but modified by addition of 0.5 to 2% acrylic or methacrylic acid to the monomer mix for enhanced colloidal stability. The precharged seed is best diluted to about 3 to 10% so that the seed should be 3 to 7% of the final latex solids. As the second stage monomer is reacted in the presence of the seed, the particle number in the reactor must not change consistent with the theory presented earlier. If in such a process the particle number decreases by partial coagulation or increases by post-nucleation, the recipe is not robust and should not be expected to have batch-to-batch reproducibility. It is the general experience that in robust industrial recipes the ratio of final-to-seed particle diameters exactly equals the cube root of the ratio between final and seed solids.

A comment on initiator and surfactant selection is in order. An anionic surfactant and an aqueous thermal initiator must be used in seed synthesis for best results. In the second stage reaction the choice of initiator or surfactant should be optimized for the composition of monomers which are gradually fed to the reactor. At this stage, redox initiators and nonionic surfactants can also be used.

It should also be noted that seeded semibatch synthesis is also beneficial for monomers which, as noted earlier, do not quantitatively comply with the assumptions underlying the mathematics of the Smith-Ewart nucleation theory. For example, this synthesis method was found to yield exceptionally monodisperse poly(vinyl acetate -butyl acrylate) latexes at 50% solids (21).

Phase Separated Latexes

Two-stage polymerization for particle size control by seeding is one of the many strategies for modifying the morphology of the latex particles. The composition of the polymer near to the center and near to the surface of the particles can be made to differ in a controlled fashion by changing the monomer composition of the feed during gradual addition. Bassett's paper (22) provides an excellent overview of this complex subject.

In the so called power feed, the change in the composition of the feed monomer is continuous. For sake of simplicity this is ignored here. The discussion below will focus on the core-shell morphology resulting from an abrupt change in the composition of the feed. Most industrial recipes are seeded requiring three stages for a core-shell synthesis. The discussion is simplified below by implicitly assuming that the first stage polymer includes the seed.

The shell thickness varies with the particle size of core-shell latexes. The properties of such latexes crucially depend upon the shell thickness so that precise particle size control by seeded synthesis is mandatory for consistent performance in paints.

If the second stage monomer mix is added in such a manner that its conversion rate is slower than the rate of addition, it will swell the latex particles. The conversion of the second stage polymer within the particles causes phase separation because the first and second stage polymers are usually incompatible. This phase separation generally results in core-shell morphology or, under special circumstances, in microdomains having complex shapes (22,23).

Whether the first stage polymer becomes the core or inverts to become the shell depends on whether it is more or less polar than the second stage. Surface thermodynamics mandate that the more polar or hydrophilic of the two polymers ends up in the shell (24,25). However, a more polar first stage polymer cannot invert if it is crosslinked (26) as in the synthesis of the opacifying latexes of the Rohm and Haas Company (27).

The rate of addition of the second stage monomer can be programmed to be slow enough so that the interior of the latex particles becomes starved of monomers. The diffusion coefficient is about 7 orders of magnitude smaller in an unswollen polymer than in a swollen one with representative values for the diffusion coefficients 10^{-13} and 10^{-6} cm^{2}sec^{-1} respectively. It would follow that under monomer-starved conditions the diffusion of unconverted monomers into particles is suppressed and the locus of polymerization is at the surface of the particles (28). The result is very sharp core-shell definition, lowered molecular weight for the shell polymer, and reduction of the chances for phase inversion.

It seems that most current product development in industry involves exploitation of the properties achievable by phase-separated latexes. This paper is confined to basic principles, without an extensive patent or literature survey. The utility of core-shell latexes is illustrated below by four examples.

A hard core can impart good print and block resistance, while the minimum film formation temperature is determined by the composition of the soft shell (29). Physical blending of hard and soft latexes cannot match the balance of properties attainable by the core-shell structure.

Often melamine or other crosslinkers cannot penetrate the interior of the particles. To obtain uniform crosslinking, it is a good strategy to crosslink the core by a multifunctional unsaturated monomer during synthesis, and incorporate the hydroxyl or other suitable groups into the shell for crosslink acceptance during film formation (30).

Special latex binders impart to automotive basecoats pseudoplastic rheology and good strike-in protection against the solvent-borne clearcoats. This is accomplished by a hydrophobic crosslinked core and an alkali-swellable hydrophilic shell (31).

Newtonian rheology is the key property of the so called "rheology controlled" latexes pioneered by the S.C. Johnson Co. (32-35). The second stage synthesis of a hydrophobic composition is carried out in the presence of a first stage "support resin", typically a copolymer of 75% styrene and 25% methacrylic acid, having a weight average molecular weight of about 8000 and being water soluble when neutralized. Presumably phase inversion takes place during synthesis. The ratio between first and second stage polymers is about 25 to 75. Addition of a fractional percentage of surfactant during synthesis reduces the particle size into the 60-100 nm range. There is no obvious explanation for the unique rheology of this commercially important class of latexes. Their utility for aqueous inks, adhesives and pigment dispersants is well established and is evolving for aqueous high gloss architectural and various industrial paints.

Polyurethane Latexes

Typical synthesis methods are shown in Table III taken from review papers (35-37). It is misleading to call this class of polymers polyurethanes. Actually they contain both urethane and urea groups in the backbones.

Industrial scale manufacturing is in reactors similar to those used for acrylic latexes, but with the difference that the stirrers must be equipped with high torque motors because the prepolymer solution is often very viscous. The reactor must also be well cooled because the chain extension reaction is quite exothermic.

The isocyanate terminated prepolymer must be dissolved in a water-miscible solvent for effective incorporation into the polymer. Acetone is occasionally used and is stripped from the system after the reaction is complete. The most often used solvent is N-methyl pyrrolidone (NMP).

Electron micrographs indicate that the particles have a microporous morphology possibly caused by the NMP. The presence of NMP may also account for the finding that the minimum film formation temperatures of polyurethane dispersions (PUDs) are lower than that of acrylics with equal hardness (38).

The particles can exhibit core-shell morphology with the shell having higher molecular weight and higher urea functionality than the core. This effect was found to be quite pronounced with isophorone diisocyanate-based PUDs (39).

Most paint grade PUDs are stabilized by carboxyl or polyethylene oxide groups grafted to the backbones. The particle sizes are in the 100-300 nm range and decrease with increasing stabilizing group content (38).

Table III. Preparation of Polyurethane Dispersions

STEP 1:   -Prepare a carboxylic NCO-terminated prepolymer in N-methyl
pyrollidone or acetone

-Add 2 moles of diol (polyester, polyether or polycarbonate, M_W: 500 - 2000)

plus   -1 mole of dihydroxy propionic acid

plus   -4 moles of diisocyanate (Note: OH/NCO = 6/8)

RESULT: Molecules ideally containing:

- 2 chain end NCO's
- 1 COOH
- 2 urethane groups within chain

STEP 2:   -Neutralize COOH with a tertiary amine

-Emulsify prepolymer into water

-Add a di-primary amine (e.g. hexamethylene diamine) at an NCO/NH_2 ratio of 1/1

RESULT: -Anionically stabilized latex with a multiplicity of urethane and urea groups in the main chain

-MW > 100,000

-Particle diameter typically 100-300 nm

The great variety of available diol, isocyanate and amine raw materials provides much freedom to adjust durability, mechanical properties and glass transition temperature.

It is easy to graft acrylic unsaturated groups or fatty acid side chains onto PUDs for UV or oxidative cure. In most practical applications the PUDs are left either uncrosslinked or are crosslinked through the carboxyl groups by polyaziridine or carbodiimide.

Recently "hybrid" acrylic-urethane latexes have been made by simultaneous polymerization of acrylic monomers and chain extension of urethane prepolymers. The resulting structure within the particles is reminiscent of interpenetrating network polymers. The mechanical properties of such hybrid films exceeded those obtainable by blending acrylic latexes with PUDs (40).

The urethane latexes are tougher than acrylics: at equal hardness they are more flexible, and at equal softness they are less tacky. Generally, the PUDs have better solvent resistance than acrylic latexes, although they are inferior in their resistance to alcohol. Of course, PUDs are much more expensive than acrylic latexes and are rarely used as sole binders. They are most cost effective in paints when blended with acrylics.

Preparation and Properties of Aqueous Polymer Dispersions

Low molecular weight (<20,000) polymers are synthesized in nonaqueous media and subsequently dispersed into water for applications in coatings. Unlike latexes, such polymers do not lacquer-dry when used as sole binders, and depend on chemical crosslinking for film formation. Alkyds, acrylics, epoxies and polyesters are the most important resins used in this manner.

The particle size of latexes is independent of the stirring rate as long as the system is kept homogeneous without causing shear flocculation. In contrast, the particle size of dispersion polymers crucially depends on the energy expended for breaking up large droplets and it makes a difference whether this happens under laminar or turbulent flow conditions (41). As a matter of practical convenience, Cowles type high speed dispersers are often used if the viscosity of the organic phase and the interfacial tension are low enough.

In particular alkyds, acrylics and polyesters can be designed for easy dispersability. For this, very low molecular weights (<3000) and incorporation of some hydrophilic groups are essential. The preferred hydrophilic group is carboxyl and dispersion polymers typically have acid values in the 30-50 range. These polymers must be neutralized to pH $\geq$ 7 for obtaining good dispersions. The ease of dispersability can be further enhanced by adding the polymer dissolved in a water-miscible organic solvent. Direct emulsification often yields 100-300 nm particle sizes at about 50% solids and less than 10% solvent contents.

In dispersion systems, which are more sophisticated than those indicated above, carboxyl groups are complemented by nonionic ethylene oxide side chains grafted to the polymer. Modification by anionic and nonionic "external" surfactants is beneficial for very hydrophobic polymers. Efficient crosslinking of the paint films can satisfactorily counteract the water sensitivity of the stabilizing groups.

Incorporation of 10,000-20,000 molecular weight epoxies into water cannot be done with dispersing equipment routinely available in a paint plant. One option is direct emulsification by homogenizers or colloid mills. Alternately, special high shear dispersers are used for the inversion process. In the inversion process water is gradually added to the hot organic phase, first creating a water-in-oil emulsion. Gradual increase of the water content causes the viscosity to rise to a maximum value. At this point the emulsion inverts to an oil-in-water type. On further dilution with water the viscosity decreases. Dispersions made in this manner typically have 1000-2000 nm particle size at 40% solids. The epoxy resins suitable for emulsification are acid modified, either by grafting a carboxylic acrylic resin or by reaction with phosphoric acid. Before emulsification the acid groups must be neutralized. Presence of water miscible organic solvents and "external" surfactants is helpful.

The rheology of the aqueous polymer dispersions depends on solids and cosolvent content. Figure 2, taken from Williams (42), shows the viscosity behavior of a chain-stopped aqueous alkyd originally prepared at 70% solids in a blend of 20 parts of butyl cellosolve and 80 parts of water. Dilution with water yields the top curve showing maxima and minima. Dilution with a blend of butyl cellosolve and water yields monotonically decreasing viscosity, characteristic of solution polymers.

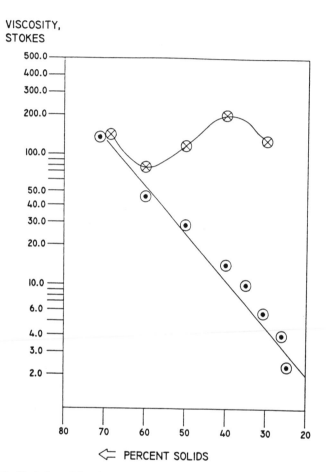

Figure 2. Variation of the viscosity of an aqueous alkyd dispersion with concentration. The dispersion was prepared at 70% solids in a butyl cellosolve/ water blend of 20/80. Dilution with the same solvent blend provides the lower curve, dilution with water alone yields the upper curve.

Hill (43-45) obtained a similar pattern of results with acrylic dispersion polymers and interpreted the irregular shape of curves obtained on dilution with water in terms of dissociation of the ionic groups on the colloidal particles.

End Uses of Aqueous Polymer Dispersions

Aqueous alkyds are successfully used in Sweden as sole binders for high gloss architectural paints (46). Ostberg (47,48) extensively studied the influence of surfactants on colloidal and shear stability. In the United States aqueous alkyds are mostly used as additives to latexes for improving chalk adhesion of architectural paints and for wetting of dirty metal substrates by industrial paints. The limited long term hydrolytic stability of aqueous alkyds puts them at a disadvantage, so that in the United States "rheology controlled" acrylic latexes have become the preferred vehicles in high gloss aqueous architectural paints.

The two most significant applications of dispersion polymers is in electrocoatings and in interior coatings for metal cans. Electrocoatings hardly require discussion here. The aqueous beer-and-beverage can liners are melamine crosslinked acrylic-epoxy graft copolymers with exceptionally good atomization and metal wetting (49-51). Epoxy emulsions crosslinked with melamine are now being used as high bake sanitary can liners. Amine-crosslinked epoxy emulsions are binders in ambient cure primers for aircraft and for industrial maintenance coatings. Generally, coatings based on dispersion resins atomize better than latexes and can achieve higher gloss values. Compared to high solids coatings, aqueous dispersion coatings can be applied at a much lower VOC.

It is noteworthy that dispersion polymer resins often have molecular weights and structures similar to resins used in high solids coatings. Recently, Jones (52) developed solventless coatings containing reactive diluents whose application properties could be improved by dilution with water. Here the high solids and aqueous dispersion technologies actually overlap.

Crosslinking Systems

The most surprising recent development is that two component aqueous urethanes became practical. The pigments are ground into hydroxyl functional aqueous polyester or acrylic polymer dispersions. The isocyanate used is either a conventional trimer of hexamethylene diisocyanate or the same modified for dispersability with grafted poly(ethylene oxide). The isocyanate must be diluted with a water miscible organic solvent to match the viscosity of the aqueous polyol dispersion before blending these two components. In the experiments described by Bock (53), carbon dioxide evolution due to reaction between water and isocyanate is initially suppressed, and can be detected only 30 minutes after the isocyanate is incorporated into the aqueous paint. The useful potlife may extend up to three hours. The main reaction between the two components occurs after the coating is applied and the water evaporates.

It now seems that high solids resins modified for aqueous dispersion application can yield aqueous two component urethane coatings with properties very similar to their high solids counterparts, but with much lower VOC. This is made plausible by the new understanding of the film formation mechanism, elucidated by Hegedus (54). In the first critical step, the hydrophobic isocyanate is emulsified by the resin and is absorbed into the resin particles. Thus the polyol and the isocyanate are blended in the organic phase before the coating is applied to the substrate.

Independently from Hegedus, Walker (55) established an analogous reaction mechanism for two component aqueous epoxy-amine coatings. Accordingly, the amine curative becomes absorbed into epoxy particles before the coating is applied.

These developments allow the use of the two highest performance ambient temperature curing systems in aqueous coatings. Other ambient cure crosslinking systems are reviewed elsewhere (56-60); here only a brief selective summary is presented.

Carboxyl functional latexes respond well to crosslinking by polyaziridines, carbodiimides and diepoxides in systems having few days pot life. Alternatively, blends of carboxyl functional latexes with amine complexes of zinc or zirconium salts are stable in the liquid state. On drying, the amine volatilizes and the liberated metal cation crosslinks the latex.

Diamines or a dicarbohydrazides very effectively crosslink latex copolymers of acetoacetoxyethyl methacrylate or diacetone acrylamide. The combination of diacetone acrylamide with dicarbohydrazide is of particular importance because it has practically infinite potlife.

Polyester or polyurethane dispersion polymers can be modified with unsaturated acrylic pendant groups. Under alkaline catalysis in a Michael reaction, these react with the activated methylene groups of multifunctional malonate or acetoacetate esters yielding crosslinks by newly formed carbon to carbon bonds. Alternative proton donor crosslinkers are diamines or dithiols which, in a pseudo Michael reaction, crosslink acryloyl functional polymers by forming a multiplicity of diamine or thioether links respectively. Acryloyl functional polymers can also be cured by UV irradiation without using external crosslinkers.

The cure systems described above are effective at temperatures of about 20°C or higher. There is a special need for an acid catalyzed cure at around 65°C for wood kitchen cabinet and office furniture coatings because acidity in the coating maintains the color of the wood. Typically, methylolated urea is used to crosslink hydroxyl functional binders in the presence of a few percent paratoluene sulfonic acid in coatings having a few hours potlife. Such coatings emit unacceptably high levels of formaldehyde. New formaldehyde free curing systems are latexes containing the dimethyl or the diethyl acetal of acrylamido butyraldehyde. For crosslinking, the diacetal groups can either self-condense to yield ring structures or react with hydroxyls (61).

Curing under high bake conditions (>120°C) continues to involve hydroxyl functional polymers and amino resin crosslinkers. Cathodic electrocoating is the only exception; for this the crosslinker of choice is a blocked isocyanate.

Conclusions

The panorama of resins and crosslinkers for aqueous coatings is awe inspiring. No other area of paint technology benefits from a comparable variety and sophistication of chemical approaches. Aqueous coatings already command a large share of the total paint market. It is certain that this share will be further increased.

Literature Cited

1. Wicks, Z.W.; Jones, F.N.; Pappas, S.P. *Organic Coatings: Science and Technology*, Volume II: *Applications Properties and Performance*; John Wiley & Sons, Inc.: New York, **1993**.
2. Gardon J.L. *Polymer Science and Engineering: The Shifting Research Frontiers* edited by a committee of the National Research Council, Stein, R.S. Chairman, National Academy Press: Washington, D.C., **1994**, pp 98-100.
3. Gardon, J.L. *Current Trends in Coating Technologies* (a) *Coatings & Adhesives-Prognosis for the Future*, Symposium of the Business and Management Division, American Chemical Society; Chicago: August **1995**, pp 71-78, (b) paper presented at the FATIPEC Meeting, Budapest **1993**.
4. Gardon, J.L.; Prane, J.W. *Nonpolluting Coatings and Coating Processes;* Plenum Press: New York, **1973**.
5. U.S. Paint Industry Data Base, National Paint and Coatings Association, Washington, D.C. **1992**.
6. Padget, J.C. *J. Coat. Technol.* **1996**, *66*, 839, 89.
7. Doren, K.; Freitag, W.; Stoye, D. *Water-Borne Coatings;* Hansen Publisher: New York, **1994**.
8. Gilbert, R.J. *Emulsion Polymerization: A Mechanistic Approach;* Academic Press: New York, **1995**.
9. Vanderhoff, J.W. *J. Polymer Sci.: Polymer Symposium* **1985**, *72*, 161.
10. Odian, G. *Emulsion Polymerization, Principles of Polymerization*, 3rd ed.; John Wiley & Son: New York, **1991**, pp 335-355.
11. Klein, A. *Latex Technology, Kirk Othmer Encyclopedia of Chemical Technology*, Vol.14, 3rd ed., **1981**, pp 82-97.
12. Gardon, J.L. *Emulsion Polymerization*, High Polymers, Volume 29, *6: Polymerization Processes*, Schildknecht, C.E.; Skeist, I. Eds.; Wiley Interscience: New York, **1977**, pp 143-196.
13. Smith, W.V.; Ewart, R.H. *J. Chem. Phys.* **1948**, *16*, 592.
14. Gardon, J. L. *J. Polymer Sci.* **1968**, *A-1*, 6, 623.
15. Ibid. 643.
16. Klevens, H.B. *J. Colloid Sci.* **1947**, *2*, 365.
17. Hansen, F. K. *Polymer Latexes: Preparation, Characterization and Application*, A.C.S. Symposium Series 492, Daniels, E.S.; Sudol E.D.; El-Asser, M.S., Eds.; **1992** pp 12-27.

18. Fitch, R.M.; Tsai, C.H. *Polymer Colloids,* Fitch, R.M. Ed.; Plenum Press: New York, **1972**, *73*, 103.
19. Hansen, F.K.; Ugelstad, J. *J. Polymer Sci.*, Polymer Chem Ed., **1979**, *17*, 3069.
20. Ugelstad, J.; El-Asser, M.S.; Vanderhoff, J.W. *J. Polymer Sci.,* Polymer Letters ed. **1973**, *11*, 503.
21. Vandezande, G.A.; Rudin, A. Monograph *Ref.17*, 134-144.
22. Bassett, D.R.; Hoy, K.L. *Emulsion Polymers and Emulsion Polymerization* Bassett, D.R.; Hamielec, A.E. Eds., A.C.S. Symposium Series 165, American Chemical Soc.: Washington, D.C., **1981,** pp 371-384.
23. Chen, Y.C.; Dimonie, V.; El-Assar, M.L. *J. Appl. Polymer Sci.* **1989**, *42*,1049.
24. Sundberg, E.J.; Sundberg, D.C. *J. Appl. Polymer Sci.* **1993**, *47*, 1277.
25. Winzor, C.L.; Sundberg, D.C. *Polymer* **1992**, *33*, 20, 4269.
26. Lee, S.; Rudin, A. monograph of ref. 6, pp 235-254.
27. Kowalski, A.; Vogel, M.; Blankenship, R.M. USP 4 468 498, **1984**.
28. Gardon, J.L. *J. Polymer Sci.*, **1973**, *11*, 241.
29. Devon, M.; Gardon, J.L.; Roberts, G.; Rudin, A. *J. Appl. Polymer Sci.* **1990**, *39*, 2119.
30. Leiner, H.H.; Gelarden, D.R.; Gardon, J.L. USP 4 174 336, **1979**.
31. Backhouse, A.J. European Patent 0 038 127, **1984**.
32. Esser, R. *International Patent Publication* **1993**, WO93/16133.
33. Morgan, L.W. USP 4 894 497, **1990**.
34. Kielbauch, R.A.; Tsaur, S.L. USP 4 839 413, **1989**.
35. Tsaur, S.L. USP 4 820 762, **1989**.
36. Rothauser, J.W.; Nachtkamp, K. *Waterborne Urethanes, Advances in Urethane Science and Technology;* Frisch, K.C.; Klempner, D. Eds., **1987**, *10*, 121-162.
37. Dietrich, D. *Progress Org. Coatings* **1981**, *9*, 281.
38. Satguru, R.; McMahon, J.; Padget, J.C.; Coogan, R.G. *J. Coat. Technol.* **1994**, *66*, 830, 47-55.
39. Lee, H.T.; Hwang, Y.T.; Chang, N.S.; Huang, C.C.T.; Li, H.C. Proceeding of the 22nd Waterborne, High-Solids and Powder Coating Symposium, The University of Southern Mississippi, **1995**, 224-233.
40. Hegedus, C.R.; Kloiber, K. Ibid., 21st Symposium, **1994**, pp144-167.
41. Walstra, P. *Principles on Emulsion Formation*, Preprints on Conference on the Preparation of Dispersions, Laven, J.; Stein, H.N. Eds., Veldhoven, The Netherlands, **1991**, pp 77-92.
42. Williams, C.J. *Formulating with Waterborne Resins*, Reichhold Chemicals: Durham, NC, **1994**.
43. Hill, L.W.; Wicks, Z.W. *Progress in Organic Coatings*, **1980**, *8*, 161.
44. Hill, L.W.; Brandenburger, L.B. Ibid., **1975**, *3*, 361.
45. Hill, L.W.; Richards, B.M. *J. Coat. Technol.* **1971**, *51*, 654, 59.
46. Rodsrud G.; Sutcliffe, J. *Alkyd Emulsions: Solvent Free Binders for the Future*, paper 14, Proceeding of the 3rd Nurnberg Congress, **1994**.

47. Ostberg, G; Bergenstahl, B.; Hulden, M. *J. Coat. Technol.* **1994,** *66,* 832, 37.
48. Ostberg, G.; Bergenstahl, B. *J. Coat. Technol.* **1994,** *66,* 838, 37.
49. Parekh, G.G.; Sharp, L.J.; Thompson, K.R., p 139 in Ref. 38.
50. Woo, J.T.K.; Toman, A. *Prog. Org. Coat.* **1993,** *21,* 371.
51. Evans, J.M. USP 4 212 781, **1980.**
52. Jones, F.N. *J. Coat. Technol.* **1996,** *68,* 852, 25.
53. Bock, M; Petzoldt, J., p 502 in Ref. 38.
54. Hegedus, C.,A.; Gilicinski, A.G.; Haney, R.J. *J. Coat. Technol.* **1996,** *68,* 852, 51.
55. Walker, F.H.; Everett, K.E.; Kamat, S., pp 88-95 in Ref. 39.
56. Warson, H. *Emulsion Polymerization,* Piirma, I.; Gardon, J.L. Eds., A.C.S. Symposium Series 24, Am. Chem. Soc., Washington, D. C. **1975,** pp 330-340.
57. Athey, Jr., R.D. *Emulsion Polymer Technology,* Marcel Dekker, Inc.: New York, **1991;** pp 93-102, 223-236.
58. Wicks, Z.W.; Jones, F.N.; Pappas, S.P. *Organic Coatings: Science and Technology,* Volume I, Film Formation, Components and Appearance; John Wiley & Sons: New York, **1992;** pp 212-228.
59. Noomen, A. *Prog. Org. Coat.* **1989,** *17,* 27.
60. Eslinger, D.R. *J. Coat. Technol.* **1995,** *67,* 850, 45.
61. Kjellgvist, K.; Rassing, J.; Wessler B. Proceedings of the XIXth International Conference in Organic Coatings, Science and Technology, Athens, **1993,** p 281.

Chapter 3

Progress in Predicting Latex-Particle Morphology and Projections for the Future

Yvon G. Durant and Donald C. Sundberg[1]

Polymer Research Group, Department of Chemical Engineering, University of New Hampshire, Durham, NH 03824

The ability to predict the structure of composite latex particles has improved markedly since the introduction of free energy analyses of the phase separated particles. Such analyses have highlighted the importance of the interfacial tensions at the polymer/water and polymer/polymer interfaces. The variation of these interfacial tensions with important parameters such as surfactant level, polymer type, monomer concentration, and initiator end groups can be utilized to predict equilibrium particle morphology in a straightforward manner. All possible structures can be displayed on a surface energy map and the morphology with the lowest free energy can be easily located. The present challenge is to be able to determine the simultaneous effect of all of the important variables on the various interfacial tensions. The future challenge is to move away from the thermodynamic equilibrium constraints of the free energy approach and to describe the dynamic changes of the internal particle phase structure resulting from polymerization reaction within the particle.

The importance of latex particle morphology to finished goods properties has been known for a long time. Two good examples of this are impact modified thermoplastics, such as ABS, and many types of water borne coatings. However it has only been in the past decade or so that there has been much progress reported in the predictive capability for latex morphology, while the need to learn what controls the morphology has been recognized for a long time. Experimental approaches to identify the important parameters controlling the morphology (1-10) have been helpful and have also resulted in the identification of particle structures such as core-shell, inverted core-shell, hemisphere, occluded, sandwich, raspberry, etc. Predictive

[1]Corresponding author

approaches (11-23) began to appear later and have principally been restricted to equilibrium conditions. Dynamic analyses are now becoming available (24, 25) but are quite limited at the present time, as expected for such a complex subject. In the present discussion we limit our descriptions to the conditions of thermodynamic equilibrium.

Torza and Mason (26) can be credited with describing the first predictive analysis about 25 years ago. They utilized the spreading coefficients concept to write inequalities which rank the total surface energy of a limited number of particle structures. They then applied this analysis to two component oil droplets dispersed in water and were very successful in predicting the correct morphologies. Nearly 15 years later our group presented a more general free energy analysis (11-13) so as to be able to deal with a wider variety of particle structures. Since that time others (15, 16, 19-22) have made contributions along the same lines. In all of these approaches it readily becomes clear that the limiting factor is the ability to understand the influence of the many experimental variables upon the various interfaces within and at the external surface of the particle. The purpose of this paper is to outline some of the advances in these areas and to show their application to experimental latices.

Free Energy Analysis of Particle Morphology

When there are no elastic forces involved in the latex particle, unlike that existing in crosslinked polymer particles, the free energy analysis amounts to the consideration of the hypothetical change in the Gibbs free energy in transitioning from two bulk phase polymers to a two component particle dispersed in water. The water may or may not contain surfactant or other components such as salts. The above transition involves the formation of interfaces within the particle and at its boundary with the water phase, and the free energy change can simply be written in terms of the interfacial tensions and areas as $\Delta G = \sum \gamma_i A_i$. An efficient way to do such calculations is to consider the change from bulk phase polymers to composite particles of every conceivable shape and to compare the free energies of all of the different particle structures. The preferred equilibrium morphology will be that with the lowest free energy. The free energy change equation for each structure is different from another only due to the different geometries of the particles if the interfacial tensions are assumed to be independent of particle structure, as is the case for most situations. This was shown clearly by Sundberg et al (13) where the free energy equations were presented for a variety of different morphologies. Equation (1) is an example of this for the transition to a core-shell (CS) particle in which the seed latex particle remains as the core and the second stage polymer forms the shell.

$$\Delta G / A_0 = \gamma_{P1/P2} + \gamma_{P2/w}\left(1 - \phi_P\right)^{-2/3} - \gamma_{P1/w} \qquad (1)$$

where $\gamma_{P1/w}$, $\gamma_{P2/w}$, and $\gamma_{P1/P2}$ are the interfacial tensions at the seed polymer/water, second stage polymer/water, and seed polymer/second stage polymer interfaces, respectively. In this equation ϕ is the volume fraction of the second stage polymer in the particle, and A_0 is the surface area oif the initial seed polymer. Similar equations can be written for any other particle structure, although they may be quite complex due to the particular geometry of the structure. Three component particles can also be described by similar equations (23).

A more elegant approach has been developed by Durant (27) in which all fully phase separated, two component particles were geometrically placed on a multi-axis plane. This "topological map" is arranged according to the angles associated with the phase boundary intersections in the particle and is pictured in Figure 1. At the far left of the plane we have a CS particle, while at the far right we have an inverted core-shell (ICS) particle. In between we have any variety of other fully phase separated particles. When the free energies associated with each of these particle structures (one can choose to do many hundreds via computer) are calculated, an energy surface can be constructed and when graphically presented, the differences in particle energies are readily apparent. A computer software, UNHLATEX™, has been developed in order to make the computations rapid and efficient. Figure 2 is a sample graphical output from the software for a particular set of interfacial tensions and at a stage ratio (second stage polymer to seed polymer) of 300%. The circle with a cross inside of it signifies the point at which the energy surface is at its minimum and indicates the preferred particle morphology. In this example it turns out to be an hemisphere particle.

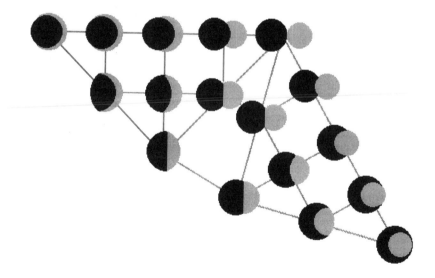

Figure 1. Topological map. Dark phases represent the seed polymer and graphases the second stage polymer.

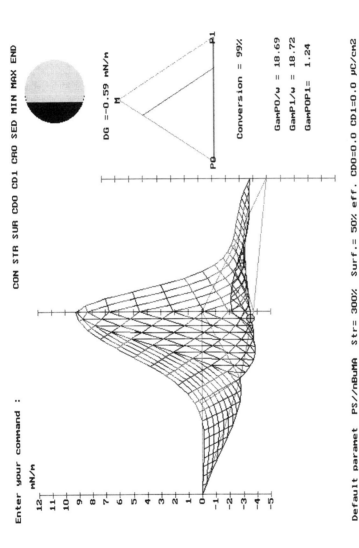

Figure 2. Free energy surface for a polystyrene seed, n-butyl methacrylate system, with a stage ratio of 300% and a concentration of SDS surfactant of 50% of the CMC.

Importance of Particle Interfaces. In all of the above calculation schemes it is readily apparent that no results can be obtained unless one has values for the various interfacial tensions. The real challenge is to be able to determine these values with some confidence for conditions of practical importance. We have spent quite a lot of energy trying to do this for some polymer and monomer systems, choosing to work predominantly with sodium dodecyl sulfate surfactant (SDS) and potassium persulfate initiator (KPS). When any of these materials are changed in the latex formulation, the effects on latex morphology are computationally seen through the changes in the associated interfacial tensions. Below we present a few experimental and computational results for selected systems.

Surfactant Effects. One of the more dramatic effects we have seen has been with a simple artificial latex formed with a copolymer of methyl methacrylate and n-butyl methacrylate and decane using two surfactants varying greatly in their ability to lower interfacial tensions at the organic/water interface. The first is Mexpectin 100S (MXP), a natural extract from citrus fruit, and the second is SDS. As reported in a previous paper (13) and shown in Figure 3, when MXP is used at 0.5 wt. % in water, the particle clearly has a CS structure. Figure 4 shows that when SDS is substituted for the MXP at the same water phase concentration, the particle morphology changes to hemispheres. Thus it is apparent that surfactants cannot necessarily be switched in latex formulations without consideration of possible changes to the particle morphology. Having determined the three interfacial tensions for this system for each of the surfactants (13), the free energy analysis predicts the correct results.

Seemingly more subtle results would be expected if the changes were only in the concentration of surfactant, not in the type of surfactant. This is a very important consideration in normal latex manufacture because the surfactant is always held below its critical micelle concentration (CMC) during the entire seeded latex polymerization. However, changes in the degree of particle surface coverage by the surfactant often occur during polymerization. Figures 5a and 5b show the free energy surfaces for a system with SDS coverages of 25% and 99% of full saturation (i.e. at the CMC) on a composite particle composed of poly(butyl methacrylate) and polystyrene. One can see a shift in the predicted morphology from an hemisphere with a flat interface to an hemisphere with a significant engulfment of the PS by the PnBMA. The variations in the polymer/water interfacial tensions with SDS concentration were modelled from experimental data developed from an interfacial tensiometer using the pendant drop method (28). The polymer/polymer interfacial tension is not considered to be dependent upon surfactant level and was taken to be 1.24 mN/m in these computations (29).

A more complex situation occurs when mixed surfactants are used, as is often the case in industry. Here there is competition at the particle/water surface and one surfactant may dominate over the other in different ways as the relative concentrations are changed. This is quite a complex situation and we know of no reported results for predicted changes in morphology. We are presently measuring the polymer/water interfacial tensions for some anionic-nonionic surfactant systems and will then apply the results to experimental latices.

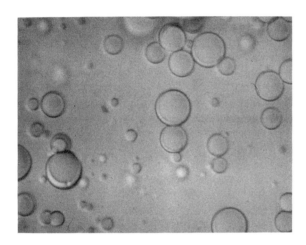

Figure 3. Optical micrograph of an emulsion of P(MMA-co-nBMA)/decane with Mexpectin 100S at 0.5 wt.% in water.

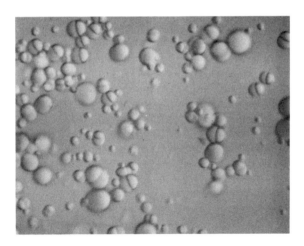

Figure 4. Optical micrograph of an emulsion of P(MMA-co-nBMA)/decane with Sodium Dodecyl Sulfate 0.5 wt.% in water.

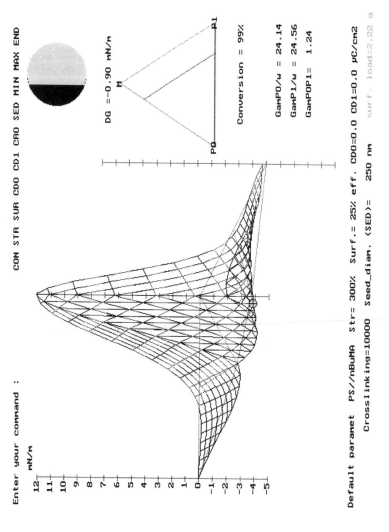

Figure 5a. Free energy surfaces for a polystyrene seed, n-butyl methacrylate system, with a stage ratio of 300% and a concentration of SDS surfactant of 25%.

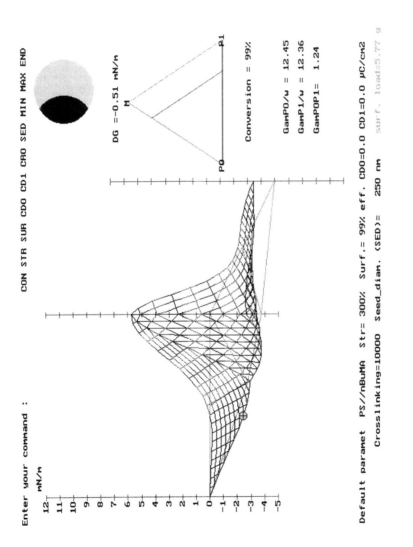

Figure 5b. Free energy surfaces for a polystyrene seed, n-butyl methacrylate system, with a stage ratio of 300% and a concentration of SDS surfactant of 99%.

Monomer Effects. It is well known the most monomers are more surface active than their polymers. Thus when there is less than full conversion in the polymerization reaction, the presence of the monomer will suppress all of the interfacial tensions. This can clearly be seen in Figure 6 where the interfacial tension between water with 0.5 wt. % SDS and a very non-polar polymer, poly(dimethyl siloxane) at 3900 molecular weight, is plotted as a function of the volume fraction of a polar monomer, MMA, in the polymer phase. These experimental results were obtained from our pendant drop interfacial tensiometer. It is clear that the behavior is quite non-linear, but it turns out that it can be modelled reasonably well from the thermodynamic treatment put forth by Prigogine and Marachal (30). Our extension of their thermodynamic treatment is presented as the solid line in Figure 6. In order to extend the model capability to other polymer/monomer systems one needs to have interfacial data for the pure polymer and pure monomer with water containing surfactant at the concentration of the experimental system which is being modelled, and also some reliable information about the Flory-Huggins interaction parameter between the polymer and monomer.

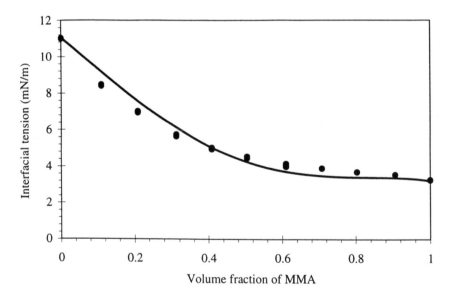

Figure 6.   Interfacial tension of a solution of Polydimethylsiloxane / methyl
methacrylate against a solution of sodium dodecyl sulfate at 0.5
wt.% in water. The continuous line represents the modified
Prigogine and Marechal model, while individual points represent
experimental data.

Applying the above information to the free energy analysis, one can make predictions for the equilibrium morphology as a function of monomer conversion during a batch reaction. Semi-batch reactions in which the monomer is fed over a period of time can also be handled as long as the monomer can be assumed to be distributed between the polymer(s) and water phases under equilibrium conditions. It should be clear that the interfacial tensions at play during the conversion history of a batch reaction will be lower than those at play during a reaction in which the monomer is "starve fed" to the latex. Models emanating from data such as those displayed in Figure 6 allow one to readily calculate conversion dependent morphology. Of course it needs to be mentioned that the monomer also affects the interfacial tension between the polymer phases. We have used the work of Broseta (31) to model the monomer effect on this interfacial tension, but have found in general that very simple expressions can be used to describe monomer concentration dependency. Obtaining particle morphology data as a function of conversion is much harder.

For the sake of illustration, Figure 7 shows the sequence of equilibrium morphologies predicted for the system described in Figure 5. These results would be for a batch reaction and the dark phase represents the seed polymer.

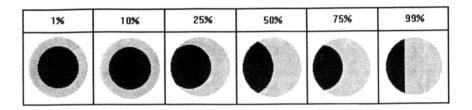

Figure 7. Sequence of predicted morphologies for a PS/PnBMA system, stage ratio 300%, SDS concentration of 25% and polymer conversion changing from 1% to 99%.

1. **Initiator End Group Effects.** The qualitative impact of having SO4 end groups on the polymer at the water interface is easy to appreciate. Its quantitative impact is another matter. Since the SO4 moiety is reasonably close to resembling the polar end of the SDS chain, one can possibly use interfacial tension data derived as a function of SDS coverage to approximate the situation for initiator derived SO4 end groups at the particle surface. Although the relationship is not direct (33), a crude model has been constructed to link the polymer/water interfacial tension with the density of SO4 end groups the particle surface. The density of the end groups can be experimentally determined by titrating the latex and determining its surface charge density, CD ($\mu C/cm^2$). That for the seed polymer we call CD0 and that for the second stage polymer we call CD1. A detailed description of this model is beyond the scope of this paper, but the presentation of

some computed results is useful. Figure 8 presents a matrix of morphology results (at full monomer conversion) derived from the UNHLATEX program for changes in the charge density at the surface of either or both of the polymers in the composite particle. Moving from left to right along the top row shows the impact of increasing the end group concentration at the second stage polymer/water interface for a seed polymer (dark phase) which was produced with a relatively low density of end groups. As the second stage polymer surface becomes more polar due to the increased SO4 concentration, the particle becomes more and more like a CS. By moving down the first row of the matrix one sees the predicted impact of starting with seed latices with more and more end groups for a fixed end group density on the second stage polymer. Here the morphology becomes an ICS. Other changes can readily be seen in the matrix.

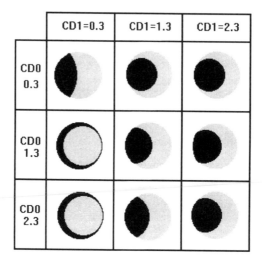

Figure 8. Matrix of morphologies for a PS/PnBMA system, stage ratio 200%, SDS concentration of 11% and polymer conversion of 99%. The charge density CD0 is for the polystyrene seed, and the charge density CD1 is for the second stage PnBMA, both are in $\mu C/cm^2$.

It needs to be stated that the above predictions only apply to the conditions of full conversion for the polymer types chosen in this example and for the SDS concentration assumed (11% of CMC). If the surfactant level was higher or if one or both of the polymers were more polar, the effect of changing the end group density is less dramatic. This is quite reasonable but underscores the need for having quantitative models which can consider simultaneous changes in a number of variables. Much more needs to be learned about this subject.

Future Considerations

There is still a great deal to do in understanding the interactive effect of all the important experimental parameters on the interfacial tensions at the particle surface and also at the polymer/polymer interface. Because of the critical dependence of the particle morphology on these interfacial tensions, there is a great need for the building of interfacial tension data bases for a wide variety of polymer systems, with various surfactants and combinations of surfactants. Reliable models need to be developed from such data so that approximations can be made for conditions that have not been fully explored by experiment. Extension needs to be made to copolymers, including the acid copolymers, and initiator end groups need to be carefully considered from both experimental and theoretical standpoints.

The most significant challenge for the future appears to be to move away from the constraints of thermodynamic equilibrium and to consider non-equilibrium structures. This must involve the dynamics of polymer phase separation caused by polymerization reactions within the latex particle. It is clear that the degree to which the morphology becomes non-equilibrium is dependent upon the speed of the polymerization reactions and also upon the degree to which the monomer is starve fed to the polymerizing latex. In either case higher viscosity leads to restricted diffusion within the particle and fully phase separated structures will be difficult to achieve during the time of the reaction. Further, it is our opinion from the work we have begun on polymer reactions within two phase latex particles that the reaction behavior can be dependent upon the instantaneous particle morphology. This leads us to suspect that both the morphology and the polymerization kinetics are interdependent, making their quantitative descriptions somewhat more challenging than may have been otherwise expected.

Acknowledgments

The work reported here has been financially supported by a variety of sponsors, including Rhone-Poulenc, the Centre National de la Recherche Scientifique (LCPP - Lyon, France), the donors of the Petroleum Research Fund of the American Chemical Society, and the University of New Hampshire for which both authors are appreciative.

References

1. Lee, D. I. *ACS Symposium Series 165,* American Chemical Society, Washington, DC, 1981; 405.
2. Lee, D.I.; Ishikawa T. *J. Polym. Sci., Polym. Chem. Edn.* 1983, 21, 147.
3. Muroi, S.; Hashimoto, H.; Hosoi K. *J. Polym. Sci., Polym. Chem. Ed.* 1983, 22, 1365.
4. Okubo, M.; Katsuta Y.; Matsumoto, T. *J. Polym. Lettl Ed.* 1980, 18, 481.
5. Okubo, M.; Ando, M.; Yamada, A.; Katsuta, Y.; Matsumoto, T. J. Polym. Sci.; Polym. Lettl Ed. 1981, 19, 143.

6. Okubo, M.; Katsuta Y.; Matsumoto, T. *J. Polym. Sci.; Polym. Lettl Ed.* 1982, 20, 45.

7. Stutman, D.R.; Klein, A.; El-Aasser, M.S.; Vanderhoff, J.W. *I&EC Product Research & Development* 1985, 24, 404.

8. Cho, I; Lee, K-.W. *J. Appl. Polym. Sci.* 1985, 30, 1903.

9. Lee, S.; Rudin, A. *Macromol. Chem. Rapid Commun.* 1989, 10, 655.

10. Lee, S.; Rudin, A. *J. Polym. Sci.: Part A: Polym. Chem.* 1992, 30, 2211.

11. Berg, J.; Sundberg, D.C.; Kronberg, B. *J. Polym. Mater. Sci. Eng.* 1986, 54, 367.

12. Berg, J.; Sundberg, D.C.; Kronberg, B.J. *Microencapsulation* 1989, 6, 327.

13. Sundberg, D. C.; Cassasa, A., J.; Pantazopoulos, J.; Muscato, M. R.; Kronberg, B.; Berg J. *J. Appl. Polym. Sci.* 1990, 41, 1425.

14. Muscato, M.R.; Sundberg, D.C. *J. Polym. Sci., Polym. Phys. Edn* 1991, 29, 102.

15. Dimonie, V. L.; EL-Aasser, M.S.; Vanderhoff, J.W. *IACES Polymeric Materials Science and Engineering* 1988, 58, 821.

16. Jonsson, J. -E. L.; Hassander, H.; Jansson, L. H.; Törnell, B. *Macromolecules* 1991, 24, 126.

17. Winzor, C.L.; Sundberg, D.C. *Polymer* 1992, 33, 3797.

18. Winzor, C.L.; Sundberg, D.C. *Polymer* 1992, 33, 4269.

19. Chen, Y.C.; Dimonie, V.L.; El-Aasser, M.S. *Macromolecules* 1991, 24, 3779.

20. Chen, Y.C.; Dimonie, V.L.; El-Aasser, M.S. *J. Appl. Polym. Sci.* 1992, 45, 487.

21. Chen, Y.C.; Dimonie, V.L.; El-Aasser, M.S. *J. Appl. Polym. Sci.* 1992, 46, 691.

22. Chen, Y.C.; Dimonie, V.C.; Shaffer, O. L.; El-Aasser, M.S. *Polymer International* 1992, 30, 185.

23. Sundberg, E. J.; Sundberg, D.C. *J. Appl. Polym. Sci.* 1993, 47, 1277.

24. Gonzalez-Ortiz, L.J. and Asua, J.M., *Macromolecules* 1995, 28, 3135.

25. Gonzalez-Ortiz, L.J. and Asua, J.M., *Macromolecules*, in press.

26. Torza, S. and Mason, S.G., *J. Colloid Interfacial Sci.* 1970, 33, 67.

27. Durant, Y.G. and Sundberg, D.C., *J. Appl. Polym. Sci.* 1995, 57, 1607.

28. Anastasiadis, S.H., Chen, J.K., Koberstein, J.T., Siegel, A.P., Sohn, J.E. and Emerson, J., *Colloid Interface Sci.,* 119 (1986) 55.

29. Durant, Y.G., Ph.D. Thesis #8094, University Claude Bernard, Lyon I, France, 1994.

30. Prigogine, I. and Marechal, J., *J. Colloid Sci.* 1952, 7, 122.

31. Broseta, D., Liebler, L. Kaddour, L.O. and Strazielle, C., *J. Chem. Phys.* 1987, 87, 7248.

Chapter 4

Particle Coalescence

Peter T. Elliott, Wylie H. Wetzel, Lin-Lin Xing, and J. Edward Glass[1]

Department of Polymers and Coatings, North Dakota State University, Fargo, ND 58105

The prior art on particle coalescence and film formation of latex particles is reviewed in the first part of this chapter. In the latter part, the particle coalescence of step-growth oligomer dispersions is discussed. The synthesis of the latter type is discussed in chapter 1, and their application properties are discussed in chapters 7 and 8. The coalescence of water-borne epoxies is discussed in one of the sections in Chapter 5. Peripheral to this chapter, the phenomena of dispersion coalescence is discussed from a uniquely different perspectives for alkyd dispersions in chapter 10 and for high clay content, paper coatings in chapter 13. The reader also is referred to chapter 14 where the importance of the drying process on film formation properties, in areas outside consumer architectural coatings, is discussed.

To achieve acceptable film properties, the resin dispersion of a water-borne coating must initially possess colloidal stability (discussed in chapters 1 and 6) and have the proper rheology during formulation, storage, and application. The disperse phase - be it a latex or a water-reducible resin - must then undergo particle coalescence and resin interpenetration after application to achieve mechanical strength.

If water-borne latex coatings are to surpass the volatile organic component (VOC) limits that high-solids failed to reach, the use of coalescing aids, glycol ethers, and fugitive neutralizing agents (i.e., low boiling amines) must be discontinued in water-borne coatings. These materials are currently needed to obtain particle stability in storage and particle coalescence after application. Without adequate particle coalescence, good touchup, scrub resistance and pigment acceptance will not occur. Freeze thaw resistance, open time, and dispersion of colorants will be lost with the removal of glycol ethers. In addition to these obvious volatile organic components, the surface stabilizers, (oligomeric acids or grafted water-soluble polymer fragments, discussed in chapter 1) that are necessary to stabilize the aqueous dispersions prior to application, have to be reconsidered, for the current ones impede particle coalescence.

[1]Corresponding author

Removal or significant reduction of the coalescing aid will only serve to enhance stabilizer impedance of particle coalescence, and a resin with a lower film formation temperature will be required. The coalescence of step-growth oligomers used in original equipment maintenance (OEM) applications has received attention only recently, but it offers some different alternatives. A review of prior art related to these areas is presented below, before considering research approaches that may facilitate significantly lower or zero VOC water-borne coatings.

LATEX COALESCENCE

There have been several descriptions of the latex particle coalescence process during the past half century. It is generally accepted that the film-forming process occurs in three stages (Figure 1): (I) evaporation of the water until the particles reach close-packing; (II) formation of particle contacts and deformation of the latex particles as the particle volume fraction goes above that of a close-packed structure; a polyhedral-foam-like structure is developed with interparticle bilayers and Plateau borders; (III) gradual coalescence by interdiffusion of polymer molecules between latex particles.

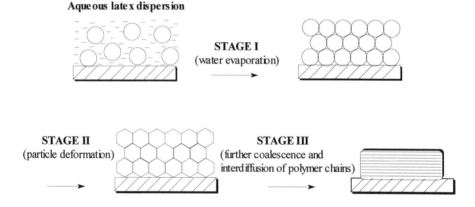

Figure 1. Three stages in the general film formation process of aqueous dispersions.

These stages are distinct if the stabilizing surfactant layer is effective, if there is no tendency to spontaneously form particle contacts or aggregates (as there is in depletion flocculation, discussed in chapter 6), and if the particles are not too hydrophilic.

Stage I. Latex films are typically prepared with thicknesses on the order of 50 μm. This is much larger than the particle diameter, and the film will comprise many layers of particles. During stage I, the water phase starts to evaporate and the concentration of the latex phase increases. As the volume fraction, ϕ, increases above that of close-packed spheres, capillary stresses develop which force the particles into contact. This type of stress can be thought of as arising from surface tension at the particle/water/vapor contact lines, at the top surface of the film. In effect, this tension pulls the film down (Figure 2), in the direction of the substrate, and leads to the "compression" of the latex particles. Sheetz first proposed (*1*) that a thin layer of coalesced particles closes the surface of the drying latex. The remaining water evaporates after diffusion through this polymer layer, and the packing of particles is compressed as if by a piston.

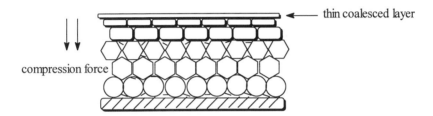
thin coalesced layer

compression force

Figure 2. "Compression" of latex particles due to surface tension forces

Stage II. Once the particles have made intimate contact, stage II of the film formation process begins. Up to this point, the particles have been stabilized by surface acid (electrostatic forces), surfactant molecules, and/or grafted water-soluble polymer fragments (steric forces). As the water evaporates and the particles pack closer together, the stabilizing forces are overcome and coalescence begins. It is at this point where descriptions of the primary forces driving coalescence have varied over the past four decades. In the early fifties, Bradford and coworkers modeled (*2,3*) film formation as a Frenkel viscous flow of contracting polymer spheres under polymer/air and polymer/water interfacial tensions. Brown complemented (*4*) this with arguments that water also contributed to the deformation process through capillary compression of the polymer assemblage by the serum/air surface tension.

As the water evaporates, all of the interfacial tensions drive the particles to coalesce despite the viscoelastic resistance of the polymer to flow. To address the coalescence of large particles, Vanderhoff and his colleagues proposed (*5*) that as the water continues to evaporate, the forces due to the water-air surface tension pushed the particles together until the stabilizing layers were ruptured, resulting in polymer-polymer contact. The pressure on the particles was then increased by forces arising from the polymer/water interfacial tension. It has also been proposed (*6*), that interfacial tension forces act along with the capillary force to cause film coalescence. Attempts have been made to verify the theories proposed. For example, the coalescence of core-shell latices (*7*) (core: St/BuA, shell: MAA/St/BuA) were studied, and the rate of coalescence decreased when the particle-water interfacial tension decreased.

Brown also observed that film formation occurs concurrent with the evaporation of water and is complete with total water evaporation. He observed that porous, incompletely coalesced films could be maintained by keeping the temperature lower than a certain critical value (below the glass transition temperature, T_g, of the resin) during the water evaporation stage, and that some voids would not coalesce when warmed to a temperature where coalescence would have occurred in the presence of water. His observation that liquid water was central to the deformation process, led several industrial laboratories, in the past decade, to use 2-hydroxyethyl methacrylate and similar comonomers in latex synthesis. In addition, the ability of coalescing aids to decrease the rate of evaporation of water has been suggested to facilitate coalescence. Similar results have also been noted (8, 9) with hydrophilic compositions where it has been observed that higher humidity, which both slows water evaporation and plasticizes the particles, will lower the MFFT. The minimum temperature at which film formation occurs is know as the Minimum Film Formation Temperature (MFFT); it generally occurs a few degrees below the T_g of the resin. If the MFFT is above the temperature range of application, the dispersion will not form a continuous film; if the MFFT is below the application temperature, a continuous film will be formed. However, if the MFFT is too far below the application temperature, a film will be formed with poor properties (e.g., mechanical, abrasion, solvent resistance, dirt pickup, etc.).

It has also been demonstrated that the hydrophilicity of the monomer is important to the morphology. For example, lower amounts of coalescing aid are needed for film formation with a methyl methacrylate/methyl acrylate (MMA/MA) porous copolymer than for a styrene/2-ethylhexyl acrylate (10) copolymer latex with a nonporous topography. Surprisingly, there had been a lack of studies on comparative compositional influences until this investigation and the one on hydroplasticization by water (9); both demonstrated the greater difficulty in film formation of a styrene/acrylate latex relative to a methacrylate/acrylate latex.

Stage III. At the end of stage II, the film is dry but interfaces between the particles still exist. Stage III corresponds to the evolution of these interfaces. The interface between particles dissipates due to interdiffusion of macromolecular chains between particles, driven by interfacial and surface tension forces. This behavior is called autohesion or further coalescence. Film properties such as mechanical strength and chemical resistance begin to develop (6). In a theoretical treatment of coalescence during the end of the second stage, it is envisaged that as the volume fraction tends to unity, a polyhedral-foam type structure is formed with the water contained in a network of bilayers, of thickness δ_0, and Plateau borders, with radius of curvature, r_{PB} (Figure 3). If the surfactant is effective, this is likely to occur only if the surfactant bilayer has been dissipated by diffusion of the surfactant into the particles or into the Plateau borders and interparticle interstices. This will occur at very low water contents, when the water of hydration of the surfactant is eliminated (11). The latter comment highlights a fact not generally considered: the colloid stabilizers that promote stability of the initial colloid particle (Figure 1, with discussion, in chapter 1). It would be harder for these types of stabilizers to disappear, and this has not been addressed in the literature. Based on the prior art, with only surfactant stabilization, it was concluded that the plateau boarders are so narrow and the water/vapor surface tension is so high that vapor penetration into the film is effectively impossible, and liquid latex particles with a surfactant layer will always form a film - independent of the particle size.

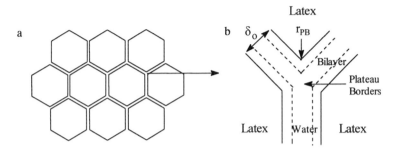

Figure 3.   (a) cross-section through a dry film of deformed particles, (b) expanded view of a Plateu border with bilayers.

MFFT.   As noted above, the MFFT has a large impact during Stages II and III. The factors effecting the MFFT are the polymer's modulus (i.e., ca. $\sim T_g$), materials that influence the modulus such as surfactants, glycol ethers, crosslinking, and the particle size of the latex. The particle size dependence is related to the pressure dependence defined in the stronger capillary force among smaller particle latices defined in the LaPlace equation (*12*).

$$\Delta P = 2\gamma/r \qquad\qquad (1)$$

where ΔP = pressure difference across the film
γ = surface or interfacial tension
r = radius of latex particle

The importance of particle size on coalescence has generally been abandoned since the mid sixties; however, there have been recent commercial studies in support of the impact of the median size of latices on the minimum filming temperature (*9,13*). The drawback is the large amount of surfactant required to synthesize truly small particle sizes. The minimum filming temperature of vinyl acetate copolymer latices was shown to increase as the average particle radius increased (*14*). This has also been shown in the work of Eckersley and Rudin (*6*).   In the latter study of methacrylate latices, the influence of the median particle size was small, but the particle size study stopped at 148 nm, and the particles contained methacrylic acid surface stabilizers.   This effect is still under investigation due to questions raised, such as particle size distribution, concentration of surface stabilizer (grafted polymers), and type of stabilizers.

Controlling the modulus of the polymer particles has been the most practiced method of effecting the MFFT. Through the use of different monomer combinations, the T_g of the particle can be adjusted above a desired MFFT. Coalescing aids are then added to the dispersion to lower the T_g of the particle still further by plasticization and perhaps by retarding the rate of water evaporation. This creates a dispersion with the advantage of a low MFFT while maintaining a relatively high T_g after evaporation of the coalescing agent and glycol ethers. The most important properties (*10*) of a coalescing aid are its non-polar solubility parameter (that reflects the relative insolubility in water) and molar volume (that reflects its ease of diffusion through the film forming matrix). The hydrophilicity of the monomers used in the dispersion synthesis can also contribute to lowering the MFFT. These hydrophilic monomers will tend to be at the particle surface where water will plasticize the particle. For example, at the same T_g a methyl methacrylate/ethyl acrylate latex will form a film much more readily than a styrene/2-ethylhexyl acrylate copolymer latex (*10*). These studies are consistent with the ability of oligomeric acid stabilizers to reach the surface of related copolymer latices (Figure 2 in chapter 1). It also has been reported that for a MMA/acrylate copolymer emulsion, the MFFT displayed a greater increase with increasing MMA when the co-monomer was less hydrophilic (*15*).

Surfactant Effects. Early studies of nonylphenol surfactants by Bradford and Vanderhoff observed (*16,17*) that as the length of the ethylene oxide segment decreased, film integration with a styrene/butadiene copolymer increased. Early studies by Vijayendran observed (*18*) surfactant absorption into latices. Other studies observed (*19*) that surfactants markedly reduced the MFFT and T_g of vinyl acrylic latices. In a recent study, a nonionic surfactant was observed (*20*) to plasticize an acrylic latex. There also have been studies of optimum surfactant concentrations. Post addition of surfactant to a surfactant-free PBMA latex observed (*7*) that the best film surface corresponded with a surfactant concentration providing full surface coverage of the latex.

STEP-GROWTH POLYURETHANE DISPERSIONS

The "water-reducible" oligomers prepared by step-growth polymerizations (discussed in chapter 1) must also undergo coalescence to form a continuous film. As noted in chapter 1, detailed studies in the open literature are essentially restricted to aqueous polyurethane dispersions (PUDs). PUDs have similar film formation characteristics to that of conventional latices, but with significant differences that are unique to PUDs. The two most common types of PUDs used in water-borne coatings are the high molecular weight aqueous dispersion and the low molecular weight two component (2K) polyisocyanate/polyurethane system.

The film formation process for PUDs is a combination of thermoplastic coalescence (as with latices) and in the 2K systems, thermoset crosslinking. It is in the areas of the stabilizers and the diffusion process were the PUDs differ most from the conventional latices. PUDs employ both anionic and nonionic stabilizers (Figure 4) in their synthesis. The nonionic stabilizers provide stability to salinity gradients but not to high temperatures. They are less frequently used because relatively high concentrations are needed for stable dispersions. The amount of dimethylolpropionic acid (DMPA, Figure 4), the most commonly used ionic stabilizer , determines the median particle size of the PUD phase and its stability. As discussed in chapter 1, the acid functionalities are not contiguous, and the dispersions are more sensitive to pH changes than conventional latices. The contiguous acid groups in the chain-growth acrylic dispersions create areas with different ionization potentials; PUDs do not benefit from this neighboring group effect, and thus display a narrower ionization range (*21*).

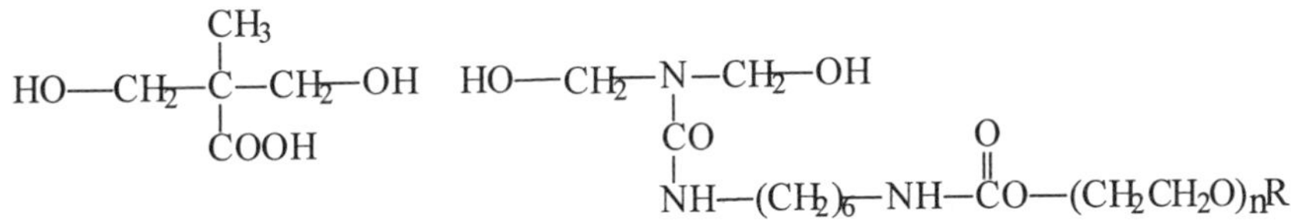

$$HO-CH_2-\underset{\underset{COOH}{|}}{\overset{\overset{CH_3}{|}}{C}}-CH_2-OH \qquad HO-CH_2-\underset{\underset{NH-(CH_2)_6-NH-\overset{\overset{O}{||}}{C}O-(CH_2CH_2O)_nR}{\underset{CO}{|}}}{N}-CH_2-OH$$

Dimethylolopropionic acid   Nonionic stabilizer
(DMPA)

Figure 4. Typical stabilizers used in PUD synthesis.

The water inside the dispersion (Figure 5) creates a more open particle, increasing the mobility of the polymer chains throughout the particle (*21*). The high molecular weight PUDs form films primarily by thermoplastic coalescence, while the 2K PUDs crosslink after the particles have coalesced to form a hard, continuous film. Diffusion of the polymer molecules, to ensure complete reaction of all functional groups, is very important in 2K PUDs.   If formulated correctly (a motherhood statement borne of proprietary needs, but often used to hide questionable experimental results), 2K PUDs can have enhanced film properties, such as solvent resistance, over more conventional high molecular weight , water-borne PUDs due to their crosslinked nature.

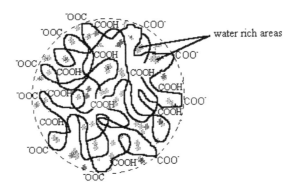

Figure 5. Representation of water swollen PUD particle

The particle size distribution of PUDs are generally broader than conventional latices. Since PUDs are dispersed mechanically, the emulsification process contributes to broader size distribution due to areas of nonuniform mixing. The particle size of the final dispersion is typically between 25-100 nm. The distribution may also be related to the statistical nature of the chain extension steps. The MFFT and polymer diffusion are directly related to the degree of swelling which is suggested to ultimately effect the final film quality. The parameters influencing the degree of particle swelling by water are: hydrophilicity of the backbone, molecular weight, DMPA concentration, and level of cosolvent (usually n-methylpyrrolidinone). Even though the particles are swollen by water prior to film formation, this appears to have little effect on the water sensitivity of the final film. The water sensitivity is determined, apparently, only by the hydrophilicity of the polymer backbone.

Two-Component Polyurethanes (2K). The formulation of a 2K water-borne polyurethane consists of combining a multi-functional polyisocyanate component, modified to be water dispersible, with a hydroxy-functional polyurethane dispersion (Scheme 1). Typical hydroxy-functional dispersions are polyurethane dispersions (I in Scheme 1) based on polyesters, polycaprolactones, acrylics, and poly(tetramethylene oxide). The reaction schemes are given in chapter 1. A common isocyanate is a modified trimer of hexamethylene diisocyanate (HDI) (III in scheme 1). The two components are mixed together, with a catalyst, prior to application. The reactivities of the functional groups dictate the length of the induction time (the time between mixing and application) in 2K waterborne polyurethanes. In spray coatings (chapter 15), formulations with induction times of less then a minute can be used, but require mixing of the two components at the spray head. During longer mixing periods the particles begin to coalesce (depending on T_g, % solids, various additives, etc.), and the isocyanate-functional groups react with both the hydroxyl-functional groups and water. Once the two reactive components in the mixed dispersion coalesce, interdiffusion can occur, thereby increasing the desired isocyanate-hydroxyl reaction (22). After application, water and co-solvent evaporate, and film formation begins as described in the previous section. Coalescence and isocyanate reactions may continue, but ideally to a lesser extent. Problems can result during this stage if the crosslinking reactions and molecular weight build-up are too rapid. Cosolvent and CO_2 (from reaction of water with isocyanate, Scheme 2) can become trapped in the film causing a decrease in the diffusion of reactive groups. These can have adverse effects on the mechanical and chemical resistive properties of the coating. The contribution of the T_g to the flexibility of the chain, in a 2K waterborne polyurethane film, becomes subordinate to the increasing modulus associated with crosslinking. As stated earlier, this is dependent on the reactivity of the functional groups.

In a recent study of 2K waterborne polyurethane systems, particle coalescence in the induction period was studied (23) using particle size analysis (capillary hydrodynamic fractionation). A dispersion of two different commercial hydroxy-functional components, one is a typical polyurethane (I) hydroxy terminated dispersion (see chapter 1); the other was an acrylic modified polyurethane (II) with

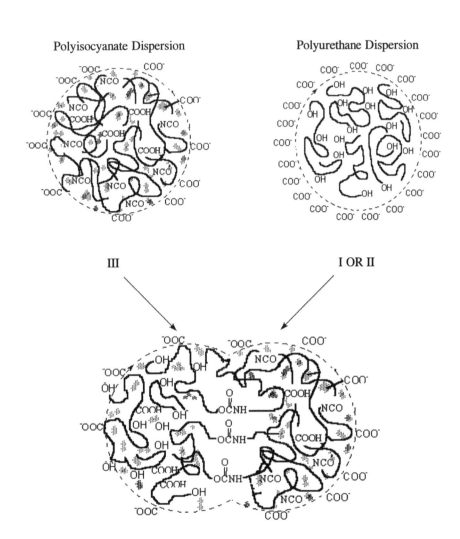

Scheme 1. Representation of water swollen polyisocyanate and polyurethane dispersion coalescence.

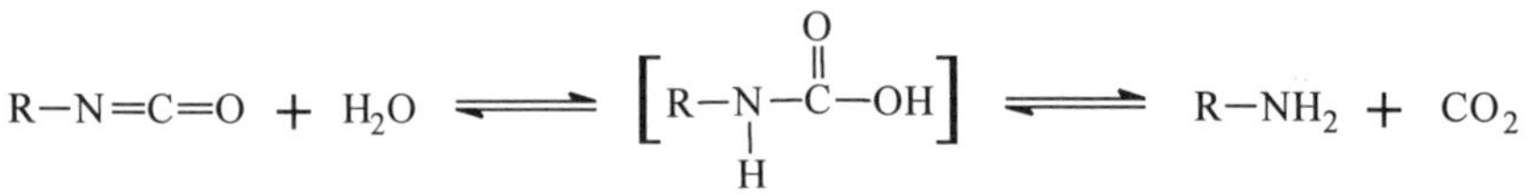

$$R-N=C=O \; + \; H_2O \; \rightleftharpoons \; \left[R-\underset{H}{\overset{}{N}}-\overset{\overset{O}{\|}}{C}-OH \right] \; \rightleftharpoons \; R-NH_2 \; + \; CO_2$$

Scheme 2. Reaction of isocyanate with water.

hydroxyls. Both were studied with a HDI (III) based polyisocyanate (23). The polyurethane dispersion (I) with hydroxy terminal groups had a median particle size of 25 nm; the acrylic-polyurethane hybrid dispersion (II) with hydroxyl functionalities had a Gaussian distribution with a peak at 66 nm. When the polyisocyanate (III) is dispersed in water a bimodal distribution with a predominance of the particles at ~60 nm and between 200 nm and 600 nm is observed. This dispersion, over a 4-5 hour period, merged into a single, broad peak at ~200 nm. This was thought to result from the reaction of the polyisocyanate dispersion with water.

When I is mixed with III the dispersion has a bimodal particle distribution at 60 nm and 200-600 nm, similar to the peaks in the original dispersed polyisocyanate. There is no evidence of a 25nm dispersion size. When dispersion II is mixed with III, a bimodal particle distribution of 50-80 nm and 100-200 nm is observed The small particle size peak arises from both the acrylic-hybrid (II) as well as the polyisocyanate (III). The lowering of the larger particles size by dispersion II, as compared to dispersion III alone and I plus III is attributed(23) to the acrylic-polyurethane hybrid system II emulsifying (i.e., "encapsulating") the polyisocyanate dispersion (III).

When monitored over a four to five hour induction period, neither the polyurethane/polyisocyanate (I with III) nor the acrylic-polyurethane hybrid/polyisocyanate (II with III) showed significant changes in their particle size distributions. From this data it was assumed that for both the polyurethane/polyisocyanate and polyurethane-acrylic hybrid/polyisocyanate 2K waterborne polyurethanes, particle coalescence or flocculation does not occur to a significant degree during the induction period. Although the above statements hold true for these particular systems, it is not known whether these results apply to other more or less reactive polyisocyanates with different hydroxy-functional dispersions.

Other tests also were run on these systems to evaluate their film forming processes. Isothermal calorimetry indicated that while reactions of isocyanate with hydroxyl groups occur almost immediately, it takes more than two hours for the isocyanate-water reactions to begin. This was supported by viscosity studies over a seven hour induction period with different levels of catalyst (up to .10%). In these systems, the viscosity of the mixed dispersions increased slightly over the first 1-2 hours and then began to drop slightly. This was related to the onset of the water-isocyanate reactions and the production of CO_2. The liberation of CO_2 and formation of amine terminal groups would account for the faster reaction rates, and the liberation of CO_2 results in a foam. Therefore, in 2K water-borne polyurethanes, it is thought that the pot life depends more on the formation of CO_2 than on the effect of viscosity build or gelation as it does in solvent-based polyurethanes. Film formation of 2K polyurethanes was further studied using IR and electrochemical impedance spectroscopy (EIS). IR analysis indicated that almost all of the water and most of the cosolvent had evaporated in 15-30 min. This time frame was also confirmed by EIS measurements.

Blends and Hybrids

The use of blending techniques (acrylic latices with polyurethane dispersions) as well as hybrid systems has been investigated as a route to obtain systems with a lower MFFT while maintaining a higher, final T_g, similar to the use of coalescing aids in latices. If this can be achieved, lower VOC's would be realized. Recent studies have examined the blending of high and low T_g PUDs (*24*). This process also is employed in conventional latices as an attempt to combine good film formation and film hardness without the need for coalescing aids. The blending of low T_g and high T_g particle dispersions in PUDs has resulted in some synergism. Some of the high T_g component is thought to be contained within the film of the low T_g component, acting somewhat like a composite for reinforcement. Scanning electron microscopy (SEM) is a fairly effective way of studying film formation, and thus the extent of particle coalescence. However, with the small particle size of the PUDs, atomic force microscopy (AFM), specifically tapping mode atomic force microscopy (TMAFM), has been found (*24*) to be a better alternative. Film formation of fourteen different blends of three PUDs with different T_g's were recently studied using TMAFM (*24*). As would be expected, the PUDs with the lower T_g's form better films than PUDs with higher T_g's. An interesting characteristic of the best film forming dispersions was an ordered, recessed, hexagonal network thought to be formed from particle packing during coalescence. The AFM studies have also proven that blending of soft (low T_g) and hard (high T_g) particle dispersions results in enhanced film formation over the hard particle dispersions by themselves, but the film formed is not as complete as observed with a disperse phase at its MFFT.

In a related study, film hardness and phase separation were characterized using dynamic mechanical thermal analysis (DMTA) and small angle X-ray scattering (SAXS) (*21*). Past results using DMTA to study solvent-borne polyurethanes have resulted in two distinct transitions which are presumed to be from soft (low T_g) and hard (high T_g) regions. With PUDs, only one distinct transition was evident with DMTA; however, when analyzed by SAXS, phase separation was evident.

Blending of water-borne polyurethanes with acrylic emulsions also has been studied (*25,26*). The addition of the PUD to an emulsion enhanced the physical properties of the acrylic polymer. Morphology studies indicated that there were distinct urethane and acrylic regions. In another study (*27*), the acrylic monomers were added to a PUD and polymerization was initiated. The polymerization took place inside the PUD particles, creating a urethane stabilized acrylic particle. The overall MFFT and T_g can be controlled with the right balance between the urethane and the acrylic.

Crosslinkable Functional Groups Within A Particle

An alternative to the use of coalescing aids and blending of acrylics has been the incorporation of crosslinkable functional groups in the polymer backbone.   This approach is focused on attempting to effectively separate the MFFT of the latex from the "T_g" of the final film. This has been attempted by the addition of either an acetoacetate functional monomer or an autooxidizable component, such as a drying oil, into the latex (28,29).   The acetoacetate functional groups are converted to enamines when the latex is neutralized with ammonia or a primary amine. An excess of neutralizer is added to ensure enamine formation.   The enamine functionalities have the advantage that they do not begin to crosslink significantly until coalescence is almost complete (for discussions of this nature, see Chapter 7).   As the enamine groups crosslink, the modulus (discussed in the literature incorrectly as the T_g) of the film increases. Therefore, a latex can be formulated to a low MFFT, yet have a hard final film.   An autooxidizable component works in the same general manner, with crosslinking occurring among the unsaturation sites. The acetoacetate/enamine latices are designed to cure at ambient temperatures and with the aid of sunlight. Thus, the addition of photoinitiators enhances the curing process.   The drying oil incorporated latices cure faster with the addition of a metal drier, such as cobalt. Both can be formulated to be stable, one package systems. In addition, the coalescing agent could be functionalized, such as dicyclopentenyloxyethyl methacrylate (6,7) and used as a crosslinker to obtain the hard film properties needed in industrial coatings.

Conclusions

The original latex compositions of the sixties evolved into coatings of very acceptable properties through evolution of their chemical compositions and synthesis techniques. This progress also was achieved in large part by the use of coalescing aids and cosolvents that facilitate good particle coalescence with good final film properties. Removal of these materials to reduce the volatile organic components of the formulation while achieving good film properties is a significant challenge.

The phenomenon of particle coalescence has been an area of study for some forty years. The recent studies described in this chapter have fine-tuned our understanding, but translation to any commercial improvements is unlikely. In the more commodity latex markets the greatest probability of success may be through the well studied core-shell approach, discussed in chapters 2 and 3, with hydrophilic comonomers in the outer shell. The disperse phase must have colloidal stability and the influence of the type of stabilizers used commercially (discussed in chapter 1) has not been reported in the open literature. Hydrophobically-modified ethoxylated urethane thickeners (30) have been observed to stabilize aqueous polyurethane dispersions (31), and their use in this mode is now generally one of the standard practices in formulating. HEURs also lower the MFFT of the formulation (10). The use of crosslinkable functional groups within each particle with the low MFFT systems is also promising.

Better film formation through the use of smaller median particle sizes also is an area worth pursuing, if the amount of surfactant in producing such small sizes can be reduced. The recent innovative studies in polyurethane aqueous dispersions described in this chapter may well provide approaches to achieving this goal; however, the PUDs described in this chapter do not address the problem of VOC, and the latitude of cosolvent displacement on film properties needs quantification. Nevertheless, there are zero VOC coatings now in the marketplace, based on the concepts discussed in this and the first three chapters. Their success lies with their market acceptance. If past history is an effective guide, their acceptance will not be immediate, and continued evolution of water-borne coating technology will be needed to accomplish this goal.

Literature Cited

1. Sheetz, D.P., *J. Appl. Polym. Sci.*, **1965**, *9*, 3759.
2. Dillion, R.E.; Matheson, L. A.; Bradford, E.B., *J. Colloid Sci.*, **1951**, *6*, 109.
3. Henson, W.A.; Tabor, D.A.; Bradford, E.B., *Ind. Eng. Chem.*, **1953**, *45*, 735.
4. Brown, G.L. *J. Polym. Sci.,* **1956**, *22*, 423.
5. Vanderhoff, J.W.; Tarkowski, H.L.; Jenkins, M.C.; and Bradford, E.B., *J. Macromol. Chem.*, **1996**, *361.*
6. Eckersley, S.T. and Rudin, A., *J. Coat. Tech.*, **1990**, *62(780)*, 89.
7. Dobler, F.; Pith, J.; Lambla, M.; and Holl, Y., L., *J. Colloid Interface Sci.*, **1992**, *152*, 1.
8. Juhue, D. and Lang, J., *Langmuir* , **1993**, *9*, 792.
9. Sperry, P.R.; Synder, B.S.; O'Dowd, M.L.; and Lesko, P.M., *Langmuir*, **1994**, *10*, 2619.
10. Alahapperuna, K. and Glass, J. E., *J. Coat. Tech.*, **1991**, *63(799)*, 69.
11. Crowley, T.L.; Sanderson, A.R.; Morrison, J.D.; Barry, M.D.; Morton-Jones, A.J.; and Rennie, A.R., *Langmuir*, **1992**, *8*, 2110.
12. Adamson, A.W., *Physical Chemistry of Surfaces*, 5th ed., John Wiley & Sons, Inc., New York, New York, 1990.
13. Shah, P.K.; Blam, A.F.; and Yang, P.Y., B.F. Goodrich, Canadian Patent, (US) 07/333,376, 4/5/1989. And references therein.
14. Nyugen, B., PhD Thesis, Univ. Waterloo (1986).
15. Tongyu, C.; Yongshen, X.; Yuncheng, S.; Fu, L.; Xing, L.; and Yuhong, H., *J. Appl. Polym. Sci.*, **1990**, *41*, 1965.
16. Bradford, E.B. and Vanderhoff, J.W., *J. Macromol. Chem.*, **1966**, *1*, 335.
17. Bradford, E.B. and Vanderhoff, J.W., *J. Macromol. Sci.-Phy.*, **1972**, *B6(4)*, 671.
18. Vijayendran, B.R., *Polymer Colloids II*; Fitch, R.M, Ed.; Plenum: 1980, p.209.

19. Vijayendran, B.R.; Bone, T.; and Sawyer, L.C., *J. Disp. Sci. Tech.*, **1982**, *3(1)*, 8.

20. Eckersley, S.T. and Rudin, A., *J. Appl. Polym. Sci.*, **1993**, *48*, 1369.

21. Satguru, R.; McMahon, J.; Padget, J.C.; and Coogan, R.G., *J. Coat. Tech.*, **1994**, *66(830)*, 47.

22. Jacobs, P.B. and Yu, P.C., *J. Coat. Tech.*, **1993**, *65(822)*, 45.

23. Hegedus, C.R.; Gilicinski, A.G.; and Haney, R.J., *J. Coat. Tech.*, **1996**, *68(852)*, 51.

24. Rynders, R.M.; Hegedus, C.R.; and Gilicinski, A.G., *J. Coat. Tech.*, **1995**, *67(845)*, 59.

25. Bakker, F., *Polymers Paint Colour Journal*, **1992**, *182(4310)*, 376.

26. Reardon, G.P., *American Paint and Coatings J.*, **1991**, *76(6)*, 50.

27. Jansse, P.L., *JOCCA*, **1989**, *72*, 478.

28. Bors, D. A.; Lavoie, A.C.; and Emmons, W.D., Rhom and Haas, European Patent, Publication No. 0 492 847 A2, Application No. (913112926), 4/12/91.

29. Bors, D.A.; Warminster, W.; Emmons, D.; and Edwards, S.S., Rhom and Haas, United States Patent, Patent No. 5,296,530, 3/22/1994.

30. Glass, J.E., Ed., *Hydrophilic Polymers: Performance with Environmental Acceptance*, Advances in Chemistry Series 248, Ch. 10,17, and 24, American Chemical Society, Washington, D.C., 1996.

31. Kaczmarski, J.P.; Fernando, R.H.; Glass, J.E., *J. Coat. Tech.*, **1993**, *65(818)*, 39.

Chapter 5

Two-Component Waterborne Epoxy Coatings

Frederick H. Walker and Michael I. Cook[1]

Air Products and Chemicals, Inc., 7201 Hamilton Boulevard,
Allentown, PA 18195-1501

The chemistry and properties of two component, waterborne epoxy coatings are reviewed. The coatings are classified by the type of epoxy resin employed. Type I utilize low molecular weight, liquid epoxy resin, and type II use pre-formed dispersions of higher molecular weight, solid epoxy resins. The physical and coatings chemistry of waterborne epoxy coatings is discussed, with particular emphasis on the film formation process. After a description of the general structure-property requirements for the coatings components and the surfactant properties of amine hardeners, a detailed description of the chemistry of the binders for both type I and type II systems is provided, based mostly on the patent literature. General principles of coatings formulation are then provided.

Since their first viable commercial introduction at least some 25 years ago, *(1)* waterborne epoxy coatings have become a commercially important technology. Growth has been driven by consumer desire for the low odor and water cleanup of coatings based on thermoplastic latices, in uses that require ambient application conditions and the performance advantages typically associated with thermosetting polymers. Various environmental and worker safety regulations have also contributed to this growth. Despite this interest, to date there appears to be no systematic published review of the various technologies employed in this market.

In this paper, after first classifying the technologies available, we will discuss important aspects of polymer and colloid chemistry pertaining to waterborne epoxy coatings. The principal chemistries now employed in this technology will then be reviewed. No attempt has been made to be encyclopedic in coverage. Instead, the discussion will emphasize those chemistries that the authors suspect to be of commercial importance, or which represent novel approaches to the technology. Much

---

[1]Current address: Air Products Nederland bv, Kanaalweg 15, Box 3193, GD Utrecht, Netherlands

of the information on the chemical technologies is drawn from the patent literature. It should be appreciated by the reader that commercial practice frequently differs to some degree from that described in the patent.

Classification of Waterborne Epoxy Technologies. For the purposes of this review, we classify waterborne epoxy coatings into two categories, based on the physical state of the base epoxy resin employed. Type I systems, which were the first to achieve commercial success, generally utilize the diglycidyl ether of bisphenol-A (DGEBA, **Ia**), which is commonly referred to as liquid epoxy resin, as the principle epoxy resin. Commercial grades used in coatings are slightly oligomerized (n ≈ 0.1) and have a viscosity at 25°C of about 12,000 m·pa·s. The diglycidyl ether of bisphenol-F (**II**, DGEBF, which also contains the *o,p'* and *o,o'* isomers in addition to the *p,p'* isomer shown) may be substituted for all or part of **Ia**. Various epoxy reactive diluents (generally glycidyl ethers of various phenols or aliphatic alcohols and diols) are used to modify the viscosity and crosslink density of the resin.

Ia, n = 0-0.1
Ib, n >2

II

In type I systems, the amine hardener is usually designed to act as the emulsifier for the epoxy resin, although sometimes the epoxy resin is pre-emulsified in water with surfactants, primarily to adjust package ratios. Thus, hardeners for type I systems are amphiphilic molecules, possessing both hydrophilic and hydrophobic sections.

Type II waterborne epoxy systems are based on much higher molecular weight epoxy resin, **Ib**. At room temperature, commercial solid epoxy resin with an epoxy equivalent weight of 450 - 550 (**Ib**, $n_{avg.}$ ≈ 2) is a solid with a melting point of 75 - 80°C. Its viscosity is orders of magnitude higher than **Ia**: about 10,500 m·pa·s at 100°C. (2) To effect a small and reproducible particle size, dispersion of such viscous materials requires specialized processing equipment, and the application of heat or the use of solvents. Thus, such dispersions are always pre-dispersed either by the manufacturer or raw material supplier. Interestingly, the amine hardeners for type II systems also tend to be amphiphilic in nature, though they no longer serve to emulsify the resin.

It is difficult to point to performance differences between the two systems that are true without exception, but the following observations generally hold. Most type I coatings contain little or no co-solvent, whereas type II require a coalescing agent. Type II coatings, like solvent-borne coatings based on solid epoxy resin, will reach a touch-dry state as soon as enough of the water and co-solvent evaporate to increase the viscosity to the required level. This so-called lacquer dry is due to the very high viscosity of the binder. *(3)* Type I coatings require a significant amount of chemical reaction before the requisite viscosity is achieved, and are generally slower drying. Typically, the useable pot life of type I systems is also shorter than for type II. The differences in resin molecular weight also result in different mechanical properties.

Comparison of Waterborne and Solvent-borne Epoxy Coatings. In clear solvent-borne epoxy formulations, the resin and amine curing agent exist in a true, isotropic solution. Epoxy resins, on the other hand, are highly water insoluble. To be used in an aqueous medium, the epoxy resin must be dispersed with the aid of a surfactant into colloidal particles, ranging in size from about 100 to several thousand nanometers. In an aqueous medium, the amphiphilic amine hardeners will partition between the aqueous and epoxy phases and the epoxy/water interface. They may also exist in self-assembling aggregates of their own.

Unlike most other aqueous dispersions used in coatings, the constituents of an epoxy coating react at room temperature. Indeed, water is an excellent catalyst for the amine/epoxy reaction, *(4)* as shown in Figure 1. The net result is that the molecular weight and chemical nature of an epoxy formulation is constantly changing once the components are mixed. In solvent-borne epoxy coatings the primary result of this reaction is an increase in viscosity that eventually indicates the end of the useable pot life of the system. In waterborne systems, the combination of colloidal phases and an ongoing reaction affect the rheology and pot-life in different ways, and also add significant complexity to the film formation process. Variability in performance also results from particle size effects, and from changes in the humidity and temperature of application.

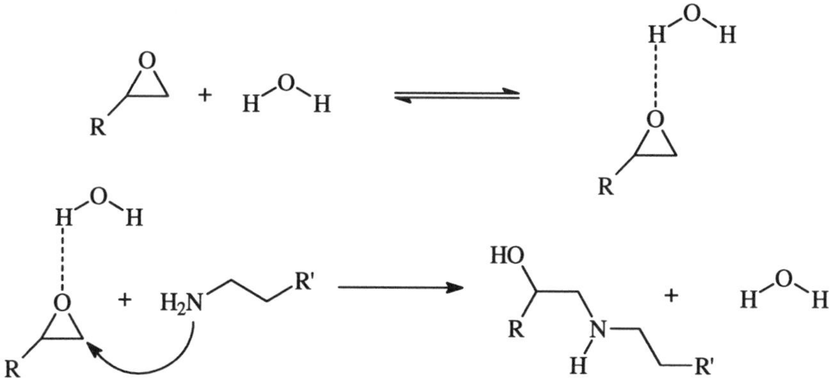

Figure 1. Water Catalyzed Epoxy Ring Opening With Amines

Rheology. Epoxy resin dispersions and emulsions, and their corresponding clear coatings, display a thixotropic or pseudoplastic rheology, typical of aqueous dispersions. *(5)* The viscosity of waterborne epoxy formulations exhibit complex trends during the pot life. Depending on specific compositions, formulations may decrease in viscosity soon after mixing, remain relatively constant throughout the pot life, or exhibit an increasing viscosity. Examples are shown in Figure 2.

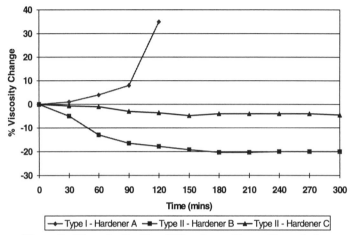

Figure 2. Viscosity Profile of Waterborne Epoxy Formulations

Of the systems depicted in Figure 2, the formulation with the increasing viscosity is a type I system, and the two other systems are of type II, made from the same epoxy dispersion, but different hardeners. An explanation for the drop in viscosity has not been reported in the literature. It may be the result of transfer of some of the amine curing agent from the aqueous phase (where its presence would increase viscosity) to the disperse phase (where its presence would have little or no effect on viscosity. *(6)* In any event, the effect is generally undesirable, since it can lead to a decrease in the sag resistance, and other inferior application properties of the coating.

Pot Life. Though sometimes the end of pot life for a waterborne epoxy system is signaled by an increase in viscosity, it is more commonly indicated by a change in some important property of the resulting film. Figure 3 shows the gloss vs. time after mixing for two representative top coat formulations.

As Wegman *(7)* has shown, the ongoing reaction increases the molecular weight and T_g of the disperse phase resins. This causes an increase in the minimum film-forming temperature (MFT) of the composition. Pot life can be increased by raising the cure temperature, or in some cases by addition of co-solvent or plasticizer to a formulation. In low gloss coatings the end of pot life may be indicated by some other property, such as a change in the humidity resistance of the film.

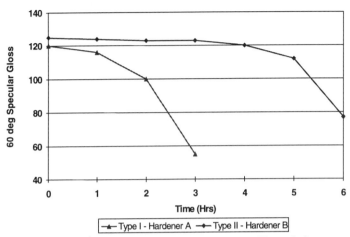

Figure 3. Gloss Profile of Waterborne Epoxy Formulations

Film Formation. Though a typical solid epoxy dispersion particle is only 500 nm in diameter, on a molecular level this is a large distance. Taking the density of epoxy resin to be 1.16 g/mL, assuming a molecular weight of 1000, and ignoring the effect of any swelling of the particle by water, it is easily calculated that there are about 4.6×10^7 molecules per 500 nm diameter particle. To achieve uniform film formation comparable to a solvent-borne epoxy formulation, it is necessary for the amine to diffuse uniformly throughout this viscous resin phase. For this reason, film formation in some waterborne epoxy systems is incomplete, and complex, heterogeneous morphologies result. Meanwhile, the ongoing chemical reactions continually raise the viscosity even higher, increasing the barrier to diffusion. At a certain degree of reaction, T_g exceeds room temperature, at which point diffusion becomes very slow. *(8)* Consequently, the morphology of films may also change as the pot life progresses.

El-Aasser and coworkers *(9)* used transmission electron microscopy to look at the structure of films formed from a solid epoxy resin dispersion and a conventional polyamide curing agent dispersion. Both dispersions were stabilized with a hexadecyltrimethylammonium bromide surfactant plus an unspecified co-emulsifier and a water-immiscible solvent, then homogenized with a Manton-Gaulin disperser, and the solvent removed by steam distillation. They had extremely small particle sizes ranging from about 30 to 200 nm, and thus are not typical of commercially available dispersions. Nevertheless, the results were quite interesting. The surface of the films prepared from freshly mixed dispersions and examined by a replication technique was shown to gradually coalesce over a period of 15 days. When a catalyst for the amine/epoxy reaction was included in the epoxy dispersion, no coalescence occurred. The films were also cross-sectioned and stained with osmium tetroxide. The osmium stains the double bonds present in the dimer acid moieties of the polyamide curing agent. The cross sections of both the catalyzed and uncatalyzed films showed heterogeneous morphologies, with epoxy domains dispersed in a polyamide matrix, although there was clearly much more coalescence in the absence of catalyst.

Walker and Schaffer *(10)* developed the use of 3-iodopropionic acid as a stain to look for the presence of amine-rich domains in microtomed cross-sections of cured waterborne epoxy films. Figures 4 and 5 show the transmission electron photomicrographs obtained from a type II system consisting of a solid epoxy dispersion with an equivalent weight of 625 and epoxy-amine-adduct curing agent. The films were prepared 0.5 hours and 2 hours after mixing, respectively, then cured for >2 weeks at ambient temperature. The dark domains at 0.5 hours are attributed to phase separation of some of the amine from the epoxy continuous phase. Two hours after mixing, more of the amine has reacted with the epoxy particles. This increases the compatibility of the amine with the epoxy, removing the thermodynamic driving force for phase separation. However, the individual particles are still present, indicating incomplete coalescence. The particles also stain darker at the particle boundaries than in the cores, suggesting a higher amine concentration at the interparticle boundaries. It was speculated that these morphologies may explain some of the difference in water resistance properties generally associated with the use of waterborne epoxy coatings when compared to their solvent-borne counterparts.

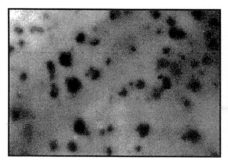 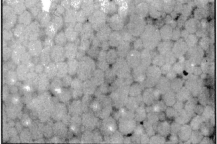

Figure 4. Cross-section of a type II   Figure 5. Cross-section of a type II
waterborne epoxy film, cast 0.5 hours   waterborne epoxy film, cast 2 hours after
after mixing. Stained with 3-   mixing. Stained with 3-iodopropionic
iodopropionic acid.   acid.
(Figures 4 and 5 reproduced with permission from reference 37. Copyright 1995 University of Southern Mississippi.)

The same technique was applied to a type I system that was developed to specifically improve the film coalescence. Liquid epoxy resin was employed because its lower viscosity would be expected to lead to higher diffusion rates, and the resin was emulsified and cured with a modified, highly epoxy-compatible cycloaliphatic amine. The film formation was much more uniform at this level of resolution. Electrochemical impedance spectroscopy showed that these films also gave a very high pore resistance, which is a measure of the resistance of the film to the penetration of ions, when immersed in a sodium chloride solution.

This points to an interesting dilemma as chemists work to improve the water resistance properties of these coatings. By employing type II resins with their higher epoxy equivalent weights, the required amount of amine curing agent is substantially reduced. Since the curing agent is generally more hydrophilic than the epoxy resin, this results in a composition that has an overall more hydrophobic balance. On the other hand, the higher molecular weight results in higher barriers to film coalescence. When type II technologies were first developed, they offered improved water resistance properties over the earlier type I systems, and were the first technology to see significant use in direct-to-metal applications. It would now appear that water resistance properties of certain type I systems is at least comparable, based on reported salt spray and EIS data. It will be interesting to see how these competing approaches compare as waterborne epoxy technology continues to advance.

Mechanical Properties of Films. The stress-strain and dynamic mechanical properties of a proprietary solid epoxy resin dispersion and curing agent combination were reported by Arora and coworkers. *(11)* Interestingly, the tensile strength, elongation, storage modulus, calculated molecular weight between crosslinks (M_c) and T_g were fairly close to those obtained for a solvent-based solid epoxy resin of about the same equivalent weight cured with the same hardener. Surprisingly, however, the values of M_c and T_g showed little change when the amine to epoxy stoichiometry was varied by 20% to either side of unity.

When these results are compared to the morphology data discussed under film formation it would be interesting to discover if the system studied by Arora has a uniform morphology, in which case mechanical properties similar to a solvent borne film would be expected. Also of interest are how the mechanical properties of heterogeneous waterborne systems compare to those formed from materials of similar molecular weight and functionality, but cast from solvent. We have conducted some informal experiments of this nature in our laboratories, and determined that systems cast from solvent had, for example, much greater impact resistance than corresponding formulations cast from water. *(12)*

Waterborne Amine Hardeners: General Structural Requirements. In many type I systems, the curing agent serves a dual role: i) as emulsifying agent for the liquid epoxy resin; and ii) as crosslinking agent. There are three generic types of curing agents, all based on polyethyleneamines, which have been in common use in traditional solvent borne epoxy coatings and which have the basic amphiphilic structure required. The first of these are amides prepared by reaction of the amine with a fatty acid, **III**, which are commonly referred to as amidoamines. The second are similar condensates prepared from dimer fatty acids, **IV**, known as polyamides. The final type, **V**, are prepared by reaction of the amine with an epoxy resin, and have come to be called amine adducts. It should be noted that **III - V** are highly idealized structures. The polyethyleneamines themselves are mixtures of linear, branched, and cyclic structures. Since the polyethyleneamines are multifunctional, unless special techniques are employed, reactions with dimeric reagents such as dimer acid or epoxy resin will result in some degree of polymerization. Even in reaction with a monofunctional fatty acid, there will be significant numbers of species formed with more than one fatty acid moiety attached to the polyethyleneamine. The fatty acids employed in both the amidoamines and polyamides are complex blends of mostly C_{18}

fatty acids with differing amounts of unsaturation. The dimerization of tall oil fatty acid yields numerous isomers besides that depicted. Finally, for both amidoamines and polyamides there is at tendency for the β-aminoamides to lose a second mole of water to form the corresponding imidazoline, **VI**.

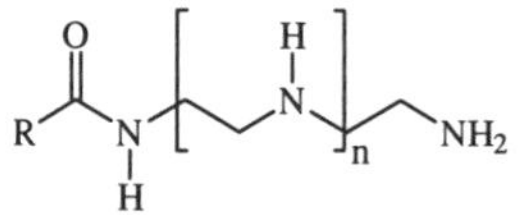

III

R = C$_{17}$H$_{29-35}$

n = 1 - 4

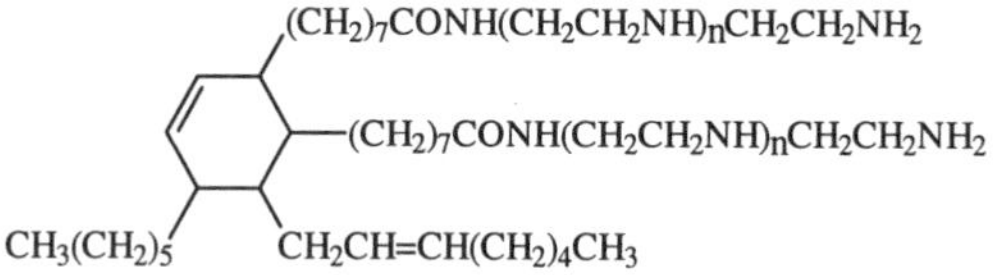

IV

n = 1-4

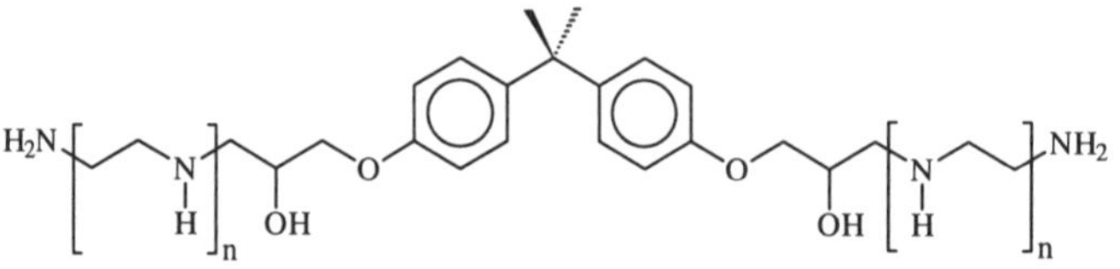

V

n = 2 - 4

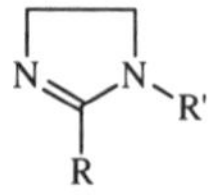

VI

R' = (CH$_2$CH$_2$)$_n$CH$_2$CH$_2$NH

Though the traditional curing agents **III** - **V** frequently can be seen to be the basis of waterborne curing agents, they must be modified before they will be acceptable both as surfactants and film crosslinkers. The specific chemistries employed are discussed in the section on type I curing agents, below. However, there are some common themes to these modifications.

To extend pot life of waterborne epoxy systems, it is usually necessary to decrease the amine reactivity. Primary amines are more reactive with epoxy functional groups than secondary amines. *(13)* By treating the hardener with a reagent that reacts preferentially with primary amines, the overall reactivity is reduced. Examples of commonly employed reagents for this purpose include monofunctional epoxides, formaldehyde, and unsaturated reagents capable of undergoing Michael addition reactions such as acrylonitrile.

The general incompatibility of epoxy resins and amines can also lead to difficulties. This incompatibility has always been a problem with traditional epoxy coatings, which have a tendency to exude amine on their surface. This can result in the formation of carbonate salts known as 'blush', *(14)* or a greasy surface layer. In waterborne formulations incompatibility may lead to other surface appearance defects such as cratering, and may also decrease the stability of the epoxy emulsion. Compatibility of a hardener is frequently enhanced by reaction with monofunctional epoxy diluents, particularly aromatic diluents, or by reaction with epoxy resin. Because polymer compatibility normally decreases with increasing molecular weight, these modifications are even more important in type II technologies due to the higher molecular weight of the epoxy dispersions. These adduction procedures have the desirable effect of also reducing reactivity.

Finally, many waterborne curing agents are treated with a small amount of an organic acid. The most commonly used is acetic acid, which because of its low molecular weight, will at least partially volatilize from a film and thus will be less detrimental to water resistance properties than some other organic acids. The pK_a of acetic acid is 4.76, while the pK_a's of the conjugate acid of organic amines are typically in the range of 8 - 11. When combined, the acid is essentially completely ionized, and the solubility of the hardener in water is increased. In addition, the HLB value of protonated hardeners is increased. Generally this will be desirable, since a surfactant must be more water soluble than oil soluble in order to obtain a stable oil-in-water emulsion. *(15)* Emulsion stability will also be enhanced by the formation of an electrical double layer and repulsion of similarly charged dispersions, as generally explained by the DLVO theory. *(16)* Finally, it has been found that addition of small amounts of acid increases the pot life. This has been attributed to a retarding effect on the amine/epoxy reaction due to amine neutralization. However, we have found that levels of acetic acid employed in waterborne epoxy coatings actually slightly decrease the thin film set time of a nonaqueous epoxy-polyamide formulation (unpublished data), and other workers have found that carboxylic acids, including acetic acid, are accelerators for epoxy/amine systems. *(17)* It seems more reasonable that the acid shifts the partition of amine more into the aqueous phase, decreasing reaction with epoxy resin by a physical separation of the reagents.

Jack et. al. *(18)* measured the critical micelle concentration (CMC) values of three polyamide curing agents, which range from 0.4 to 1 g L^{-1}. These authors also reported that the polyamides lower the epoxy resin-water interfacial tension from 10 to 2 dyne cm^{-2}.

In contrast to the above work, in a recent publication Verkholantsev *(19)* measured the surface tension of a number of commercial waterborne hardeners, including amindoamines, **III**, polyamides, **IV**, and adducts, **V**. He found that they all exhibited a cloud point as the solutions were reduced with water, somewhere between a solids content of 6 and 25%. Surface tension was found to be constant, even when diluted below the cloud point. Since no inflection in the curve was found, it was concluded that none of these hardeners exhibit a CMC. The author suggested that these hardeners should be referred to as "noncolloid" (i.e. nonmicellar) surfactants.

Verkholantsev also conducted some interesting experiments that led to his assertion that type I hardeners can 'self-emulsify' epoxy resin. Some confusion can arise reading this paper over the term 'self-emulsification'. In normal usage within the coatings industry, self-emulsification simply refers to the fact that no additional surfactant must be added to the system to emulsify the epoxy resin; i.e., the curing agent is a sufficient emulsifier. This author is instead referring to the ability of the hardener to spontaneously form microemulsions of the resin; i.e., thermodynamically stable dispersions. The assertion for a spontaneous process was based on two experiments. In the first, an aqueous mixture of hardener and epoxy resin were rapidly mixed for two minutes. The mixtures were divided into two portions. One was stirred, and the other was left at rest. After the initial mixing, a coarse (> 1000 nm) unimodal dispersion was obtained. After resting for one half hour, the dispersion had turned into a bimodal distribution with a small particle size fraction of 155 nm diameter. In the second type of experiment, an aqueous solution of a polyamide hardener was carefully spread over a layer of epoxy resin in a petri dish. In one to two hours, a milky-white interfacial layer forms, and within 6 - 20 hours, a crosslinked film is formed that is durable enough to be collected and washed. The amount of film formed varied inversely with the viscosity of the epoxy resin, and film thicknesses ranging from 60 to 350 μm were obtained. Were the film to form by Fickian diffusion of the hardener into the resin rather than the proposed self-emulsification process, and assuming a diffusion constant of 10^{-12} cm/s^2, it was calculated that the process should take approximately 230 days. Confirmation of the hypotheses of Verholantsev would constitute an interesting area for further study.

Jack et. al. *(18)* also studied the film formation process in these type I systems by monitoring electrical resistance and water loss concurrently. Resistance was constant until the solids content reached about 80%, then gradually started to rise, while the water evaporation rate began to slow. This was taken as the beginning of the rupture of water as the continuous phase, and the starting point for conversion from an oil-in-water to a water-in-oil emulsion. Between about 90 and 95% solids, there was an abrupt increase in resistance, and water evaporation became very slow. This signaled the end of any conductive continuous phase, and the point where water must migrate through polymer or pores in the coating to evaporate.

Type I Waterborne Epoxy Technologies

Polyamide Curing Agents. The first commercially available waterborne curing agents were essentially unmodified polyamides, **IV,** partially neutralized with carboxylic acids. *(20,21,22)* Although polyamides of this type readily emulsify a liquid epoxy resin, the quality of the cured coatings was clearly inferior to their conventional solvent-borne polyamide counterparts. They were slow to cure, extremely sensitive to the cure environment, and more often than not gave soft, low gloss coatings with poor chemical and corrosion resistance. The lack of performance was probably due to poor system compatibility and film formation. In the mid 1970s the performance of the polyamides was significantly improved. This was accomplished via the partial adduction of the polyamide with either an aromatic mono glycidyl ether, *(23)* or with the diglycidyl ether of bisphenol A, the latter approach being the preferred route. Although both liquid and solid epoxy resins could be used, partial adduction with solid epoxy resins with an average molecular weight ≥900 gave the greatest improvement in compatibility and emulsion quality. Partial neutralization with acetic acid was required to maintain water solubility. The above modifications resulted in products with improved water stability and resin compatibility. The resulting coatings offered improved dry speed, gloss, hardness and chemical resistance.

In the early 1970s, Richardson *(24)* developed a curing agent for 0 VOC systems based on a combination of modified polyamide and polyamine adducts. It was the first product to gain wide commercial acceptance, and found widespread use in topcoats in hospitals, schools and industrial warehouses and as sealers for concrete. Even today these are among the largest markets for waterborne epoxy coatings. Another application area where waterborne epoxy systems offered significant performance advantages over existing technology was coatings for nuclear power plants, since these coatings proved relatively easy to decontaminate. *(25)*

Despite the widespread introduction and application of modified polyamides, the relatively short pot life of waterborne systems was always considered a shortcoming. A conventional solvent based system has a typical pot life of 6-8 hours, compared to only 1 hour for the modified waterborne polyamides. Richardson *(26)* reacted the polyamide with carbon dioxide, and achieved pot lives up to 6 hours without adverse effect on the dry speed and coating performance.

A patent published by Moes and Small *(27)* describes a method of preparing a polyamide based waterborne curing agent from a blend of a C_{18} dimer acid-tetraethylenepentamine polyamide and a mono epoxide-polyethyleneamine adduct. The addition of the mono epoxy adduct was shown to significantly reduce the dry time.

Polyamine Epoxy Adducts. Despite the success of the polyamide curing agents, the products suffered from several inherent weaknesses. These included, a) poor color, which made it difficult for formulators to develop high gloss white enamels and pastel paints; b) high initial viscosity resulting in low film build at application viscosity; and c) poor water resistance and anti-corrosive properties. The poor color appears to be an unavoidable result of the high temperatures required for polyamide production. Many developments have focused on resolving several of the above issues and have led to the introduction of new curing agents based on polyamine adduct technology.

Use of cycloaliphatic amine epoxy adducts for waterborne epoxy systems was described by Neffgen and Allewelt. *(28)* They were the first class of waterborne curing agents not to exhibit flash rusting on steel, thus opening up the potential for the use of these products as anti-corrosive coatings. No chemical basis for the improvement was given. Excellent anti-corrosive properties have been claimed. *(29)*

Cornforth and Darwen *(30)* combined epoxides with differing levels of hydrophobicity to yield epoxy-amine adducts with an appropriate balance of performance properties. A more hydrophilic adduct was prepared by reacting a polyamine, preferably a high molecular weight polyethylene polyamine with a mixture of mono- and polyepoxides. The latter are preferably a combination of aliphatic and aromatic polyepoxides. The preferred aliphatic di-epoxides are those having an epoxy equivalent weight in the region 120-140, for example hexanediol diglycidyl ether. The aromatic polyepoxides include the polygylcidyl ethers of bisphenol-A **Ia, Ib** or bisphenol-F **II**. Preferably the levels of epoxides used ensure that from about 20-40% of the available primary amine functionality in the polyethylene polyamine are allowed to react with the monoepoxide and from about 5-65% of the available primary amine groups are reacted with the polyepoxide. A more hydrophobic adduct was prepared by reacting a low molecular weight polyethylene polyamine with an aromatic mono glycidyl ether, preferably phenyl or cresyl glycidyl ether. To facilitate dispersibility of the polyamine adduct blend, the mixture is finally treated with formaldehyde.

An important property claimed for this technology is reduced viscosity in aqueous media when compared to commercial waterborne polyamide hardeners. This results in higher formulated volume solids, in turn leading to higher film build per application. Commercial products have been formulated with both liquid resins (type I) and more recently with solid epoxy resin dispersions (type II). Coatings were shown to have rapid hardness development and excellent chemical and stain resistance in enamel topcoats.*(31)* Another important property was long term retention of gloss at ambient storage conditions, which has been shown to slowly deteriorate in some waterborne epoxy systems.

Miscellaneous Curing Agents: Toshiaki and co-workers *(32)* recently described a new approach for the synthesis of waterborne epoxy curing agents. The patent discloses an aqueous epoxy resin composition that is composed of an amidoamine curing agent obtained by the reaction of a fatty acid with polyamine, **VII**. The patent claims that the coatings obtained possess excellent gloss and surface appearance. Hardness development and adhesion to steel substrates are also said to be improved, which makes them ideal candidate materials for a wide variety of industrial applications.

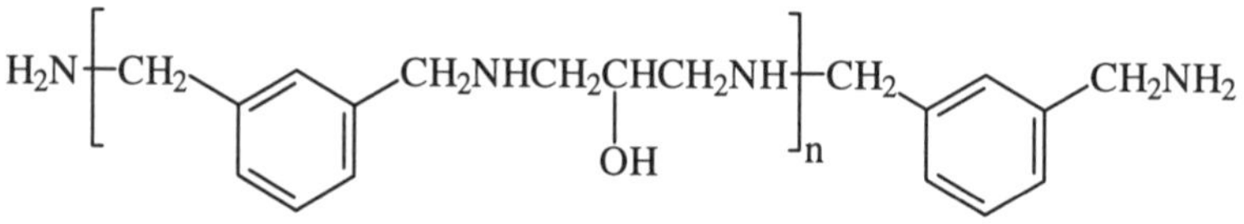

VII

Another novel approach is the use of alkyl phenol Mannich condensates as developed by Speranza and Lin. *(33)* Phenol formaldehyde Mannich condensates are widely used to cure solvent based liquid epoxy resin systems, but are not normally water soluble. This problem is solved by forming Mannich condensates from polyoxyalkylene diamines to yield **VIII**. The polyoxyalkylene diamines used in the invention can be either polyethylene oxide or polypropylene oxide based; nonyl phenol is the preferred alkyl substituted phenolic compound. These products are claimed to form excellent stable emulsions with liquid epoxy resins and cure to give extremely tough, high gloss films.

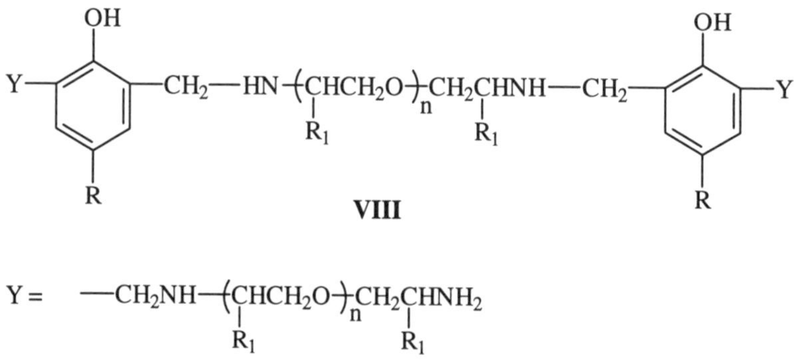

VIII

$Y = $ $-CH_2NH-(CHCH_2O)_n CH_2CHNH_2$
         R_1 R_1

$R = C_9H_{19}$, $R_1 = H, CH_3$

There are patents describing the modification of traditional solvent-borne curing agents by neutralization with carbon-based organic acids. Albers *(34)* demonstrates the use of nitroparaffins such as 2-nitropropane and nitroethane with polyamines, including aliphatic and cycloaliphatic amine-epoxy adducts and polyamides. The nitroparaffin increases the solubility of the polyamine and increases colloidal stability, but is believed to be more volatile than carboxylic acids, yielding improved water resistance. A similar approach to this technology has also been patented by Hoefs *(35,36)* where the hydrophobic polyamine is solubilized in water via addition of alternative carbon acids such as phenyl sulphonyl propan-2-one, methyl cyanoacetate and ethyl formate.

Due to the increased hydrophobicity of the cured films, these systems behave more like traditional solvent-borne coatings and have been recommended for use on a variety of metal substrates as both topcoat and anti-corrosive primers. Many of the paint formulations developed conform with US Military Coating Specifications, making them the first water-based coatings to achieve this certification.

Whereas most type I curing agents employ an amine moiety as the hydrophilic portion of the emulsifier, Walker, Everett and Kamat *(37)* took the approach of covalently bonding a water soluble polymer (WSP) to a large excess of a highly water-insoluble mixture of polycycloaliphatic polyamines (PCPA), as shown in Figure 6. The system was designed to yield a more homogeneous film structure than for type II systems as discussed in the above section on film formation. By employing a highly compatible and hydrophobic amine and liquid epoxy resin, driving forces for phase separation could be reduced. The amine structure was also chosen because of its lower reactivity with epoxy resin, which led to a pot life of about 4 hours in a clearcoat formulation. The electrochemical impedance and homogeneous film formation obtained are also discussed in the film formation section (above).

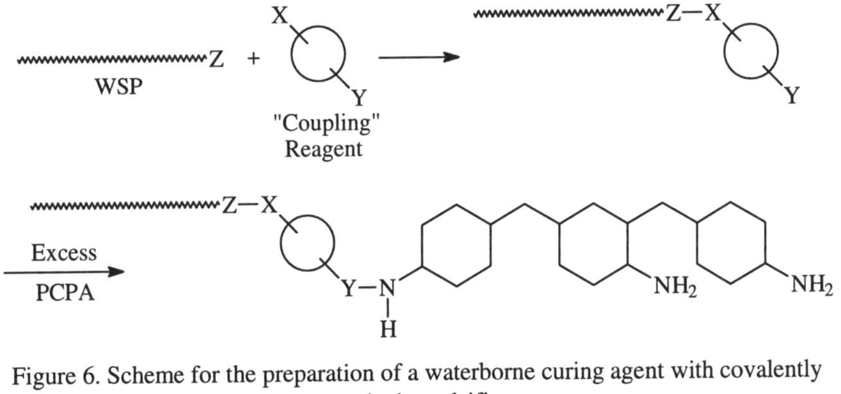

Figure 6. Scheme for the preparation of a waterborne curing agent with covalently attached emulsifier.

Recently, type I systems have been developed that exhibit some of the lacquer dry characteristics of type II technologies. This has been accomplished by the development of curing agents that contain higher molecular weight entities in the form of a colloidal dispersion.

Klippstein *(38)* described a process for preparing a curing agent dispersion by adding a solution of solid epoxy resin in glycol ether to an aqueous solution of a commercial curing agent. When equal weights of liquid epoxy resin were substituted for the higher molecular weight resin, the system gelled. When compared to the original hardener from which the product was derived, phase 4 thin film set times (B-K recorder) for clear coats were reduced from 12 hr. to 5 hr., and there was also a much more rapid buildup of hardness and solvent resistance. This was accomplished with no loss in pot life.

Nederhoff *(39)* patented a microgel dispersion prepared in a two-step process (Figure 7). In step 1, an advanced epoxy resin with an equivalent weight of 300 to 900 is treated with 0.1 to 0.3 eq. of the carboxylate salt of N-methylmorpholine in glycol ether solution. This intermediate is added to a polyamine such as meta-xylylene diamine (MXDA) or isophorone diamine (IPDA). In a later paper, *(40)* Takas and Naderhoff described a commercial curing agent that has a particle size of 0.066 μm, and an AHEW of 560 based on solids. This curing agent is typically formulated with a pre-dispersed liquid epoxy resin, and reaches a tack free state in a primer formulation in as little as 10 min. At the same time, gloss pot life was >8 hr.

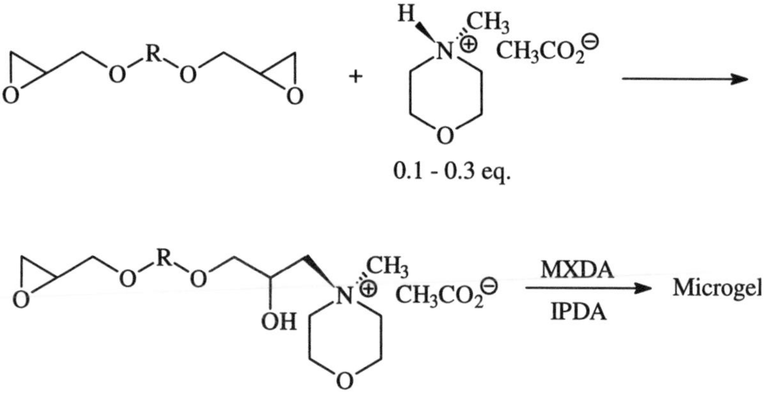

0.1 - 0.3 eq.

Figure 7. Preparation of Microgel Dispersion Curing Agent

The authors measured the particle size of a clearcoat system through the pot life and found it to be stable, in contrast to an unidentified competitive system that more than doubled. They also found that when the curing agent was combined with an epoxy resin emulsion with a particle size of 1.73 μm, that the resulting product had a particle size of 0.76 μm, with no visible presence of any particles in the 1 μm or larger size range. It was then postulated that the epoxy resin is transported to the surface of the microgels, and that the only reactions would be intraparticle, as opposed to the interparticle crosslinking that caused the increased particle size of the competitive system. It should be noted, however, that sufficient intraparticle crosslinking will limit pot life by gloss, since it will cause an increase in the MFT and T_g, as discussed in the above section on film formation. The rapid lacquer dry reported would appear to be due to the very large viscosity contribution that would be anticipated from the microgel.

Type II Waterborne Epoxy Technologies

Epoxy Resin Dispersions. The impetus behind the development of type II technologies was specifically to improve dry speed and water resistance. This led researchers to change their strategy toward the development of high molecular weight solid epoxy resin dispersions. An article published by Kurnik and Roy *(41)* discusses some of these early developments. The advantages of this approach are that coatings would reach the initial tack free state much faster, since they would rely more on the lacquer dry effect of the solid resin during the early period following application, rather than any real degree of crosslinking. The higher molecular weight resins also help to improve flexibility and increase the level of hydrophobicity.

A process for the preparation of a solid epoxy dispersion is described by Becker. *(42)* A common problem associated with the preparation of solid epoxy resin dispersions with an external nonionic surfactant such as the block polymers of polyethylene glycol and polypropylene glycol or the nonyl phenol ethoxylates is a broad particle size distribution (1-3 microns), leading to poor settling stability. To improve this, a nonionic dispersant in the form of a copolymer derived from polyalkylene glycols and **Ia** was used to disperse a solid epoxy resin. An example dispersant was prepared from a mixture of liquid epoxy resin (EEW 190) and polyethylene glycol (PEG 3000) with a BF_3 catalyst. This dispersant was then used to form a dispersion of an organic solution of a solid epoxy resin having an EEW of 470 with vigorous stirring ($\geq$ 1000 rpm). Further water dilution yields a low viscosity aqueous dispersion with a particle size < 1.0 μm.

More recently Dreischhoff and co-workers *(43)* have made a more flexible composition similar to the above, but where a polypropylene glycol diglycidyl ether is included in the advancement reaction used to prepare the dispersed epoxy resin. Workers from the same company later reported that films formulated from an experimental flexibilized epoxy dispersion had good hardness development and a relatively high T_g, yet they exhibited excellent adhesion to plastic substrates such as EPDM rubber and passed a reverse impact test at -13°F on this substrate with >90% of the energy uptake of uncoated controls. *(44)*

Elmore and Cecil *(45)* developed a modified solid epoxy resin dispersion by reacting a mixture of a liquid epoxy resin (EEW 190), bisphenol-A and a polyoxyalkylene block copolymer surfactant. Chemically the surfactant is the diglycidyl ether of a high molecular weight (>4,000) block copolymer of polyethylene glycol and polypropylene glycol. The above reaction is carried out in the presence of a catalyst at process temperatures in excess of 280°F. Water and a glycol ether are added under vigorous agitation to complete the dispersion process. The surfactant is covalently attached to the epoxy resin and is thus non-fugitive, which in turn leads to coatings with a higher degree of water resistance.

Williams, Burt and Golden, *(46)* describe a process for the manufacture of a high molecular weight, polyfunctional epoxy resin dispersions. These products were specifically developed to give coatings with improved corrosion resistance and flexibility when formulated with waterborne polyamine adduct curing agents. The first stage of manufacture utilizes the dispersant technology previously described by

Elmore. After reaction of a liquid epoxy resin, bisphenol-A and PEG-diglycidyl ether mixture, the resin undergoes a second processing step, where it is modified by a post advancement reaction with an epoxy novolac resin. The preferred novolac is one where the average functionality is ≥3.

Recent innovations include the development of self-emulsifiable novolac and semi-solid type epoxy resins. Klein *(47)* has published extensive evaluation data on the performance of the semi-solid epoxy resin with both modified polyamide and polyamine adduct curing agents. Advantages claimed using this resin are low VOC, good corrosion resistance and adhesion to a variety of substrates. Klein *(48)* also described a composition that consists of an epoxy terminated epoxy ester dispersion, stabilized with an external surfactant. The epoxy ester is prepared from the reaction of a diepoxide (preferably with a 300 - 500 EEW) with dimer acid, optionally in the presence of additional diphenol and monophenol, the latter to act as a chain terminator. The preferred surfactants are nonylphenol terminated block copolymers of propylene oxide and ethylene oxide. Coatings cured with a waterborne polyamide hardener had >160 in-lb reverse impact.

Polyamine Curing Agents. Shimp and co-workers, *(49)* were one of the first groups to successfully develop amine hardeners for use with the type II solid epoxy resin dispersion technology. Their approach was to prepare a polyamine adduct via a two stage reaction process. Stage 1 involves reacting a bisphenol-A diglycidyl ether with a polyethyleneamine. Diglycidyl ethers having an EEW in the range of 180-900, are recommended, the actual resin used depends upon the desired physical and mechanical properties of the final coatings. When the adduction reaction is complete, unreacted amine is removed by vacuum distillation. This is claimed to offer several performance advantages including reduced water sensitivity and improved system stability. Stage 2 involves reducing primary amine functionality by end capping the polyamine adduct with a combination of aliphatic and aromatic mono glycidyl ethers. In a detailed study, the inventors have demonstrated that at least 25 mole-% of the end capping agents employed must be aliphatic in nature, otherwise compatibility of the epoxy resin-curing agent dispersion is compromised and the aesthetics of the cured coating deteriorate. The preferred aliphatic end capping agents are those obtained from a long chain monohydric alcohol having 11-15 carbon atoms. Typical aromatic glycidyl ethers include phenyl or cresyl glycidyl ether. Finally the curing agents are rendered water compatible via partial neutralization with acetic acid. The chemical structure of this type of polyamine adduct is given in Figure 8.

The end capped polyamine adducts have been shown to exhibit excellent compatibility with several of the epoxy resin dispersions previously described. In addition, they are also compatible with either standard or self-emulsifiable liquid epoxy resins. As a result curing agents can be used to develop coatings with a variety of differing properties, including pot lives ranging from 2 to 8 hours. Compared to waterborne polyamide hardeners, color of the modified polyamine adducts is significantly improved, allowing for the development of high gloss white enamels and pastel shade decorative paints.

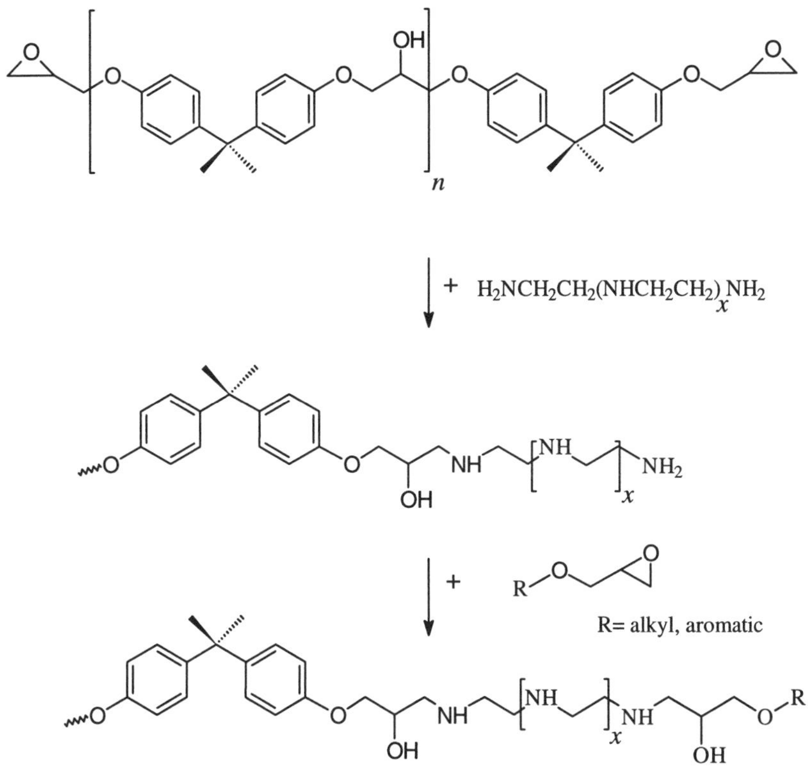

Figure 8. Synthesis of Modified Polyamine Adduct Curing Agent

DeGooyer *(50)* describes a process for the manufacture of a polyfunctional curing agent based on an epoxy novolac. The curing agent is prepared by reacting a novolac resin having a functionality in the range 3-7.5, with an excess of a polyamine. Again the preferred class of amine used for the waterborne curing agents is the polyethyleneamines, in particular, triethylenetetramine. At least 3 moles of amine per epoxy group are employed during adduct synthesis to reduce the degree of oligomerization and control curing agent viscosity. After adduction and subsequent removal of excess amine, the isolated adduct is end capped with either a mono carboxylic acid or a mono epoxide. The preferred end capping agent is the aromatic mono epoxide, cresyl glycidyl ether. As with the end capped adducts described in the Shimp patent, products based on this technology also require addition of acetic acid to aid water solubility. In general, the end capped epoxy-amine adduct is reacted with sufficient acid to neutralize from about 2 to 20% of the available amine. The proposed structure of the curing agent **IX** is given below.

IX

$n = 1\text{-}4; \ x > 1 < 5$

DeGooyer, *(51)* has also developed a modified polyamine adduct offering coatings with improved adhesion and flexibility. This is important for many direct to steel applications where such properties are considered essential for coating durability. Flexibility was achieved by incorporating a long chain ethylene oxide or propylene oxide moiety into the backbone of the curing agent. The approach first involved synthesis of an epoxy functional co-polymer by advancing an aromatic bisphenol-A diglycidyl ether (EEW 190) with an aliphatic polyether diepoxide. The preferred reagent was a polyethylene glycol diglycidyl ether with an average molecular weight of ~400. The reaction is conducted at elevated temperatures using a catalyst such as ethyl triphenylphosphonium iodide, until an epoxy resin is formed with an epoxy equivalent weight in the 600-900 range. Curing agents are produced by adducting the above resin with a polyalkyleneamine, followed by end capping all the primary amine groups with either a mono carboxylic acid or a mono epoxide. The amine hardeners described in the above patent are claimed to possess excellent dispersion stability and yield coatings exhibiting good hardness and chemical resistance. As expected the coatings were also shown to have improved flexibility over cold rolled steel. Measurements of reverse impact resistance exceeding 100 inch. lbs are reported.

Becker and Karasmanns *(52)* developed a unique curing agent. The products described in the patent were among the first curing agents with a nonionic emulsifier built into the polymer backbone. Incorporation of the emulsifier is achieved by advancing a liquid epoxy resin with a mixture of bisphenol-A and a high molecular weight polyethylene glycol (PEG 1000). The advancement is carried out in the presence of a BF$_3$-amine catalyst to yield the modified epoxy resin **X**.

Polyamine adducts are then prepared by reacting **X** with an excess of a diamine, the preferred material being *meta*-xylylene diamine. The inventors have also used the technique of end capping as a means of extending the pot life. After adduction, the primary amine concentration in the polyamine adduct is reduced via Michael addition of acrylonitrile, to yield the end capped polyamine adduct **XI**.

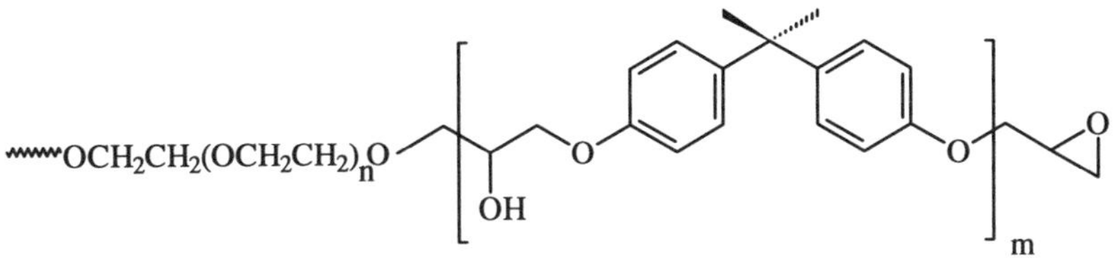

X

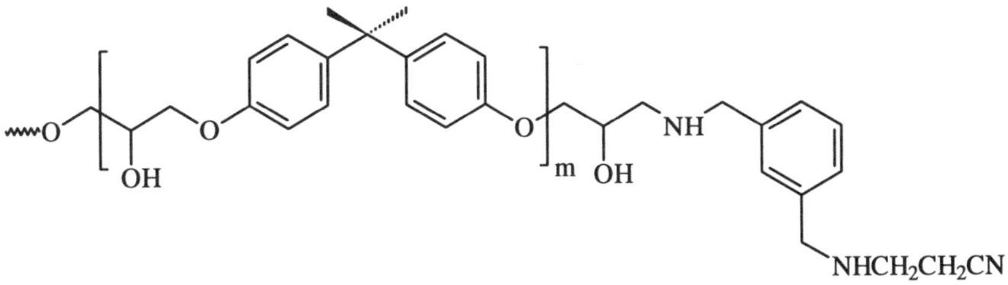

XI

By incorporating the emulsifier into the curing agent backbone, this product does not require the use of neutralizing agents to increase water solubility. More recently researchers within the same organization have patented curing agents where the polyethylene glycol modified epoxy-amine adduct was end capped with an aliphatic mono epoxide. *(53)*

The use of end capping technology as a means of reducing primary amine functionality appears to be a theme most common with all the polyamine adducts if long pot life and good system compatibility are required. Recently two new products using this strategy have been developed. Chou, Shah and Jewell, *(54)* patented a polyamine-liquid epoxy resin adduct, end capped with an excess of aromatic glycidyl ether. In this patent a combination of a mono alkylene diamine (e.g., hexamethylene diamine) and a polyethyleneamine are claimed to provide the best balance of performance properties. The preferred ratio of mono alkylene diamine to polyethyleneamine is from about 5:1 to about 3:1. Products are free of neutralizing acids and resulting coatings do not exhibit flash rusting when applied over untreated steel. Flash rusting is a phenomenon occasionally observed with acetic acid containing curing agents. Walker, Cook and Dubowik, *(55)* report a new modified polyamine adduct which was specifically designed to give improved coating performance with the solid epoxy resin dispersions. An important property of this curing agent is that it exhibits uniform viscosity throughout the pot life. Formulated coatings have been shown not to experience the sharp viscosity decrease commonly observed with many products used in type II epoxy coating systems. Primer formulations applied to grit blasted steel gave rather good performance, with little deterioration in over 2000 hours accelerated corrosion and humidity resistance.

Formulation Guidelines

When formulating with waterborne epoxy resin systems, special attention must be paid to the choice of pigments, wetting agents, defoamers and co-solvents, and their likely effect upon the stability and performance properties of a formulated paint. In recent years several papers have been published highlighting formulation guidelines. Some of the important factors to be considered are discussed below.

Resin Stoichiometry. One of the most important formulating tools that can be utilized with two pack waterborne epoxy systems is varying the resin stoichiometry to achieve a whole range of performance properties, as reviewed by Jackson, *(56)* and by Galgocci and Weinmann. *(57)*
With solvent borne two pack epoxy coatings, it is conventional formulating practice to combine the amine curing agent and epoxy resin at a stoichiometric ratio of about 1:1, where the loading of curing agent is calculated based on the epoxy equivalent weight (EEW) of the resin and the amine hydrogen equivalent weight (AHEW) of the curing agent. Typically, the curing agent usage is expressed as parts per hundred resin, in accordance with Equation 1.

$$PHR = 100 \times [AHEW]/[EEW] \qquad (1)$$

For waterborne type I technologies the 1:1 stoichiometry or a slight epoxy excess also tends to yield optimum performance. With type II technologies studies have shown that varying the stoichiometry from the theoretical loading can offer several performance advantages. Formulation with excess solid epoxy resin dispersion can dramatically improve water and corrosion resistance properties. Salt fog and humidity resistance of anti-corrosive primers are significantly improved if resin levels as high as 60-90% above the theoretical level are used. Some of the effect is probably due to the fact that the amines are more hydrophilic than the epoxy resin, so that by reducing the amount of amine employed, the overall hydrophobicity of the composition is therefore increased. In addition, at higher amine levels, the nonuniform film morphology discussed under film formation (above) may result in high concentrations of amine in localized domains, through which the transport of water is accelerated relative to other parts of the film. *(9,37)*
In certain applications it can be more beneficial to employ conventional stoichiometry, or even an excess of curing agent. This is of particular importance when solvent or stain resistance of the cured coating is required. Other advantages gained using high levels of curing agent include an improvement in the abrasion resistance, adhesion and dry speed. Table I highlights some of the performance parameters that are improved with varying resin and curing agent stoichiometry.

Effect of co-solvents. Coating properties, particularly of type II systems, are frequently improved if additional co-solvents are added during formulation. Cosolvents can affect several basic properties including film formation, pot life, gloss and ultimate barrier protection. Sometimes a combination of water soluble (e.g., glycol ethers) and water insoluble solvents (e.g., benzyl alcohol) are required for optimum performance properties. *(55)* Coating properties are enhanced due to the effect co-solvents have on both the minimum film formation temperature and film coalescence.

Table I. Performance Parameters Improved With Varying Stoichiometry

Increase in Epoxy Resin	Increase in Curing Agent
Pot Life	Dry Speed
Acid Resistance	Solvent Resistance
Water Resistance	Stain Resistance
Alkali Resistance	Flexibility
Salt Fog Resistance	Higher Gloss
Humidity Resistance	Abrasion Resistance

Recommended Pigments for Waterborne Coatings. In developing paint formulations other considerations include the type of pigmentation packages that are acceptable for waterborne systems. For white gloss enamels, titanium dioxide is used as the main pigment and talc is often used as an extender. Usually a silica or alumina surface-treated grade of TiO_2 is required for dispersion stability and optimum gloss development. Due to increasing environmental concerns the use of lead and chromate based corrosion-inhibiting pigments is no longer considered acceptable. In an extensive study conducted by Jackson *(56)*, anti-corrosive pigments that function well in waterborne epoxy primers are those that are relatively inert and do not release large amounts of water soluble salts. Inert pigments are required to prevent adverse ionic interactions with the strong cationic character of the water miscible amine functional curing agent. Pigments shown to offer the best salt fog, prohesion and humidity resistance include calcium strontium phosphosilicate, modified zinc phosphates, and more recently calcium strontium zinc phosphosilicate. Typically the loading of the anti-corrosive pigment required to obtain maximum corrosion protection is in the range 0.5-1.0 lbs/gal.

In choosing the extender pigments, good results are often obtained if a combination of particle sizes and shapes are employed, which may improve barrier protection. It is usually important that extender pigments with low water soluble content be employed. The use of pigments with low oil and water absorption minimize adverse effects on vehicle-volatile demand and coating viscosity. Typical pigments used include calcium metasilicate, barium sulphate, silica-alumina ceramic spheres, and wet ground mica.

Literature Cited

1. Richardson F. *Pigment and Resin Tech.*, **1973**, 5, 41-43.
2. McAdams, L.V.; Gannon, J.A. In *Enc. Pol. Sci Eng.*, Kroshcwitz, J.I., Ed. 1986, Vol. 6, p 339.

3. Wicks, Z.W. ; Jones, F.N.; Papas, S.P. *Organic Coatings: Science and Technology,* John Wiley and Sons, Inc.: New York, NY, 1992, Vol. 1, p 35.

4. Rozenberg, B.A. *Adv. Polym. Sci.* **1986,** *75,* 113.

5. Wegmann, A. *Spec. Publ. - R. Soc. Chem.* **1995,** *165,* 33.

6. Wicks, Z.W.; Jones, F.N.; Pappas, S.P. *op. cit.,* Vol 1, p 64.

7. Wegman, A. *J. Coatings Tech.,* **1993,** *65* (827), 27.

8. Gillham, J.K.; Wisanrakkit, G.*J. Coatings Tech.* **1990,** *62* (783), 35.

9. El-Aasser, M.S.; Vanderhoff, J.W.; Mirsa, S.C.; Manson, J.A. *J. Coatings Tech.* **1977,** *49* (645), 71.

10. Walker, F. H.; Schaffer, O.; in Provder, T., Ed. *Film Formation,* ACS Symposium Series, American Chemical Society: Washington D.C., in press.

11. Arora, K.S.; Chou, J.L.; Devore, D.I.; Papalos, J. *Proc. XXII Waterborne, High-Solids and Powder Ctgs. Symp.,* **1995,** 129.

12. Dubowik, D.A.; Lindenmuth, D.L. Unpublished results.

13. (a) Gillham, J.K.; Xiaorong, W. *J. Appl. Polym. Sci.* **1991,** *43,* 2267-2277. (b) Dusek, K.; Bleha, M. Sunak, S. *J. Polym, Sci.* A-1, **1977,** *15,* 2393. (c) Charlesworth, J. *J. Polym. Sci. Polym. Chem. Ed..* **1980,** *18,* 621.

14. Wicks, Z. W.; Jones, F.N.; Pappas, S. P. *op. cit.,* Vol 1, pp 173-174.

15. Evans, D.F.; Wennerstrom, H. *The Colloidal Domain,* VCH Publishers: New York, NY, 1994, pp 490-494.

16. Ross, S.; Morrison, I.D. *Colloidal Systems and Interfaces,* John Wiley and Sons: New York, NY, 1988, pp 205-263.

17. Bowen, D.O.; Whiteside, R.C., Jr. In *Epoxy Resins;* Lee, H., Ed.; Advances in Chemistry Series 92; American Chemical Society: Washington D.C., 1970, pp 48-59.

18. Jack, V.L.; Ashitkov, V.A.; Mnatsakanov, S.S.; Kalaus, E.E.; Shaltyko, L.G. *Materials and Structures.* **1990,** *23,* 442-448.

19. Verkholantsev, V.V. *J. Coatings Tech.,* **1996,** *68* (853), 49.

20. Bulgier, L. UK Patent 1,108,558, 1965.

21. Afford, J.A. US Patent 4,013,601, 1977.

22. Ashjian, H., *et. al.* DE Patent 2519390, 1975.

23. General Mills Inc. UK Patent 1,242,783, 1971.

24. Richardson, F.B., *Waterborne Coatings, Surface Coatings 3, Elsevier Scientific Publishing:* New York, 1990, pp 229-254.

25. Richardson, F.B. *Polymers Paint Colour Journal,* **1982,** 714-715.

26. Richardson, F.B. US Patent 4,526,721, 1985.

27. Moes, N.S.; Small, M.P. US Patent 4,089,826, 1978.

28. Neffgen, B., Allewelt, K.H. *Pitture e Vernici,* **1985,** *91* (5), 33-43.

29. Nelson, I. *Polymers Paint Colour Journal,* **1982,** 251.

30. Darwen, S.P.,Cornforth, D.A. US Patent 5,246,984, 1993.

31. Dubowik, D.A., *Proceedings of the Epoxy Resin Formulators conference,* **1995.**

32. Toshiaki, N., *et. al.* European Patent 548,493 A1, 1993.

33. Speranza, G.P., Lin, J.J. US Patent 5,098,986, 1992.

34. Albers, R.A. US Patent 4,495,317, 1985.

35. Hoefs, C.A.M. US Patent 4,598,108, 1986.

36. Hoefs, C.A.M. US Patent 4,737,530, 1988.

37. Walker, F.H.; Everett, K.E.; Kamat, S. *Proc. XXII Waterborne, High-Solids and Powder Ctgs. Symp.,* **1995,** 88.

38. Klippstein, A. World Patent WO93/21250, 1993.
39. Naderhoff, B.A. US Patent 5,204,385, 1993.
40. Takas, T.P.; Naderhoff, B.A. *Proc, XXII Waterborne, High-Solids and Powder Ctgs. Symp.,* **1995**, 110.
41. Kurnik, W.J.; Roy, G. A. *American Paint & Coating Journal,* **1989**, pp 36-44.
42. Becker, W. US Patent 4,415,682, 1983.
43. Dreischoff, H. D.; Geisler, J.P.; Godau, C.; Hoenel, M. US Patent 5,236,974, 1993.
44. Rutkowski, B.S.; Godau, C.; Pfeil, A.; Geisler, J.P.; Becker, P. *Proc. XXII Waterborne, High-Solids and Powder Ctgs. Symp.,* **1995**, 100.
45. Elmore, J.D.; Cecil, J.C. US Patent 4,315,044, 1982.
46. Williams, P. R.; Burt, R.V.; Golden, R. US Patent 4,608,406, 1986.
47. Klein, D.H. *Pitture e Vernici Europe,* **1994**, 7-23.
48. Klein, D.H. European Patent 491,550 A2, 1992.
49. Shimp, D.A., US Patent 4,246,148, 1981.
50. DeGooyer, W.J. US Patent 4,539,347, 1985.
51. DeGooyer, W.J. US Patent 4,608,405, 1986.
52. Becker, W.; Karasmann, H. US Patent 4,197,389, 1980.
53. Dreischoff, D., *et al.*, Eur Pat. 610787 A2, 1994.
54. Chou, J.L.; Shah, S.; Jewell, B.G.; Moon, M.A. World Patent WO95/01386, 1995
55. Walker, F.H., Cook, M.I., Dubowik, D.A., *Proc XXII Waterborne, High-Solids and Powder Ctgs. Symp.,* **1996**, 289.
56. Jackson, M.A., *Journal of Protective Coatings and Linings,* **1990**, pp 54-64.
57. Galgocci, E.C.; Weinmann, D.J. *Proc, XXII Waterborne, High-Solids and Powder Ctgs. Sypm.,* **1995**, 119.

Chapter 6

Particle Interactions and Dispersion Rheology

J. W. Goodwin[1] and R. W. Hughes[2]

[1]Colloid and Condensed Phase Sector and [2]Bristol Colloid Centre, School of Chemistry, University of Bristol, Cantock's Close, Bristol BS8 1TS, United Kingdom

The rheological properties of dispersions are markedly dependent on the particle-particle interactions that occur in the systems. This paper briefly reviews the rheological measurements that are relevant to coatings and then the types of particle interactions that are important. The depletion interaction, which occurs in many coating systems on the introducion of polymeric thickeners, is re-interpreted in some detail. The rheological behaviour is then discussed with reference to the colloidal interactions. The high and low shear limits of the viscosity, shear thinning behaviour, high frequency elasticity, relaxation spectrum and strain melting are examined.

The rheological requirements for colloidal dispersions, such as those used in the coatings field, can be extremely broad. They may range from viscoelastic solids under storage to low or moderate viscosity fluids under application conditions. In addition the rate of change from one state to the other is usually important. To achieve the desired rheological response, often a compromise, the loading of colloidal particles as well as that of "rheology modifiers" is adjusted.The latter species may be either soluble polymers or particles. A key to understanding the concentration dependence of the rheology is in the understanding of the interparticle forces and how these effect the rheology of concentrated dispersions.

1. Rheological Characterisation.

The range of rheological response require of coatings can vary enormously, even for the same coating, depending on the timescale of the performance. For example, the dispersion may be required to behave as a *soft solid* over a range of times $10^8 > t > 1$s and as a *mobile liquid* for $10^2 > t > 10^6$ s. The factors which are available to control the rheology are:-
  i. particle concentration,
  ii. particle size,
  iii. particle size distribution,
  iv. particle shape,
  v. stabiliser/dispersant type,
  vi. chemical environment,
  vii. soluble polymers.

Each of these can effect the interaction energy that acts between the particles in the dispersion and therefore changes the rheology.

The viscosity - shear rate behaviour is a characterisation measurement normally carried out routinely.This is because the application window is of great importance. For example, if a decorative coating is considered, we can sketch an application window as in Figure 1. Here the viscous response is of a power law fluid with a time dependence which is the result of concentration increase as solvent is lost by evaporation as well as any contribution from polymeric thickeners. In addition to this application response, the long time mechanical stability must be assessable so that problems of sedimentation can be predicted. This takes the performance to a very long timescale (i.e. very low shear rates) which is outside the window sketched in Figure 1.

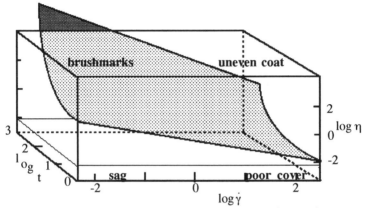

Figure1. Typical performance window for a decorative coating.

The characterisation should enable us to describe the viscosity from measurements, as shown in Figure 2, of the **shear stress - shear rate - time** carried out as a function of volume fraction. This enables us to record measurements of

i.   the zero shear viscosity, $\eta(0)$, and the static yield stress, σ_y;

ii. the high shear (plastic viscosity), $\eta(\infty)$, and the dynamic (Bingham) yield σ_B;

iii. the power law index and consistency coefficient, $\eta = K\dot{\gamma}^{-n}$;

iv. the thixotropic recovery time, τ_T.

The latter quantity is best measured by following the build-up of the storage modulus after the cessation of a high-shear preconditioning period.

A selection of these parameters should provide enough information for adequate quality control, but all will be required if performance is to be optimised and, in addition, the **viscoelastic behaviour** will be required so that measurement of the dynamic moduli as a function of frequency, see Figure 3, and as a function of strain will have to be carried out. This will result in values of :-

v.   the high frequency storage modulus, $G(\infty)$,

vi.   the zero frequency viscosity, $\eta^*(\omega \to 0)$ $(\cong \eta(0)?)$,

vii. the characteristic or relaxation time, τ,

viii. the static or zero frequency modulus, $G(0)$,

ix.   the critical strain γ_c, (the limit of linear behaviour),

x.   the melting strain γ_m (the strain at which the storage modulus of a soft solid has fallen to a value equal to that of the loss modulus.

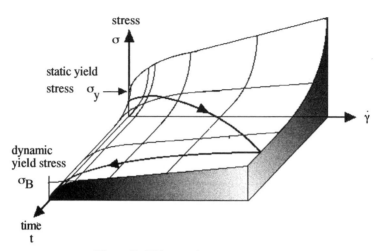

Figure 2. Thixotropic response.

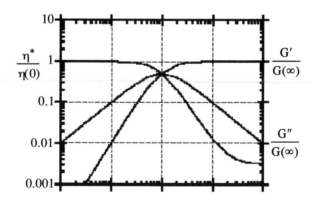

Figure 3. The frequency response of a Maxwell viscoelastic fluid.

2. Colloidal Interactions.

There are four colloidal interactions that we will consider at this time as being the major ones that can be modified to control the rheology of our dispersions. There are two forces giving rise to a repulsive interaction: **a**. electrostatic and **b**. steric. The two which we will consider which give rise to attractive interactions are: **c**. dispersion forces and **d**. depletion forces. All of these are documented in the usual colloid texts e.g. (*1* ,*2*).

Electrostatic Repulsion. The charge on the surface of a colloidal particle can arise from ionogenic groups chemically bound to the particle surface (e.g. latex particles), physically adsorbed species (as is the case with surfactants), or due to isomorphous substitution in a crystal lattice (e.g. clay particles). The balancing charge of counterions is distributed in the solution phase and screens the particle charge. When two particles approach, the counterion clouds overlap and repulsion occurs between the particles giving the pair potential for spherical particles as :-

$$\frac{V_e}{kT} = \frac{4\pi\varepsilon a^2\psi^2}{kTr}\exp(-\kappa h) \qquad \text{for} \quad \kappa a < 5$$

$$\frac{V_e}{kT} = \frac{2\pi\varepsilon a\psi^2}{kT}\{\ln[1 + \exp(-\kappa h)]\} \qquad \text{for} \quad \kappa a > 5$$

2.1

where kT is the thermal energy, a is the particle radius, ε is the permittivity of the medium, ψ is the surface potential, h is the surface-surface separation between the particles with a centre-centre separation of r and κ is the Debye-Huckel potential decay parameter :-

$$\left.\frac{V_s}{kT}\right|_r \propto \frac{a}{v_s}\delta^2(0.5-\chi)\kappa^{-1} = \sqrt{\frac{\varepsilon kT}{2e^2 N_A I}} \qquad \text{with I = ionic strength and } N_A \text{ is}$$

Avogadro's Number.

Note however that with concentrated systems of charged particles the ionic strength must include the counterions of the particles in addition to to the background electrolyte concentration (2). The former term can sometimes dominate the calculation of κ.

The expressions given above can give rise to a long range repulsion which is clearly sensitive to the chemical environment and is hence controllable in principle.

Steric Repulsion. The adsorption of non-ionic surfactants or polymers is often used to provide a repulsive interaction between particles. Block copolymers are generally better than homopolymers as the lyophobic block(s) provide strong anchoring of the stabiliser at the surface. Also high surface densities are desirable to maximise the interaction between layers as they approach. The origin of the repulsion is the change in free energy of the adsorbed polymer as the layers of thickness δ start to mix when the surfaces are 2δ apart (3). The repulsive force may be computed from the change in the osmotic pressure and the pair potential determined :-

$$\left.\frac{V_s}{kT}\right|_r \propto \frac{a}{v_s}\delta^2(0.5-\chi)$$

2.2

where χ is the Flory-Huggins polymer-solvent interaction parameter and v_s is the partial molar volume of the solvent. The proportionality constant include a volume term describing the overlap and a concentration profile of the outer sections of the polymer layers.This gives a short range interaction which commences at a surface-surface separtion of the particles of h=2δ and then increases very rapidly. Control of this interaction is only possible by changing the temperature or adding solvent components which modify the χ-parameter.

Dispersion Forces. These interactions are a function of the dielectric properties of the particles (and solvent) and arise from the coupling of the oscillating atomic dipoles. The approximate form of the interaction energy for close separations (I) is :-

$$\frac{V_a}{kT} = -\frac{Aa}{12\,hkT} \qquad\qquad 2.3$$

where A is the net Hamaker constant for the particle/solvent combination.

Depletion Interaction. Weak attraction can arise between particles when soluble non-adsorbing polymers are added to a dispersion. This can be observed experimentally(4).At a critical concentration of polymer the system can undergo a phase transition. The system separates into two phases, a polymer rich phase and a particle rich phase. The particle rich phase is often observed as a sediment. A further increase in polymer concentration changes the relative composition of the phases. Some systems display a second critical polymer concentration at which a single phase is reformed.

The mechanism which gives rise to phase separation is thought to be a reversible weak aggregation process. The particles can be supposed to aggregate and after an initiation period begin to sediment. The aggregation process is termed depletion flocculation. A system in thermodynamic equilibrium will occupy its lowest free energy state. Thus the phase diagram for the system can be found by calculating the Gibbs free energy as a function of composition(5). However the initial aggregation process can be thought of as being due to a pair interaction acting between the particles. The physical origin of this depletion pair potential can be considered as due to a volume exclusion effect.

Asakura and Oosawa(6) used a simple geometric argument to calculate the depletion energy. Consider an isolated colloidal particle in a solvent containing polymer. The polymer possess an osmotic pressure which exerts a force normal to the surface of the particle. The time averaged force acting normal to any point on the surface integrated over that surface gives a zero net force. If the centre of mass of the polymer coil approaches the surface of the particle closer than the radius of gyration the coil must distort changing its internal and configurational energy. If the coil is modelled as a rigid hard sphere the radius of gyration gives a minimum distance of approach. At this point an arbitrary mathematical construction can be used which supposes that there is a *depletion layer* associated with a particle. This is a layer surrounding the particle with a thickness equal to the radius of gyration. It is depleted of polymer. If two colloidal particles now approach the depletion layers overlap. The total volume of the layers reduces with a consequent change in polymer concentration in the bulk solution. The resulting increase in pressure acts on the particles pushing them together. Vrij(7) gives the more general expression for the geometric exclusion term and gave interaction energy variation with distance as,

$$\frac{V_d}{kT} = -\frac{4\pi}{3}\left(a+R_g\right)^3\left(1 - \frac{3r}{4\left(a+R_g\right)} + \frac{r^3}{16\left(a+R_g\right)^3}\right)\frac{\Pi}{kT} \qquad\qquad 2.4$$

where R_g is the radius of gyration, r=(2a+h) is the centre to centre separation and Π is the osmotic pressure.The attractive force increases monotonically with reducing surface separation. In treating the coils as rigid hard spheres the configurational contribution of the polymer is oversimplified. This problem was tackled by Joanny *et al* (8) giving rise to essentially similar solutions. Despite the elegance of their approach and the utility of the AO equation the depletion layer model is not without its drawbacks. The notion that there is a layer surrounding the particle depleted of polymer is essentially modelling the properties of a hard sphere dispersion in the presence of a curved wall. It is attempting to describe the interfacial region. The calculation of an excluded volume between the particles does not require this assumption. At this point a previously unpublished model will be introduced to calculate the depletion energy.

The Depletion Potential Revisited.   The colloidal particles and polymer molecules are in continuous motion due to the Brownian forces. On occasion the surfaces of a pair of colloidal particles will approach closer than twice the typical radius of gyration of the polymer. All the polymer molecules are now excluded from between the particles. This gives an imbalance in osmotic pressure. The pressure is lowered between the particles. The difference in osmotic pressure acts to push the particles together and flocculate the system.

Consider two hard sphere colloidal particles in a solvent and polymer mixture. Imagine the system is contained in a box of volume v_T. The total volume occupied by the particles is $2v_p$ where v_p is the volume of a particle. If the particles are well spaced so that the polymer coils can move freely about the particles there is no net attraction between the particles due to the soluble polymer. However as the particles approach to a distance of $2x_m$ (Figure 4) polymer molecules can no longer pass freely through the gap between them. A volume is excluded to the polymer and the polymer concentration

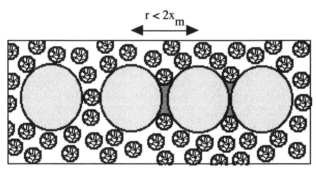

Figure 4.   Depletion volume between two spherical particles.

in the box has risen. The increase in the number density of the polymer molecules $\rho(x)$ as the particles approach closer than $2x_m$ is given by,

$$\rho(x) = \frac{N}{v_T - 2v_p - v_e(x)} \qquad x \le x_m \qquad 2.5$$

where $v_e(x)$ is the excluded volume as a function of separation and N is the number of polymer molecules in the box. Now the number density of polymer molecules in the box when the particles are separated by a distance greater than $2x_m$ is given by,

$$\rho = \frac{N}{v_T - 2v_p} \qquad x \ge x_m \qquad 2.6$$

Combining these two equations gives an expression describing the variation in polymer concentration with distance of separation of a pair of colloidal hard spheres.

$$\rho(x) = \rho \frac{1-\phi}{1-\phi\left(1 + \frac{v_e(x)}{2v_p}\right)} = \rho \frac{1-\phi}{1-\phi g(x)} \qquad x \le x_m \qquad 2.7$$

The function $g(x)$ is defined in terms of the particle radius a and the polymer radius of gyration R_g.

$$g(x) = 1 + \frac{3f(x)}{4R_g^3} \qquad\qquad 2.8$$

The term $f(x)$ is directly related to the excluded volume $v_e(x)$. The osmotic pressure of the polymer varies as a function of distance of separation between the particles. So for example using the van't Hoff law :-

$$\Pi(x) = \rho(x)kT = \; = \rho\frac{1-\phi}{1-\phi g(x)}kT \qquad x \leq x_m \qquad 2.9$$

The calculation of the excluded volume and $g(x)$ is complicated but not mathematically difficult if it is assumed the polymer coils can be represented by rigid hard spheres. Before presenting the result, inspection of Figure 4 shows an interesting result. In Figure 4 when $x=x_m$ the depleted volume is zero. As the particles approach to a distance given by x_1 the depleted volume shown as the shaded area reaches a maximum value and as $x \to 0$ the volume reduces again. Thus the volume is **not** a monotonically increasing function and so neither is the depletion energy. This result significantly differs from that of the AO equation.

   If the polymer coil is represented by a hard sphere the volume excluded can be calculated by a simple three dimensional integration.

$$v_e(x) = 2\pi f(x) = 2\pi\left[f_1(x) - f_2(x) - f_3(x) - f_4(x) - f_5(x)\right] \quad x \leq x_m$$
$$v_e(x) = 0 \quad x > x_m \qquad\qquad 2.10$$

with $h=2x$ and each function given as,

$$f_1(x) = (x + L(x))\left(R_g^2 + \left(R_g + y(x)\right)^2\right) \qquad\qquad 2.10a$$

$$f_2(x) = \frac{1}{3}\left\{(x + L(x))^3 - (L(x))^3\right\} \qquad\qquad 2.10b$$

$$f_3(x) = a(L(x))^2 \qquad\qquad 2.10c$$

$$f_4(x) = (x + L(x))\left(R_g + y(x)\right)\sqrt{R_g^2 - (x + L(x))^2} \qquad\qquad 2.10d$$

$$f_5(x) = R_g^2\left(R_g + y(x)\right)\sin^{-1}\left(\frac{x + L(x)}{R_g}\right) \qquad\qquad 2.10e$$

$$L(x) = a\left(1 - \sqrt{1 - \sin^2\left(\tan^{-1}\left(\frac{y(x) + r}{x + a}\right)\right)}\right) \qquad\qquad 2.10f$$

$$y(x) = -R_g + \sqrt{R_g^2 - \left(x^2 + 2a\left(x - R_g\right)\right)} \qquad\qquad 2.10g$$

This formal calculation of the excluded volume could be extended to allow for changes in the internal degrees of freedom of the polymer coils although this will prove geometrically more complex than that for the depletion layer approach. The closest approach of the surfaces of a pair of colloidal particles before the particle experiences a depletion force is no longer $2R_g$ but is given by,

$$\frac{h}{2R_g} = -\bar{a} + \sqrt{\bar{a}^2 + 2\bar{a}} \qquad \text{where } \bar{a} = \frac{a}{R_g}. \qquad\qquad 2.11$$

Calculation of the energy of interaction is straight forward. The depletion pressure will be resolved into a 'pair' interaction energy as :-

$$\frac{V_d}{kT} = -\frac{1}{kT} \int_{v_e(0)}^{v_e(h/2)} \Pi(h/2) dv_e \qquad \text{for } h > 0$$

$$\frac{V_d}{kT} = +\infty \qquad \text{for } h \le 0 \qquad\qquad 2.12$$

where V_d represents the depletion energy required to move a particle at a surface separation h to infinity. This formulation is designed to retain consistency with the DLVO formulation of the pair interaction energy (9,10). The energy distance curve is calculated using Equation 2.12 and an osmotic pressure equation appropriate for the polymer. A comparison of the depletion layer and the depletion volume models is shown in Figure 5 with a=10R$_g$.

Retaining the spirit of the hard sphere model a Carnahan-Starling model(11) is appropriate for comparison with the AO model.

$$\Pi(x) = \rho(x)kT \frac{1 + \phi_p(x) + \phi_p(x)^2 - \phi_p(x)^3}{\left(1 - \phi_p(x)\right)^3} \qquad\qquad 2.13$$

with $\phi_p(x) = \rho(x)v_{pol}$ where v_{pol} is the volume of the polymer coil found from the radius of gyration. The curve was obtained by numerical integration of Equation 2.12. There is a significant difference in functional form of the potentials, the depletion volume model realising a potential with a minimum away from the particle surface and a change in curvature of the function. The AO function is similar in magnitude but monotonically decreasing. This difference is very important for rheological measurements particularly elasticity which is very sensitive to potential and in particular the second derivative (or curvature) of the potential. A simpler approximation can be used to replace Equation 2.12.

Approximate form of the equation. The integral Equation 2.12 can be reduced to a simpler analytical approximation. Provided the depletion volume is relatively small compared to the particle size the osmotic pressure change can be approximated as a linear perturbation. This can be tested by plotting Π against v_e and the linearity can be determined from the correlation coefficient. The exact applicability of this approximation will depend on R$_g$, a and the form of the equation used for the osmotic pressure. It has proved to be a good approximation for many of the models investigated. If the linear approximation proves appropriate the osmotic pressure is now given by :-

$$\Pi(x) = mv_e(x) + \Pi_{eqlbm} \qquad\qquad 2.14$$

where Π_{eqlbm} is the equilibrium osmotic pressure of the polymer and m the slope of a Π versus v_e plot. Integration is now trivial giving,

$$\frac{V_d}{kT} = \frac{m}{kT} \frac{\left(v_e(0)\right)^2 - \left(v_e(h/2)\right)^2}{2} + \frac{\Pi_{eqlbm}}{kT}\left(v_e(0) - v_e(h/2)\right) \text{ for } h > 0$$

$$\frac{V_d}{kT} = +\infty \qquad \text{for } h \le 0 \qquad\qquad 2.15$$

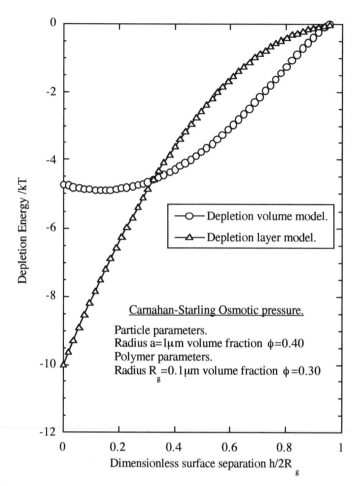

Figure 5. Calculated depletion potential curves from Equations 2.3 and 2.12

Before particle contact the equation is purely algebraic in form requiring no numerical integration.

The Total Interaction. The linear combination of the above pair potentials yields curve as illustrated in Figure 6. The figure illustrates two typical situations that can be expected to occur with practical colloidal dispersions. Conditions for A were a = 100nm, [NaCl] = 10^3 M, and for B were a = 500nm, [NaCl] = 10^{-1} M, and a steric barrier of 35nm. Both systems had an electrical potential of 75mV and a net Hamaker constant of 7×10^{-20} J. The stable system has a long exponential repulsive tail whilst the weakly flocculated system has an attractive well at distances close to the point where the steric stabiliser layers just start to interact.

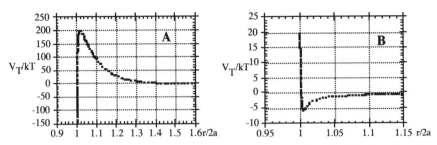

Figure 6. Pair Potentials. A. Electrostatically stable and B. Weakly flocculated.

3. Concentrated systems.

As a system becomes concentrated, the particles become part of a space filling structure. Computer experiments (*12*) with hard sphere systems indicate that a liquid/solid transition occurs at $\varphi \sim 0.5$, with freezing occuring at $\varphi = 0.495$, melting at $\varphi = 0.54$, and coexistence of liquid with solid between these two limits. This is basically an excluded volume effect. The presence of either repulsive or attractive interparticle forces lowers the phase transition boundary.

The structure can be described through the pair distribution function, $g(r)$, which gives the volume fraction of particles in a spherical shell, radius r, relative to the global average. Peaks in the $g(r)$ - r plot indicate structure in the suspension. Figure 7 illustrates a typical example.

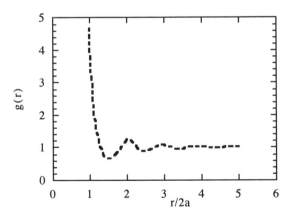

Figure 7. The pair distribution function for hard spheres at $\varphi = 0.45$.

The internal energy, U. of the dispersion of N particles is given by the energy of each particle :-

$$U = \frac{3NkT}{2} + \frac{N}{2}\int_0^\infty 4\pi r^2 g(r)\rho_p V(r)dr \qquad 3.1$$

where ρ_p is the number density of particles in the system. For unit volume of dispersion we therefore have :-

$$E = \frac{3\rho_p kT}{2} + 2\pi\rho_p^2 \int_0^\infty r^2 g(r)V(r)dr \qquad 3.2$$

$$\begin{array}{cc} \text{Hard} & \text{Colloid} \\ \text{sphere} & \text{interaction} \end{array}$$

To illustrate how we should think about the internal energy and the structure we can consider a strongly interacting repulsive system. These systems form a microcrystalline structure with face-centred-cubic order (*13*). This will enable the calculation of the high frequency storage modulus from a lattice model (*14*) as:-

$$4\pi r^2 g(r)\rho_p \Rightarrow N_1\delta(r - R) \quad \text{and} \quad R = 2a\left(\frac{\varphi_m}{\varphi}\right)^{\frac{1}{3}}$$

i.e. as we have a lattice g(r) becomes a δ-function and N_1 is the coordination number of the lattice. For an fcc structure N_1 is 12 and φ_m is 0.74. Now the high frequency modulus is (*14*) :-

$$G(\infty) = \frac{3\varphi_m N_1}{32R} \frac{\partial^2 V(R)}{\partial R^2}$$

$$= \frac{\pi}{32}\left\{9\varphi^2\frac{\partial^2 E}{\partial\varphi^2} - 6\frac{\partial E}{\partial\varphi} + 6E\right\} \qquad 3.3$$

and we may write the reduced internal energy as :-

$$\overline{E} = \frac{Ea^3}{kT} = \frac{3\varphi}{8\pi}\left(3 + \frac{N_1 V(R)}{kT}\right) \qquad 3.4$$

Figure 8 shows a phase diagram for dispersions up to a volume fraction of 0.6. The hard sphere line forms the boundary between the electrostatically stabilised systems and the weakly flocculated systems. Broadly, at a given volume fraction, the further away the internal energy from the hard sphere line, the more rigid the structure. the means the more viscoelastic and non-Newtonian the dispersions become. For strongly interacting electrostatically stabilised particles, Russel (*15*) suggested an effective hard sphere model based on the balance of Brownian interaction with the electrostatic interaction. The model is given in more detail below where it is used to compare with experimental values of the low shear viscosity.

Figure 9 compares the prediction from Russel's phase boundary with the experimental observations recorded in Figure 8. Russel included coexistence in his model but our experiments just indicated from rheological behaviour wether liquid or solid behaviour was found. A reasonably good qualitiative agreement is shown in the figure between the model and the experiments.

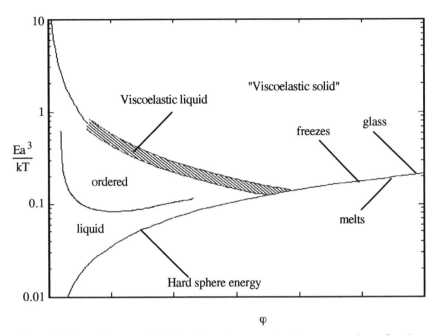

Figure 8. Phase diagram plotted as internal energy density versus volume fraction.

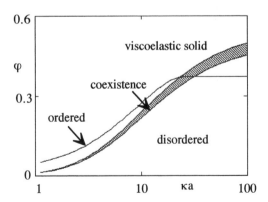

Figure 9. Phase boundary for polymer latex with a = 105nm, ψ = 80mV. The solid line is from rheological experiments and the hatched area is calculated from Russel's model (2).

4.The Viscosity Limits.

Above volume fractions of 0.1, the estimation of the viscosity of dispersions using rigorous hydrodynamic analysis is an unsolved problem. Computer simulations have been used for hard sphere sytems with some success (16) whilst experimental data is available for sterically stabilised systems which are thought to approximate to hard spheres (*17*). The Doughery-Krieger equation (*18*) provides a reasonably good description over most of the concentration range:-

$$\eta = \eta_0 \left(1 - \frac{\varphi}{\varphi_m}\right)^{-[\eta]\varphi_m} \qquad\qquad 4.1$$

where η_0 is the viscosity of the continuous phase, $[\eta]$ is the intrinsic viscosity with a value of 2.5 for hard spheres, and φ_m is the maximum volume fraction at which flow can occur. There are many alternative expressions but this one has some merit in that it comes from a free volume argument (*19*) and fits the data quite well.

For hard spheres, two limiting conditions for φ_m become apparent (*20*). Under very low shear or quiescient conditions where the colloid and Brownian forces dominate any imposed hydrodynamic force, we can expect flow to occur up to $\varphi_m = 0.54$ (*20*), although the model may become poor at $\varphi > 0.5$ due to the coexisting solid phase. At high shear, the imposed hydrodynamic field will dominate the Brownian colloid interactions and linearised structures are reported from simulation (*21*) and from scattering studies (*22*). This gives a high shear limit of $\varphi_m = 0.605$ (*20*). These limiting values give equations that should describe the high and low shear limits of the viscosity:-

$$\eta(0) = \eta_0 \left(1 - \frac{\varphi}{0.54}\right)^{-1.35} ; \qquad\qquad \eta(\infty) = \eta_0 \left(1 - \frac{\varphi}{0.605}\right)^{-1.51} \qquad 4.2a,b$$

Experimental studies on near hard sphere systems were fitted to similar equations(*23*) :-

$$\eta(0) = \eta_0 \left(1 - \frac{\varphi}{0.63}\right)^{-2} ; \qquad\qquad \eta(\infty) = \eta_0 \left(1 - \frac{\varphi}{0.71}\right)^{-2} \qquad 4.3a,b$$

Figure 10 shows curves generated from all four equations up to a volume fraction of 0.5. Above this value, as mentioned above, the models may be poor due to phase coexistence, but also experiments are difficult with slip and poor flow profiles in the instruments becoming an increasing problem.

When the colloidal particles are charge stabilised the viscosity is increased over that found with hard spheres. Three electroviscous effects have been identified (*24*). The primary electroviscous effect increases the intrinsic viscosity but is only important at $\kappa a < 1$ (*25*), whilst the tertiary electroviscous effect refers to soluble polyelectrolyte species and so does not concern us here. The secondary electroviscous effect is due to the interaction between charged particles and has been analysed by Russel (*2*). An important feature of this treatment is that of defining the excluded volume of the particles under quiescent conditions such that the onset of a phase transition occurs at $0.495(2a/d_0)^3$. Here d_0 is the effective hard-sphere diameter of the particles estimated from equating the Brownian and electrostatic interactions so that :-

$$d_o = \frac{1}{\kappa} \ln\left(\alpha / \ln\left(\alpha / \ln(\alpha/....)\right)\right) \qquad\qquad 4.4$$

and $\qquad \alpha = \dfrac{4\pi\varepsilon\psi^2\kappa a^2}{kT} \exp(2\kappa a).$

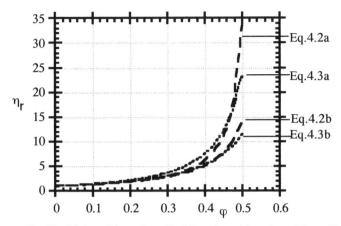

Figure 10. The high and low shear limiting viscosities plotted from Equations 4.2 and 4.3.

The equation for the low shear limiting viscosity now becomes :-

$$\eta_r(0) = \left(1 - \frac{\varphi}{0.54\left(\frac{2a}{d_0}\right)^3}\right)^{-1.35\left(\frac{2a}{d_0}\right)^3} \qquad 4.5$$

A plot of this equation is shown in Figure 11 calculated for a polystyrene latex with a diameter of 170 nm.

Also shown in the figure are the experimental data obtained from creep compliance measurements and the calculated high shear limiting viscosity. It should be noted that viscoelastic liquid-like behaviour is limited to the volume fraction range of 0.1 to 0.21 as it is only within this range that both $\eta(0)$ and $\eta(\infty)$ are observable. Viscoelastic solid-like behaviour should be expected between volume fractions of 0.21 and 0.61 with strain-melting being observed. In this example, the liquid-solid phase transition was estimated to commence at $\varphi = 0.19$ and to be complete at 0.21, hence in this region the coexisting liquid phase should remain at $\varphi \sim 0.19$ but decrease in proportion to the solid phase. The precise description of the value of $\eta(\varphi)$ in this region is complex but the calculated curve shows the same qualitative behaviour as the experimental data.

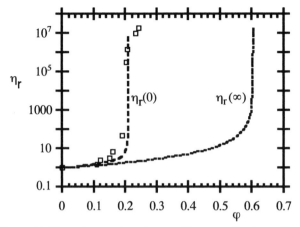

Figure 11. The relative viscosity as a function of volume fraction. Curves are calculated, the points are experimental data.

5. Shear Thinning Behaviour.

Cross (26) developed an expression to describe the shear thinning behaviour of both polymer solutions and dispersions of particles. The equation was written in terms of the limiting viscosities and two material parameters; a consistency and a power law index :-

$$\eta_r = \eta(\infty) + \frac{(\eta_r(0) - \eta(\infty))}{1 + A\dot{\gamma}^n}. \qquad 5.1$$

The index was shown to have a value $1 > n > 2/3$ as the distribution broadened from a monodisperse system. Papir and Krieger (27) gave a similar equation but written in terms of the reduced stress :-

$$\eta_r = \eta_r(\infty) + \frac{(\eta_r(0) - \eta_r(\infty))}{1 + (B\sigma_r)^n} \qquad 5.2$$

where the reduced stress is :-

$$\sigma_r = \frac{(\eta\dot{\gamma})a^3}{kT}. \qquad 5.3$$

This latter form has the advantage that the use of the suspension viscosity in the equation for the critical stress implies the same mean field approximation assumed in the derivation of the Krieger equation and hence includes an approximation for the multi-body hydrodynamic interactions. The reduced stress compares the diffusional time to the characteristic shear time and the shear thinning should occur around a value of $(B\sigma_r)^n \approx 1$, i.e. $B = \sigma_r^{-1}$ and the value of n defines the width of the stress range. For a simple monodisperse system, i.e. one with a single relaxation time, a range of three orders of magnitude in reduced stress covers the range (to within 2% of each limiting viscosity). Note that this is identical to the width of the frequency response to a simple Maxwell liquid in forced oscillation experiments. Choi and Krieger (*28*) carried out an extensive series of experiments   with monodisperse sterically stabilised poly (methyl methacrylate) dispersions in silicone oils. The particles were stabilised with a 7nm layer of poly(dimethyl siloxane) and the range of diameters used was 170nm to 620nm. The data were fitted with a B value of 5.70 and a plot is shown in Figure 12. The experimental data are shown as points and were obtained from data over a wide range of background medium viscosities. This work indicated that B is independent of volume fraction (although this was not the case with other workers, see Russel (*2*) for example). The poor fit at high reduced stress levels is due to flow instabilities in the viscometer. Values of B ranging from 2 up to 50 can be found in the literature and the value is   indicative of the colloidal interactions between the particles. B becomes larger as the interactions become softer. Figure 9 shows a plot for an electostatically stabilised system and here the fit is with a B-value of 21. The data were obtained with a monodisperse poly(styrene) latex with a particle diameter of 170nm, a volume fraction of $\varphi = 0.2$ and a univalent ion concentration of 10^3 M.

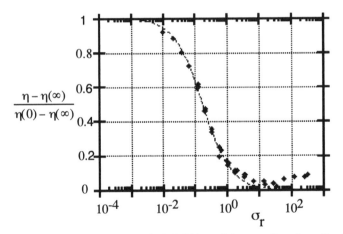

Figure 12. Shear thinning of p(MMA) particles as a function of reduced stress. Redrawn from Choi and Krieger (*28*).

For a 'true' hard sphere system the constant B might be expected to be independent of particle size or concentration and be given by the balance between the characteristic time scale of diffusive movement and that due to the shearing motion. A simple method for extracting a value of B for this limiting case is to consider the mathematical properties of the curve and the microstructural properties of the system. Shear thinning occurs when the shearing field disrupts the equilibrium microstructure and the Brownian motion of the system. The balance of the diffusive processes to the shear processes is described by the Peclet number, Pe. Under the condition of high Peclet number the structure is disrupted and the system flows with a low viscosity. Under the condition of a Peclet number much less than unity, diffusive motion dominates and the rate of directional motion due to the application of a stress is controlled by Brownian diffusion. The diffusion is controlled by the particle size, the thermal energy and the viscosity. This is the viscosity that the particle is subjected to

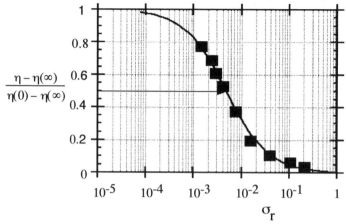

Figure 13. Shear thinning response for a polystyrene latex at $\varphi = 0.2$ and $[NaCl]=10^{3}$.

by the surrounding medium. However, in the case of a dense suspension a particle will diffuse against a background viscosity determined by the zero shear rate viscosity of the suspension. The Peclet number for a dense suspension should therefore be given by :-

$$Pe = \frac{6\pi a^{3}\eta(0)\dot{\gamma}}{kT} \tag{5.4}$$

At the shear rate where the Peclet number is equal to unity, the diffusive forces and the shear forces balance. The shear rate where this occurs marks the onset of shear thinning. Thus at Pe=1 there is a critical stress where there is a transition between a constant low shear viscosity and a shear thinning response. The transition is not clearly defined by a singular point but is marked by the transition in a continuous curve as is shown by figures 12 and 13. In order to define a critical stress where this transition occurs the point for the onset of shear thinning needs to be defined. A method which is consistent with the shape of the curve is to define the stress at the onset of shear thinning as the point where the tangent to the viscosity at the midpoint is equal to the low shear rate viscosity. The equation for the tangent T at the midpoint when n=1 is given by,

$$T = \frac{-(\eta(0) - \eta(\infty))}{4} \ln(\sigma_r) + \eta_c \tag{5.5}$$

where η_c is a critical viscosity.

$$\eta_c = \eta(\infty)\left\{\frac{1}{2} + \frac{\ln(B)}{4}\right\} + \eta(0)\left\{\frac{1}{2} - \frac{\ln(B)}{4}\right\} \tag{5.6}$$

So by setting T=$\eta(0)$ the value for B can be derived. This gives B=2.55 which, interestingly, is nearly the intrinsic viscosity for hard spheres $[\eta]=2.5$. Such a plot is illustrated in Figure 14 below.

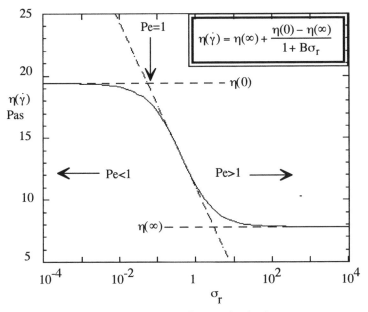

Figure 14. Shear thinning for monodisperse hard spheres.

6. Viscoelastic Behaviour.

The structure produced by the particle - particle interactions in concentrated dispersions provides a means of storing energy in linear viscoelastic measurements. The most common measurement is of the storage and loss moduli (G' and G" respectively) as a function of frequency. Usually a broader time dependence will be observed than that illustrated in Figure 3 for a simple Maxwell fluid. The polydispersity of size and molecular weight of the components of the microstrucure combine to broaden the response curve. It is useful to consider the response as a spectrum of behaviour and use integral equations of the following form for the moduli :-

the loss modulus

$$G'' = \int_{-\infty}^{+\infty} H(\ln\tau) \frac{(\omega\tau)}{1+(\omega\tau)^2} d(\ln\tau) \; ; \qquad \text{6.1a}$$

or dynamic viscosity

$$\eta' = \int_{-\infty}^{+\infty} H(\ln\tau) \frac{\tau}{1+(\omega\tau)^2} d(\ln\tau) \; ; \qquad \text{6.1b}$$

the storage modulus

$$G' = \int_{-\infty}^{+\infty} H(\ln\tau) \frac{(\omega\tau)^2}{1+(\omega\tau)^2} d(\ln\tau) \quad \text{fluid;} \qquad \text{6.1c}$$

$$G' = G(0) + \int_{-\infty}^{+\infty} H(\ln\tau) \frac{(\omega\tau)^2}{1+(\omega\tau)^2} d(\ln\tau) \quad \text{solid.} \qquad \text{6.1d}$$

The relaxation spectrum, $H(\ln\tau)$, is the key to the material behaviour and contains the high frequency modulus corresponding to each relaxation process and the probability of that process occurring.In order to characterise the relaxation spectrum as fully as possible, experiments should be carried out over as wide a frequency range a possible. There are usually two problems in inverting the oscillation data to obtain the spectrum, namely the range of frequency and the precision over that range. For example, the loss modulus data may be used to determine the spectrum and, in many cases, the "zero order" approximation will be all that the data will justify :-

$$H(\ln \tau) \approx \frac{2}{\pi} G''(\omega)\big|_{\omega = 1/\tau} \; . \qquad\qquad 6.2$$

It is the relaxation spectrum that is the basis for our characterisation and enables us to compare systems. Two parameters for example that are of particular interest are the high frequency elasticity and the zero shear viscosity. These are readily estimated from the spectrum :-

$$G'\big|_{\omega \to \infty} = G(\infty) = \int_{-\infty}^{+\infty} H(\ln \tau) \, d(\ln\tau); \qquad\qquad 6.3a$$

and

$$\eta'\big|_{\omega \to 0} = \eta(0) = \int_{-\infty}^{+\infty} \tau H(\ln \tau) \, d(\ln\tau) \quad \text{(for a fluid).} \qquad 6.3b$$

(Of course for a solid $\eta(0)$ is infinity.) The high frequency plateau elasticity is made up of contributions from particle - particle, polymer - polymer and polymer - particle interactions in real coating systems. The most obvious point here is that the slowly relaxing components, such as high molecular weight polymer additives, can make a major contribution to the low shear viscosity. This is the case if their relaxation time is very long, even though their occurrence probability (concentration) is low, because the product relationship in the above equation favours the long time processes.

The relaxation spectrum is the structural elasticity and can be modified by formulation changes. In order to illustrate the origins it is instructive to look at some simple systems and to start with the high frequency plateau modulus. At this plateau, the frequency is high enough so that the work carried out in deforming the structure is recoverable. This means that interactions between particles can be readily modelled as can polymer interactions. Let us consider the following four cases of model systems :-
  i.   monodisperse, charged stabilised latices,
  ii. monodisperse, weakly flocculated latices,
  iii. monodisperse, depletion flocculated latices,
  iv. associative thickeners of the hydrophobically modified ethoxylated
        urethanes (HEURs).

i. Elasticity of Charge Stabilised Latices. These systems, when monodisperse, form highly organised structures with bcc or fcc symmetry with N_1 nearest neighbours. The local structure can be thought of in terms of a lattice and, as we are considering the high frequency limit such that no relaxation can occur, it is relatively simple to model the elasticity (*14, 29-31*) using a simple lattice the result, which was given earlier is :-

$$G(\infty) = \frac{3\varphi_m N_1}{32R} \frac{\partial^2 V(R)}{\partial R^2} \qquad\qquad 6.4$$

The equation can be used for either packing type although the bcc arrangement usually only occurs at the lower part of the volume fraction range. The important feature here is the curvature of the pair potential as shown in Figure 6A. As the volume fraction is increased, the mean separation between particles decreases at a constant value of N_1. A comparison of experiment and model is shown in Figure 15 for a polystyrene latex in lithium chloride solution of concentration 10^{-4} M (*32*). The curves were calculated from the titrated particle charge and a good fit was obtained with 75% of the titrated charge being in the diffuse layer.

ii. Weakly Flocculated Latices. These are the types of system with a pair potential of the type illustrated in Figure 6B. The potential well is large compared to kT but small compared to the hydrodynamic forces that are exerted on the particles at moderate to low shear. This type of curve can be produced by large sterically stabilised particles at high electrolyte concentration (*33*) as well as by depletion flocculation. In this case the structural problem is a little more complex as the particles will tend to stick where they "contact" (i.e. in the energy minimum). Consider the situation in which the system is poured into a can or a rheometer. The particles form a liquid-like structure in which the separation between nearest neighbours is constant at the minimum but the number is a function of volume fraction. (Note the contrast here with the strongly repulsive systems in the previous paragraph.) The modulus again depends on the curvature of the pair potential between the nearest neighbours which are at the minimum position. The important feature is the structural arrangement. It was shown that the elasticity could be modelled successfully with (*33*) :-

$$G(\infty) \cong \frac{2\pi}{15}(\rho_p)^2 \int_0^\infty g(r)\left(r^4 \frac{\partial V(r)}{\partial r} \right) \partial r \qquad 6.5$$

The fit of equation 6.5 to experimental data is also illustrated in Figure 15 for a polystyrene latex of 1μm diameter, sterically stabilised with a non-ionic ethoxylate surfactant in 0.5 M sodium chloride solution (*33*).

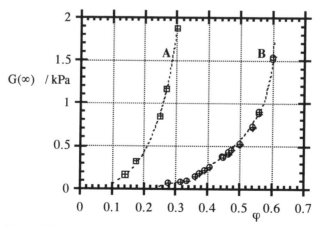

Figure 15. High frequency modulus for latex dispersions. Experimental points and calculated curves. A. charged stabilised latex and B. weakly flocculated latex.

Consider now the structure as a function of time. Initially, on the cessation of stirring, the particles stick where they touch. This is not necessarily the lowest energy configuration as that would require dense close packing. Over a long period particles will be able to locally concentrate if there is sufficient local space. This means that at moderate concentrations ($\varphi\sim0.30$), these local dense regions can be formed but at the expense of the overall structure giving rise to locally weak regions. Finally, after an induction period, the structure will spontaneously consolidate. the length of the induction period will be a strong function of volume fraction and will become to large to concern us at high volume fractions. This does have important formulation implications however as if a system is set up at moderate concentration and to low an attractive interaction, it may appear to be stable for several days but suddenly consolidate at a later date. For a *uniform network*, a first estimate of the static yield stress is given by the maximum interparticle force and the number of particle interactions per unit area (Russel *2*) :-

$$\sigma_y \approx \left(\frac{\varphi}{a}\right)^2 \left(\frac{\partial V(r)}{\partial r}\right)_{max}. \qquad\qquad 6.6$$

However, this will only prove useful in a predictive sense if the fraction of the structure which is weakened can be estimated.

    iii. Depletion Flocculated Latices. The viscoelastic response of a floc formed from a hard sphere system will depend on a complex interplay of the hydrodynamic forces, depletion forces and the spatial arrangement of particles. In the absence of depletion forces the notion that hard sphere systems posses an elastic modulus is a contentious one. Many quasi-hard sphere systems can only easily be measured at very high concentrations using current instrumentation. As a consequence it is difficult to establish if the systems are in their equilibrium state. It is possible to calculate an elasticity of a weakly attractive system using an integral equation describing the distribution of particles in space(*33*) as shown above. This calculation was extended to apply to hard spheres by assuming a liquid-like order to be maintained above the phase transition of $\phi=0.494$. This approach essentially allows for a hard sphere system to posses a glass-like order, although with the limitation of the absence of a detailed simulation of the structure. For a system undergoing depletion flocculation the range of interaction is increased from that of the hard sphere potential. The local order will extend over longer distances. Additionally depletion systems need not be at thermodynamic equilibrium and as a consequence models based on equilibrium approaches are not necessarily appropriate to predicting structural order. Thus the high frequency shear modulus is difficult to calculate for a system interacting through a depletion potential. It will be sensitive to the rate of change of force between the particles. For weakly attractive systems with moderate particle size hydrodynamic interactions do not significantly influence the elasticity of the floc(*34*).

    At high concentrations the particles will form dense close packed flocs with the number, of nearest neighbours becoming insensitive to concentration.

$$G(\infty) = \frac{\alpha}{R_{min}} \frac{d^2 V_d}{dh^2}\bigg|_{h=h_{min}} \qquad\qquad 6.7$$

where α is a constant and R_{min} is the centre to centre separation at the minimum in the energy distance curve. The constant α is a function of volume fraction and the packing in the floc. It is related to the number of nearest neighbours of particles. The second derivative of energy is insensitive to the volume fraction as the particles will occupy the minimum in the interparticle energy curve in the floc regardless of volume fraction. A number of authors have shown that the high frequency elasticity of a floc can be modelled by a power law(*34,35*). Equation 6.7 can be rewritten as:-

$$G(\infty) \propto \phi^m \qquad\qquad 6.8$$

In order to verify the depletion potential experimentally values for A and m are required. Russel and Sontag(*34*) obtained a value for m=2.5 for freshly flocculated samples, Mewis and Aerschot(*35*) obtained m=2.4 for 'dried' fumed silica samples. *The limitation of the depletion layer approach of AO* is that the second derivative of the potential is constant with separation and therefore $G(\infty) \propto -\pi a\Pi$. This is clearly physically unrealistic.

In order to verify the relative merits of the potentials a quasi-hard sphere system of polymethylmethacrylate (PMMA) latex coated with a chemically grafted layer of poly(12-hydroxystearic acid) (PHS) was prepared in dodecane(*36*). The four particle sizes investigated with radii of 125nm, 184nm, 281nm and 419nm. The polymer cis-polyisoprene (PIP from Polymer Laboratories Ltd.) was added to the dispersion. No evidence was found for adsorption of the polymer and so it can be supposed to be a free, soluble, polymer. The polymers used had a narrow molecular weight distribution (M_W/M_n<1.1). Their radius of gyration was found from capillary viscometry giving an R_g=12.2nm for M_W=86000 Dalton (P86) and R_g=6.1nm for a molecular weight M_W=28300 Dalton (P28). This is in good agreement with a polymer chain in near θ conditions. This system was the subject of a very extensive study which is to be published(*36*). Some of the findings are presented below.

The Dependence of $G(\infty)$ on the Volume Fraction. The high frequency limit of the shear modulus of the flocculated dispersions mixed with the polymer were measured using a Rank Bros. shearometer. The operating frequency of this instrument (about 1200rad s⁻¹) is lower than the characteristic frequency for polymer relaxation process so configurational changes of the PIP did not contribute directly to the elasticity. Figure 16 shows the three data sets for the shear modulus of the dispersion as a function of volume fraction for different concentrations of polymer. The curves conform to Equation 6.8 with m=2.4-2.6. An interesting feature to note is the similarity of the elasticity of the 86000 Dalton and 28000 Dalton samples. The depletion energy is a function of the concentration of the polymer but it is not sensitive to the volume fraction of the particles. Thus the curvature of the lines reflects subtle changes in interparticle spacing and the floc density.

The Dependence of $G(\infty)$ on 86000 Dalton PIP. The shear modulus was measured at a fixed volume fraction for two different particle sizes with P86. This data is presented in Figure 17. The shear modulus was calculated using Equations 6.7 and 6.8 with α=0.833 for a face centred cubic packing. The solid lines on this plot give the results of these calculations. The osmotic pressure was calculated from a Virial expansion :-

$$\frac{\Pi}{\rho kT} = 1 + B\rho kT \qquad\qquad 6.9$$

where B was used as a fitting parameter. The coefficient B=(3-3.9)x10⁻⁴Pa⁻¹. This indicates that the osmotic pressure is only weakly dependant on ρ^2.

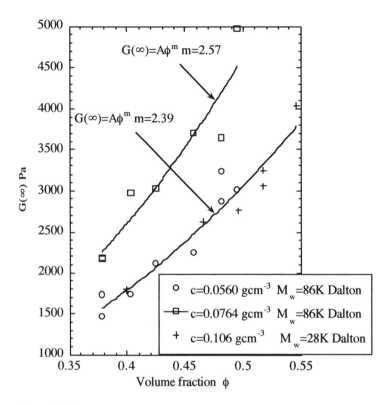

Figure 16. High frequency elasticity as a function of volume fraction and   polymer molecular weight.

The dependence of $G(\infty)$ on 28000 Dalton PIP. The shear modulus was measured at a fixed volume fraction for three different particle sizes with P28. This data is presented in Figure 18. The shear modulus was calculated using Equations (2.15) and (6.7) with $\alpha=0.833$. This data gave a very surprising result in that the osmotic pressure did not readily conform to a Virial expansion. The low concentration osmotic pressure of van't Hoff's law over estimated the shear modulus. The shear modulus was well fitted when an osmotic pressure of the form:-

$$\Pi = A(\rho - \rho')kT \qquad\qquad 6.10$$

was used. Here ρ' represents a critical concentration for the onset of the osmotic pressure in the presence of the particles. The values are shown in table 1.

Table 1. The variation in ρ', a and A for polymer P28

Radius a /nm	ρ' / molecules m^{-3} 10^{24}	A
183	0.3	0.40
281	0.9	0.28
419	1.4	0.25

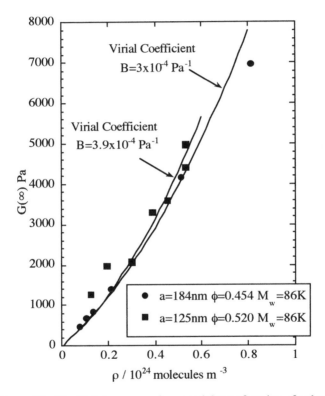

Figure 17. The high frequency shear modulus as function of polymer concentration for 86×10^3 Dalton molecular weight.

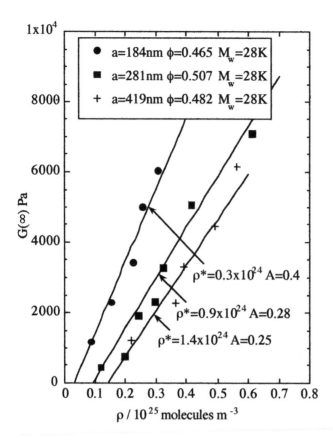

Figure 18. The high frequency shear modulus as a function of polymer concentration for 28×10^3 Dalton molecular weight.

Discussion of the Results. The majority of the data reported for both P28 and P86 are at concentrations above their respective ρ^* where:-

$$\rho^* = \frac{3\phi^*}{4\pi R_g^3} \qquad\qquad 6.11$$

with $\phi^*=0.74$. The polymer P28 has a calculated $\rho^*=0.78\times10^{24}$ molecules m^{-3} and the polymer P86 has a calculated $\rho^*=0.76\times10^{22}$ molecules m^{-3}. The osmotic pressure in this region is normally in excess of that given by the van't Hoff law. Both systems give osmotic pressures less than would be expected. As the concentration is increased above ρ^* for a polymer in a good solvent the radius of gyration tends to reduce. There is a corresponding increase in Π. Doi and Edwards(37) use a scaling argument to suggest the osmotic pressure in the semi-dilute region increases as,

$$\Pi = \rho kT \left(\frac{\rho}{\rho *} \right)^{1/(3\nu-1)} \qquad 6.12$$

and the radius of gyration reduces as,

$$R_g = R_g^0 \left(\frac{\rho}{\rho *} \right)^{(1-2\nu)/(1-3\nu))} \qquad 6.13$$

where R_g^0 is the radius of gyration at infinite dilution. This argument can be easily applied to the current depletion potential. It would imply for both the model presented here and the AO model that the range of interaction reduces since the radius of gyration reduces. Additionally the osmotic pressure increases. However the notion the range of interaction falls, fails to take account of the fact that the depletion potential was developed as an infinite dilution model. Once the system is above $\rho *$ many body forces need to be accounted for. It is unlikely that the displacement of polymer coils from between the colloidal particles can be achieved without displacing those in the solution around it. In a dilute system there is no net pressure change until all polymer coils are excluded from the gap between the particles. In a concentrated system the structural order will be displaced from equilibrium as the pair distribution function of the polymer particles is significantly perturbed by approaching colloidal particles. The range of interaction will be proportional to the structural order of the polymer coils. There will be a deepening of the energy well but it will be spread over a longer range.

It is important to remember the polymer coils are not rigid hard spheres. They posses a degree of flexibility and hence configurational entropy not present in the model presented. Once the system becomes concentrated this can become significantly disturbed and thus significantly modify the effective potential operating between a pair of particles. The notion that lower molecular weight polymers give rise to surprising viscoelastic behaviour is demonstrated by work of Reynolds and Reid(38). Their studies indicate that added high molecular weight polymers cause depletion flocculation. Phase separation is readily observed with polyisobutylene and PHS coated PMMA (38). However as the molecular weight is reduced to a region where the radius of gyration of the polymer is comparable to the adsorbed layer thickness phase separation is not observed. The viscosity is much larger than might be expected from simple depletion arguments.

The experimental study here was modelled with a more sophisticated potential than the AO model. This study has further highlighted differences between the low and high molecular responses of the PIP/PMMA/PHS system. Low molecular weight systems at high concentrations result in high number densities of polymer coils. These systems are tending to a state where the polymer coils loose there identity and the segment density tends toward a uniform spatial distribution. The viscoelasticity of dispersions in this polymer phase requires further experiments and modelling before the processes that are occurring are fully understood.

iv.   HEUR Associative Thickeners.   The hydrophobically modified ethoxylated urethane thickeners are frequently used as thickeners. The molecules consist of an ethylene oxide chain with paraffinic end-caps. Detailed discussion of the rheology of this type of system has been reported by Jenkins (39) and Annable et al (40). The hydrophobic end groups can form "micellar-like" aggregates which produce a cross-linked network. This network is easily disrupted by shear forces but readily reforms and gives rise to useful viscoelastic behaviour. Stress relaxation occurs by the hydrophobic groups temporarily leaving aggregates. The aggregation number is of order 10 and hence is relatively large. As a result the characteristic timescale is approximately the same for each aggregate and will depend on the chain length and temperature. This results in Maxwell fluid behaviour, and it has been shown that as mixed chain length aggregates are not found, two different length end-caps result in a spectrum with two sharp peaks (40).

The analysis of the high frequency plateau elasticity of polymer networks, G_N, has been given by many authors (e.g. 41,-43) and in particular for HEURs (39,40). The network modulus can be measured at frequency high enough so that neglible relaxation occurs and is given by :-

$$G_N \propto \rho kT \qquad\qquad 6.14$$

Here ρ is the number of elastically effective chains. For a fully developed network of HEUR polymers this might be expected to be simply the number of chains. However, a range of network defects can occur. A closed loop can be formed if both end-caps are in the same aggregate and this chain will not add the network elasticity except possibly at very high concentrations where it could "lock in" an entanglement. A end-cap may be free from aggregation, especially if the end-capping efficiency was less than 100%, and this will also reduce the proportionality constant to below unity. On the other hand, chain entanglements lead to an increase in the number of elastically effective links. Hence, the expected behaviour would be to firstly observe a percolation threshold where the network would just span the volume. This network could be expected to contain many defects of the closed loop type. At a concentration around ρ^*, a network close to being fully developed should be seen and above this concentration entanglements should enhance the network elasticity. The data illustrated in Figure 19 (44) was obtained on a 25,000 Dalton HEUR research sample kindly supplied by Rohm and Haas. The end-caps were C_{18}, and with rms chain dimension of ~10nm, the

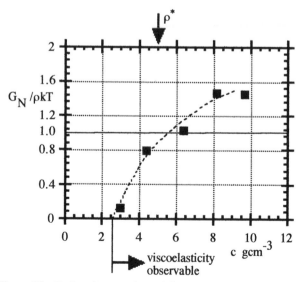

Figure 19. Reduced network modulus as a function of HEUR concentration.

value of ρ^* corresponded to $H(\ln\tau) \propto \tau^{-1/2} c^* \sim 5.2 \%$ polymer. It can be seen that the value of the network modulus reduced by the value of ρkT reaches a value of unity at only slightly above the value of ρ^* calculate from the dense random packing of spherical coils. The percolation threshold appeared at ~2.5 % polymer. At $2.5 < c < 5.5$ the system is made up of clusters and elastically effective links. Above 5.5% the network is fully developed and, in addition, entanglements are enhanceing the stiffness of the system.

7.The Relaxation spectrum.

The relaxation spectrum is a function of the range of structural diffusion times that are contributing to relaxation of the stress. The example of a strongly interacting, charge stabilised latex has been analysed successfully in terms of a model based on the "graininess" of the structure (45). The lattice model given above gives a good description of the measured high frequency modulus for monodisperse dispersions.

Dispersion

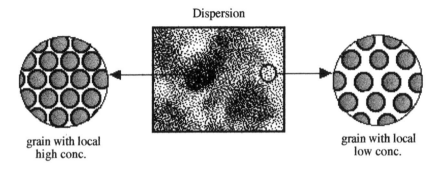

grain with local high conc.

grain with local low conc.

Figure 20. Schematic of the local concentration fluctuations giving rise to the relaxation spectrum.

The Maxwell relaxation time is given by the ratio :-

$$\tau_M = \frac{\eta(0)}{G(\infty)}$$

and the low shear viscosity is related to the diffusive motion in the system. Now a uniform concentration would imply a single relaxation time whereas we observe a small spread. The spectral width should reflect the temporal fluctuations in the local moduli of the "grains" from the mean state $(\Delta G(\tau))$:-

$$H(\ln\tau) = \Delta G(\tau)P(\Delta E) \qquad 7.1$$

where $P(\Delta E)$ is the probability of a local relaxation time which is related to the internal energy difference from the mean state. Hence the spectrum can be written as :-

$$H(\ln\tau) = \frac{G(\tau) - G(\infty)}{Q}\exp\frac{-\Delta E}{kT} \qquad 7.2$$

Here Q is the partition function and it was shown (45) that as the integral under the spectrum yields the high frequency modulus, Q was given by :-

$$Q = \int_{-\infty}^{+\infty} \left\{ \frac{G(\tau) - G(\infty)}{G(\infty)} \right\} e^{\frac{-\Delta E}{kT}} d\ln(D_{ref}\tau) \qquad 7.3$$

with

$$D(\tau) = D_{ref} e^{\frac{-\Delta E}{kT}}$$

The diffusion time of the reference state then fixes the position of the peak of the spectrum which shows a sharp symetrical peak in good agreement with experiment. An example is shown in Figure 21 for a poly(vinylidene fluoride) latex (PVDF) in sodium chloride solution.

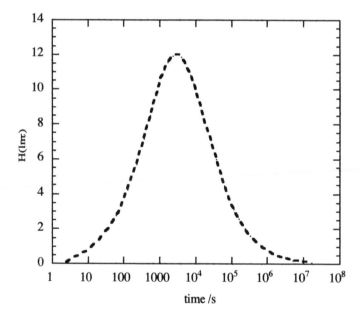

Figure 21.   Relaxation spectrum for a PVDF latex with a = 110 nm, ψ = 80 mV in a 2×10^{-3} M [NaCl] at a $\varphi = 0.38$.

Although the peak in the above example appears to be quite sharp, it should be noted that most of the relaxation occurs over 5 decades in time. This system is very close to the phase boundary and had a very high low shear limiting viscosity. it is interesting to compare this behaviour with what can be expected for a polymer solution. Ferry(46) illustrates this and gives a relationship for chains with Rouse dynamics of :-

$$H(\ln \tau) \propto \tau^{-1/2}$$

This clearly will result in a distinctly different spectral form and is due to the flexibility of the structural elements. The addition of polymeric rheology modifiers to particulates will result in a composite spectrum with a complex form.

8. Strain Melting.

The structures have been discussed above in terms of their linear viscoelastic response. This can only occur at low stresses and strains. At high stresses and strains the responses become non-linear . The range of linearity is related to the curvature of the pair potential. This is readily understood when it is remembered that we are testing the rate of change of the force-distance curve at the mean separation. If the strains are small enough for this to be approximately linear, the elastic moduli will also be linear with strain. Clearly, with sharp curvatures this can only be true for very small strains. Polymer solutions and highly isometric particulates tend to be the most tolerant to strain. The most usual experiment is to run a series of oscillation experiments at a constant frequency but progressively increasing the strain (or the applied stress). Figure 22 shows some typical data obtained for a colloidally stable system (*47*). If the onset of

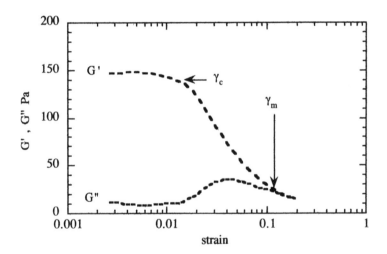

Figure 22. A strain sweep at 1 Hz for a dispersion of 300nm diameter silica particle in 10^{-3} M [NaCl} at a $\varphi = 0.44$.

non-linear response is seen at a value of strain γ_c then :-

    i.   polymer gels $\gamma_c > 0.1$;

    ii.   stable particulate gels $0.01 < \gamma_c < 0.04$;

    iii. weakly flocculated particle gel $0.001 < \gamma_c < 0.01$;

    iv. coagulated particle gels $\gamma_c << 0.001$;

and a linear region may not be attainable for the last category of system. The melting strain was taken as the value reached, γ_m, which resulted in an equality of the storage and loss moduli (*47*). This enabled a range of systems to be compared systematically in a way analogous to comparing melting temperatures. the model used was that of the "grain" structure used above, but with the disordered inter-grain zones becoming progressively larger as the melting took place.

Literature Cited.

1. Hunter, R.J. *Foundatons of Colloid Science* ; Clarendon Press: Oxford, UK 1987; Vol.1.

2. Russel,W.B.;Saville,D.A.; Schowalter,W.R. *Colloidal Dispersions* ; Cambridge University Press: Cambridge, UK, 1989.

3. Napper,D.H. *Polymeric Stabilisation of Colloidal Dispersions* ; Academic Press: London, UK, 1983.

4. Sperry, P. R., Hopfenberg, H. B. and Thomas, N. L. *J. Coll. Intf. Sci..* **1981**, *82*, 62.

5. Gast,A.P., Hall,C.K. and Russel, W.B. *J. Coll. Intf. Sci.* **1983**, *96*, 251

6. Asakura, S., and Oosawa, F. *J. Chem. Phys.* **1954**, *22*, 1255, &*J. Polym. Sci.* **1958**, *37*, 183

7. Vrij.A. *Pure and Appl. Chem.* **1975**, *48* , 471.

8. Joanny,J.F., Leibler,L. and de Gennes, P.G. *J. Poly. Sci.: Poly. Phys. Ed.* **1979**,*17*,1073

9. Verwey, E. J. W. and Overbeek, J. Th. G. *Theory of Stability of Lyophobic Colloids,* Elservier: Amsterdam, 1948.

10. Derjaguin, B., Landau, L. *Acta. Physcochim., URSS* , **1941**, *14*, 633.

11. Carnahan, N.F. and Starling, K.E. *J. Chem. Phys.* **1969**,*51,* 635

12. Alder, B.J. ;Wainwright, T.E. *Phys. Rev.* **1962**, *127*, 359.

13. Hiltner, P.A.; Krieger, I.M.*J.Am.Chem.Soc.* **1968**,*90*,3114.

14. Goodwin, J.W.; Hughes, R.W. *Adv. Colloid Interface Sci.* **1992**, *42*, 303.

15. Russel, W.B. *J.Rheol.* **1980**, *24*, 287.

16. Bossis, G.; Brady, J.F. *J.Chem.Phys.* **1989**, *91*, 1866.

17. van der Werff, J.C.; de Kruif, C.G. *J.Rheology*, **1989**, *33*, 421.

18. Krieger, I.M. *Adv.Colloid Interface Sci.* **1972**, *3*, 111.

19. Ball, R.C.; Richmond, P. *Phys.Chem.Liq.* **1980**, *9*, 99.

20. Goodwin, J.W.; Ottewill, R.H. *J.Chem.Soc. Faraday Trans* . **1991**, *87*, 357.

21. Melrose, J.R.; Heyes, D.M. *J.Chem.Phys.* **1993**, *98*, 5873

22. Hoffman, R. *J.Coll. Interface Sci.* **1974**, *46*, 491.

23. de Kruif, C.G.; van Iersel, E.M.F.; Vrij, A.; Russel, W.B. *J.Chem.Phys.* **1986**, *83*, 4717.

24. Conway, B.E.; Dobry-Duclaux, A.in *Rheology, Theory and Applications*, Eirich,F, Ed.;Academic Press, NY, 1960, Vol.3.

25. Watterson, I.G.; White, L.R. *J.Chem. Soc. Faraday 2*,**1981**, *77*, 1115.

26. Cross, M.M. *J.Coll.Sci.*. **1965**, *20*. 147.

27. Papir, V.S.; Krieger, I.M. *J.Coll.Interface Sci.* **1970**, *24*,126.

28. Choi, G.M.; Krieger, I.M. *J.Coll.Interface Sci.* **1986**, *113*, 94.

29. Goodwin, J.W.; Khidher, A.M. in *Colloid and Interface Science*, Kirker, M.Ed.; Academic Press, NY, 1977, Vol.IV.

30. Buscall, R.; Goodwin, J.W.; Hawkins, M.; Ottewill, R.H. *J.Chem.Soc.Farady Trans. 1*, **1982**, *78*, 2873, 2889.

31. Joanny, J.F. *J.Coll. Interface Sci.* **1979**, *71*, 622.

32. Keeping, S. Ph.D.Thesis, Bristol, 1989.

33. Goodwin, J. W., Hughes, R. W., Partridge, S. J., Zukoski, C. F. *J. Chem. Phys.* **1986**, *85*, 559.

34. Russel, B. W. and Sonntag, R. C. *J. Coll. Intf. Sci.*. **1987**, , 485.

35. Mewis, J.; van der Aerschot, E. *Coll. Surf.* **1992**, *69*, 15.

36 To be published, Goodwin,J.W. , Hughes,R.W. , Kwaambwa, H.M. and Reynolds, P.A.

37. Doi, M., Edwards, S. F. *Theory of Polymer Dynamics,* Claredon Press: Oxford, 1986.

38. Reynolds,P.A. and Reid, C.A. *Langmuir* . **1991**,*7*, 89.

39. Jenkins, R.D.; Silebi, C.A.; El-Asser, M.S. *ACS Symp.Series* , **1991**, *462*, 222.

40. Annable, T,; Buscall, R.; Ettelaie, R.; Whittlestone, D. *J.Rheol.* **1993**, *37*, 695.

41. Green, M.S.; Tobolsky, A.V. *J.Chem. Phys.* **1946**, *14*, 80.

42. Yamamoto, M. *J. Phys.Soc.Japan* **1956**, *11*, 413

43. Tanaka, F.; Edwards, S.F. *J.Non-Newtonian Fluid Mech.*. **1992**, *43*, 247, 273, 289.

44. Lam, S. Ph.D. Thesis, Bristol, 1993.

45. Goodwin, J.W.; Hughes, R.H. *J.Chem.Phys.* **1991**, *95*, 6124.

46. Ferry, J.D. in *The Viscoelastic Properties of Polymers*, Wiley, NY, 1980.

47. Bradbury, A.; Goodwin, J.W.; Hughes, R.W. *Langmuir,* **1992**, *12*, 2863

Chapter 7

Waterborne Radiation-Curable Coatings

Kurt A. Wood

Rohm and Haas Company, 727 Norristown Road,
Spring House, PA 19477

Radiation cured coatings constitute a relatively small, but rapidly growing segment of the total coatings market. They are attractive for many applications because they generally have a very rapid cure, require little or no heating of the coating or substrate, achieve very highly crosslinked films, and do so while emitting little or no volatile organic compounds (VOC) into the environment. The overall North American growth rate for radiation cured inks and coatings has been in the 10% range[1], with current usage at about 30,000 tons per year. The market share for radiation curable coatings is even larger in other parts of the world, notably Europe.

Radiation-curable coatings can be cured either by ultraviolet (UV) or electron beam (EB) radiation[2]. In terms of the apparatus required to achieve curing, electron beam curing units are considerably more elaborate and expensive, and this has tended to limit EB curing to certain high-throughput applications where the economics are most favorable. Ultraviolet curing lamps, on the other hand, do not generally require a great deal of space or capital investment, and can often simply be attached or inserted into the back end of an existing coating line.

Most radiation-curable coatings are cured via free radical addition polymerization, although some specialized applications make use of a cationic cure mechanism. Radiation-curing formulations are often supplied at or near 100% non-volatiles. A representative "conventional" UV curable coating of this sort might include: 1) some sort of oligomeric species, e.g. an acrylated polyurethane, epoxy, or polyester; 2) one or more, generally several, reactive diluents (often multifunctional or monofunctional acrylates or methacrylates), which have the dual function of lowering the formulation viscosity and contributing to the balance of properties of the coating; 3) a photoinitiator package, at 1-6% on total non-volatiles, that will induce free radical cure.

The high solids feature of conventional radiation-curable coatings is useful for many applications, e.g. for high build, "wet look" topcoats and for fillers where no shrinkage of the coating is desired. However, other applications require low solids either for aesthetic or practical reasons. Low solids formulations are generally preferred for spray applications, for low build curtain coat applications, for formulations incorporating a flatting agent to achieve low gloss, and for "open pore" type applications that bring out the natural beauty of many kinds of wood. For radiation-curable coatings at low solids, it is necessary that the desired film shrinkage occurs, i.e. that the solvent is substantially driven off, prior to curing.

Low solids radiation-curable coatings can be prepared either by diluting high solids conventional systems with large quantities of organic solvents, or by redesigning the oligomers and other formulation ingredients to make them water-soluble or dispersible. Formulations with high levels of organic solvents have the usual solvent disadvantages of VOCs, flammability, and a tendency to give finishes with a "plastic look". Consequently there is increasing interest in waterborne systems, which not only readily achieve low solids, but also maintain the performance and VOC advantages of conventional UV coatings. In Europe, which has a more advanced radiation curing market, there are perhaps twenty different commercial or experimental waterborne systems which have been developed by various suppliers during the past few years. The advantages of some of these systems have been discussed in a number of recent talks and papers[3,4,5,6].

Waterborne Radiation-Curable Systems Based upon Condensation Polymers

The oligomeric components used in conventional radiation-curable formulations are generally condensation polymers. A radiation-curable polyester oligomer, for example, might be prepared by first synthesizing a low molecular weight, hydroxy functional polyester, and then reacting the hydroxyl groups with acrylic acid[2]. The oligomer contributes to the final coating properties both directly, i.e. through the inherent properties imparted by its composition and internal structure, and indirectly, through the crosslinked network that it helps form through radiation curing. The oligomers contain, on the average, several radiation-curable groups per molecule. Esters of acrylic or methacrylic acid are often used due to their high reactivity and relative insensitivity to oxygen inhibition[7]. Allyl and vinyl ether groups may also be used. For free radical polymerization, the reactivity of various functional groups has been generally found to decrease in the order:

acrylate > methacrylate > allyl > vinyl ether

The reactivity of some functional groups can be enhanced by designing systems incorporating two or more different functional groups which preferentially copolymerize[8,9].

For systems curing by a cationic mechanism, vinyl ethers and epoxides tend to be the most reactive functional groups. However, almost nothing has been described in the literature in the way of *waterborne* cationic curing systems, which is not surprising given that water as a nucleophile inhibits cationic cure.

A number of waterborne radiation curable systems based on condensation polymers have been described in recent years. One major class of systems is based upon polyester and polyether technology[10,11,12,13,14,15]. The unsaturated oligomers used in these systems are very similar in structure to those used in conventional radiation curable coatings. The unsaturation may either be incorporated into the oligomer backbone-- for instance through the use of maleic anhydride or an unsaturated diacid in the oligomer synthesis-- or it may be pendant, for instance by capping the oligomer with acrylate, allyl, or vinyl ether groups. As with conventional radiation curing coatings, each kind of unsaturated group has particular advantages and disadvantages in terms of cure speed, types of photoinitiators and reactive diluents that work efficiently with it, toxicity considerations, and so forth.

In the absence of any low molecular weight diluent, the unsaturated oligomers generally have a very high viscosity at room temperature (often being solids). The factors determining the viscosity include the oligomer composition, molecular weight, and degree of branching-- resin design parameters which are typically set to certain values in order to achieve specific coating properties. To make a waterborne coating from this sort of oligomer, two practical considerations must be addressed. First, the oligomer must be dispersed into water (or in some cases dissolved)-- a process which generally requires considerable shear and usually some means of reducing the oligomer viscosity during the dispersion process. Second, if a dispersion is made, it must be stable enough that it will have an adequate shelf life and survive the stresses of formulation.

Oligomers can often be designed to be soluble in water by neutralization with an amine or permanent base[13,16]. However, truly water-soluble resins have only a limited utility, since their hydrophilic character remains even after film formation and curing, resulting in water sensitivity problems in the cured film. Water-soluble oligomers are of interest in certain graphic arts applications, where soluble resins have rheological advantages, and where water resistance requirements are often not too stringent.

Water soluble resins are also important in photoresist technology, in what might be called water-developable systems. Such a system might be based upon a high acid-number acrylated oligomer, which is soluble in aqueous base when uncured, but insoluble after UV curing. After exposure of selected regions of such a coating to UV light, the coating can be developed in aqueous base, and the unexposed regions are dissolved away[17].

Water-thinnable resins have also been described[18]. Like conventional radiation-curable oligomers, these are designed to work at very high non-volatile levels (90% or even higher); however, they use water rather than some organic solvent or reactive diluent as a viscosity suppressant. In this way such systems may be able to keep high solids and low VOCs, while avoiding the use of reactive diluents. Conceptually these water-thinnable systems are closer to conventional radiation-curable coatings than to true waterborne coatings.

In most cases it is preferable to keep the hydrophilic content in the oligomer as low as possible and make the oligomer water dispersible rather than water soluble. This may be done through the incorporation of non-ionic or ionic, surfactant-like groups into the oligomer[19,20]. As such, it may be supplied already dispersed into water by the manufacturer, or it may be supplied at 100% solids and the dispersion step left to the formulator. Another way to get stability in water is to add emulsifiers to make the resin dispersible[11,21]. These emulsifiers can themselves contain unsaturated functional groups that can participate in the radiation curing[22,23,24]. To achieve the dispersion, the polymer viscosity may need to be reduced, either by adding it to water as a hot melt, or by manufacturing it in a viscosity-reducing solvent which could then be stripped off after the dispersion into water. This viscosity-reducing solvent could be a reactive diluent, such as an acrylic monomer.

Alongside polyesters and polyethers, the second major class of condensation polymers used for waterborne radiation-curable coatings are the polyurethanes[25,26,27,28,29]. These can be fully synthesized and then dispersed into water in the same way as polyesters[30], but more often the technology used to make polyurethane dispersions[31] (PUDs) or acrylic-urethane graft copolymers is employed[32,33]. For instance, a hydroxy functional acrylic monomer such as hydroxyethyl methacrylate may be combined with other polyols to make an isocyanate-functional prepolymer, which may then be neutralized with triethylamine, dispersed in water, and chain extended with a primary diamine. In this approach the dispersion into water is accomplished prior to building the polymer up to its full molecular weight. The manufacturer is then able to avoid many of the in-process viscosity problems associated with making high molecular weight condensation polymers. At the same time the manufacturer has considerable latitude in polymer design, in terms of being able to select the molecular weight over a wide range, the crosslink density, and other polymer properties.

Most waterborne urethane systems have anionic stabilization, which is often achieved by incorporating acid diols such as dimethylolpropionic acid (DMPA) into the prepolymer. Systems with nonionic[10] and cationic[34,35] stabilization have also been described.

Some of the waterborne systems described in the literature contain multi-functional acrylates, or MFAs, e.g. ethylene glycol diacrylate (EGDA) or trimethylolpropane triacrylate (TMPTA). These same MFAs are also commonly used in conventional UV formulations to increase the film crosslink density. The multifunctional acrylates in waterborne UV systems may be post-added intentionally to boost final properties or to enhance the polymer dispersibility, or they may be by-products of the oligomer manufacturing process. However, often systems that are free from multifunctional acrylates are preferred, because certain MFAs have been identified as irritants and sensitizers[36,37]. From a worker health and safety stand-point, MFA-free systems may be simpler to use than MFA-containing systems for certain applications, for instance in manual spray applications. In fact waterborne systems generally have a major advantage over conventional radiation-curable coatings in this respect: They can take advantage of the fast cure afforded by acrylated systems, without the traditional complications associated with the presence of multifunctional acrylates in the formulation.

Waterborne Radiation-Curable Systems Based upon Addition Polymers

A second type of waterborne radiation-curable coating is based upon addition polymers. Most of the systems described in the literature are based upon acrylic emulsions[5,6,38], but systems based on other addition polymers such as styrene-acrylics[39] and chlorinated polyolefins[40] have also been described. Unlike the normal situation for condensation polymers, waterborne addition polymers are typically polymerized *after* the dispersion into water of the polymer precursors has taken place. In this way, high molecular weight, linear polymers are easily made; there are no practical viscosity limitations associated with the molecular weight build since the polymerization takes place in dispersed micelles. Molecular weight can be controlled with various chain transfer agents, while internal crosslinking can be achieved by using multifunctional crosslinking agents. Other superstructural features, such as multiple phases in one particle, are also possible by various advanced techniques[6,41,42].

This ability to make use of the full range of waterborne polymer design techniques is another advantage that waterborne radiation-curable systems have over conventional systems. Because the build-up of the molecular weight for acrylic emulsion polymers and PUDs takes place in an aqueous dispersion-- rather than in a bulk phase-- it is possible to design waterborne systems with unique features, unattainable with conventional radiation-curing technology. For instance, wood coatings binders are now available commercially that are not only tack-free before curing, but are actually hard enough after physical drying (i.e. loss of water) that they can be sanded and compounded prior to cure[3,43,44]. This is an extremely difficult property to achieve with conventional systems, because of the plasticizing effect of the monomers they contain, and the low molecular weight of most oligomers. At the same time these waterborne binders retain the conventional radiation-cure benefits of near-zero VOCs and high performance after UV cure.

Radiation-curable binders are often made from addition polymers by post-functionalization[45]. In other words, an addition polymer is first made, which contains some sort of functional group but no residual unsaturation that could undergo radiation curing. In a second step, the functional group is used as an attachment point for an unsaturated monomer. For instance, Pears *et al.* describe a radiation-curable acrylic emulsion in which an amine functional latex is prepared, which is then functionalized by reacting the amine group with acetoacetoxy ethyl methacrylate. This reaction forms an enamine group and the result is a methacrylate functional latex. (The same reaction can also be used to make a methacrylate-functional PUD.[46])

Like some of the PUD-based systems, acrylic-based systems with pendant unsaturation are now available which are free of MFAs. Other acrylic systems are designed to work in conjunction with MFAs or other reactive diluents[47]. Systems have also been described in which the addition polymer has little or no attached unsaturation, but is nevertheless able to be grafted into a network of photopolymerizing acrylic monomers, under typical conditions for radiation curing[5,48]. Even in the absence of grafting, an interesting balance of properties can often be obtained for photopolymerized films of acrylic polymers to which MFAs have been added[3]. Such films might be expected to form an interpenetrating network of two continuous phases under appropriate conditions. As another kind of approach, other workers are investigating blends of dispersions, e.g. two acrylic emulsion systems, or an acrylic blended with a polyurethane dispersion, where one of the blend components may be non-functional[44,49].

Formulation of Waterborne Radiation-Curable Systems

Photoinitiators are not normally included in electron beam curing formulations, but are essential components of nearly every waterborne UV-curable formulation. Often considerable formulation skill is required to optimize the photoinitiator package to particular application requirements. In principle one wants to match the UV lamps and photoinitiator(s), such that spectral regions exist with both high lamp output and efficient photoinitiation from light absorption at that wavelength. Generally this is not too difficult for thin, clear films, but the situation is far from trivial where pigmented systems are concerned, or for the case of thick films where efficient photoinitiator absorption may greatly reduce the amount of light which penetrates deeply into the film. Photoinitiators must also be matched to the system in terms of chemistry, since the different classes of photoinitiator are most effective when used with specific functional groups[50]. The formulator must also consider the possibility of inhibition from atmospheric oxygen, or from other substances in the formulation or in the substrate (e.g. wood tannins).

The degree of cure will depend on the amount of photoinitiator used, the UV intensity, the total UV dose received (a function of the intensity and the line speed), and sometimes also the temperature at which cure is accomplished. Since photoinitiators are relatively expensive, there is usually a strong cost incentive to keep the levels of these materials to a minimum. Other disadvantages of high photoinitiator levels are a tendency to yellowing, and in some specialized applications, health and safety considerations pertaining to unreacted photoinitiators or their photofragments[51].

For waterborne systems, another complication arises because many photoinitiators are supplied in solid form. Unlike conventional UV formulations where photoinitiators are often dissolved in the monomer, in waterborne systems a different photoinitiator dispersion strategy may be necessary[52]. Many published waterborne formulations use Darocur 1173 (Ciba-Geigy) or for pigmented coatings, Darocur 4265, since these are liquids which are relatively easy to disperse directly into a waterborne coating. Alternatively, liquid photoinitiators may be pre-emulisified[53], or solid photoinitiators may be introduced as a solution in some sort of cosolvent (although this may introduce VOCs), be predispersed in the resin itself in the case of water-reducible systems, or be included in a grind in a pigmented formulation. Recently predispersed polymeric photoinitiators have also been described. With their high molecular weights, these also have the advantage of low volatility. They therefore minimize any potential problems associated with the precure loss of photoinitiator while water is being removed from the coating. Loss of photoinitiator has occasionally been reported as a problem with some of the lower molecular weight photoinitiators, particularly when aggressive force bake conditions are used.

In the case of pigmented coatings and inks, the formulator must carefully tailor both the photoinitiator and the UV lamps, to take advantage of whatever spectral window for cure is allowed by the pigments and resins that are used[54]. Often it is advantageous in terms of cure to replace traditional rutile TiO_2 pigments with other pigments such as anatase TiO_2 or zinc sulfide, which have a larger window of UV transmission.

Besides photoinitiators, waterborne radiation-curable binders can for the most part be used with the full range of formulation additives available for waterborne coatings, for instance defoamers, dispersants, wetting agents, and flatting agents. Coalescents are generally not required, and they are often either not used at all, or only in low amounts, in order to maintain the low VOC advantage of the radiation cure technology. Some other kinds of traditional coatings additives used to enhance final properties, e.g. waxes to improve mar resistance, may also be unnecessary due to the high performance qualities of the films attained through radiation curing.

Drying and Heating Requirements

As mentioned in the Introduction, waterborne radiation-curable coatings are of particular interest in certain kinds of applications in which low solids, low application viscosity, or low gloss are important. As such, they are akin to other kinds of waterborne coatings, while providing what is often a superior level of performance. Also like other kinds of waterborne coatings, they share in the particular advantages and disadvantages that water has as a solvent. Removal of the water from the coating is a particular issue that must be addressed: it is almost always necessary that the water be substantially out of the coating prior to the radiation curing step. The need to remove water is often considered to be the major drawback to this class of coatings; unlike solvent-borne coatings, one cannot do much in the way of formulation to increase the dry speed. Heat, of course, will increase the water evaporation rate, but heating can be expensive, and can cause other problems such as volatilization of the photoinitiator, and blistering of the coating prior to or during cure due to the volatilization of water from the substrate. Heating may also not always even be possible in certain applications due to line configurations or the heat sensitivity of some substrates. (In fact, one advantage of radiation-curable coating is that high performance coatings are possible without the heat required for many high performance thermoset systems.)

A second issue requiring consideration is that molecular mobility effects may sometimes influence the extent of cure that can be achieved with waterborne radiation-curable systems[3,38,55,56]. These effect arise because polymer chains are essentially locked into place at temperatures below the effective glass transition temperature T_g. In systems undergoing crosslinking, this effective glass transition temperature depends both on the inherent rigidity and molecular weight of the polymers and other components in the system, and on the extent of conversion of the crosslinking process, which can dramatically reduce the molecular mobility. A radiation-curable film can consequently arrive at a state of "vitrification", which has been studied extensively in classical thermosetting systems, such as epoxy-amine systems[57]. In vitrification, the reacting system eventually reaches the point where no further crosslinking is possible, because the molecular mobility is so greatly reduced that the reactants cannot move to reach each other. At this point the system's temperature matches its effective T_g. The only way to drive the reaction further to completion is to restore the molecular mobility by raising the temperature back above the effective T_g.

Before cure, conventional radiation-curable materials generally have very high molecular mobilities, due to the low molecular weights of the oligomers used, and the high levels of MFAs and other monomers present. Differences in the ultimate degree of conversion can be considerable, depending on factors such as the resin type and the average number of functional groups on the monomers used[55]. However, in such cases, a considerable reduction in the molecular mobility always will imply a considerable degree of crosslinking, so that the onset of vitrification may have little effect on the coating properties that will be achieved in practice. Waterborne radiation-curable systems, on the other hand, may already have low molecular mobilities at ambient temperature, prior to cure-- especially if they are designed to give films with a measure of pre-cure hardness. Pre-cure hardness is difficult to attain with conventional radiation-curable coatings, but confers a number of practical advantages. If a film can be made tack-free before cure, for instance, this can significantly simplify the handling requirements for uncured coated objects, and minimize dirt and dust pickup should there be a delay between the coating and curing steps.

Because of the higher molecular weights of many waterborne systems, and the low or negligible monomer content, these systems can be affected by vitrification, and consequently slow cure speeds, even at low degrees of conversion, and for materials of low functionality. As a result, the degree of cure with waterborne systems, and the film properties obtained, may depend to a surprising degree on the thermal history of the coating, during and immediately after radiation curing[3]. On the other hand, because of the higher molecular weights of these polymers, *uncured* film properties are also often far superior to those attainable with conventional radiation curing formulations. To achieve a comparable level of final performance, a waterborne system may require a lower level of radiation-induced crosslinking. This may in turn result in reduced film shrinkage during cure, relative to conventional radcure systems.

Applications of Waterborne Radiation-Curable Coatings

A major applications target for waterborne UV formulations has been the wood coatings area, with a number of recent talks and papers describing waterborne UV formulations based on acrylic, polyester, and urethane resins. Hybrid systems have also been described which contain nitrocellulose[58,59], although the nitrocellulose itself is not modified to make it radiation curable. Waterborne wood coating binders are probably not the right choice for fillers or for high build, "wet look", topcoats-- for these applications, one generally wants the coating non-volatile level to be as high as possible. However, many sealers and topcoats are supplied at lower solids. By using lower solids formulations, there are not only economic advantages in many cases, but it is easier to achieve a natural, non-"plastic" look with many kinds of wood. Flatting agents are also easily incorporated into the formulation, and it is often easier to get the right kind of rheological profile for spray and curtain coat applications. Increased grain-raising, relative to solvent-based coatings, can be a problem at times, but often this can be minimized through optimization of the sealer system. With some of the newer waterborne radiation-curable binders, sanding and other wood industry finishing operations, such as compounding and buffing, can be performed before UV cure due to the inherent hardness of the polymers being used. In multiple step operations such as are common in the U.S. wood finishing industry, this opens the possibility of "repair" of coating defects before the film is hardened to its final state; it also simplifies the line requirements since UV cure need only be done once, as a final step, rather than after the application of each coat. It may also be possible with a waterborne system to avoid having to rub and buff the finish after the cure, because the gloss is not compromised by film shrinkage during cure.

"Dual cure" systems have also been developed in which a radiation curing mechanism is supplemented by some other sort of conventional curing mechanism[60].

Other potential application areas for waterborne UV curable coatings are coil and metal decorator coatings[47]. An acrylic latex binder for flooring curing by a cationic mechanism has also been described[61].

A number of graphic arts applications have been identified as particularly amenable to the use of waterborne radiation curable binders, including flexographic and gravure inks, screen inks, intaglio inks, ink-jet textile printing inks, and overprint varnishes[62,63,64]. Systems for making printing plates have also been described[65,66]. A major benefit of water in some of these applications is that very low viscosity formulations can easily be obtained. Waterborne systems can also have VOC, flammability, and toxicity advantages over conventional technology. Both soluble and dispersed polymers have been preferred for various graphic arts applications[67].

Waterborne radiation curable systems have been described, which are useful as laminating or pressure-sensitive adhesives. Besides systems based on urethanes[68], a polyamide system has been described, which is functionalized using the Michael addition reaction of residual amine groups on the resin with acrylate groups of a multifunctional acrylate such as TMPTA[69].

Literature Cited

[1] Lawson, K. *Proceedings of Radtech '94 North America*; RadTech International North America: Northbrook IL, 1994; Vol. I, pp. 298-306.

[2] *UV & EB Curing Formulation for Printing Inks, Coatings & Paints;* Holman, R.; Oldring, P., Eds.; SITA-- Technology: London, England, 1988.

[3] Wood, K. *Proceedings of Radtech '94 North America*; RadTech International North America: Northbrook IL, 1994; Vol. I, pp. 360-369.

[4] Stenson, P. *Modern Paint and Coatings* **June 1990,** *80(6)*, pp. 44-48.

[5] Wood, K. *Polymers Paint Colour Journal* **February 10, 1993,** *183(4322),* pp. 34-37.

[6] Padget, J. C. *Journal of Coatings Technology,* **December 1994,** *66(839)*, pp. 89-105.

[7] O'Hara, K. J. In *Radiation Curing of Polymers*; Randell, D. R., Ed.; Royal Society of Chemistry: Cambridge, England, 1987; Special Publication No. 64.

[8] Meij, E.; Rietberg, J. *Proceedings 3rd Nürnberg Congress,* March 1995, Paper 45.

[9] Schouten, J. J.; Noren, G. K.; Lapin, S. C. *Proceedings of Radtech '92 North America*; RadTech International North America: Northbrook IL, 1992; Vol. I, pp. 167-172.

[10] Cavalieri, R; Farronato, S.; Settin, C. *Proceedings of RadTech Europe '89;* RadTech Europe: Fribourg, Switzerland, 1989; pp. 679-688.

[11] Buethe, I. et al., *US 4,287,039* (Sept. 1, 1981).

[12] Arnoldus, R. et al. *Waterborne Coatings: Are They Meeting the Challenge?;* Proceedings of 12th International Conference on Waterborne Paints, Milan, Italy; Paint Research Association: Teddington, Middlesex, England, 1992; Paper 8.

[13] Philips, M. et al., *Polymers Paint Colour Journal,* **10 February 1993,** *183(4322),* pp. 38, 40.

[14] Dvorchak, M. J.; Riberi, B.H. *J. Coatings Technology* **May 1992,** *64(808),* pp. 43-49.

[15] Kressdorf, B.;Luehmann, E.; Dannhorn, W.;Hoppe, L. *DE 4,219,76* (December 22,1993).

[16] Corrado, G., *EP 558,788* (September 8, 1993).

[17] Behr, A.H. et al., *EP 546,768* (June 16, 1993).

[18] Ravjist, J.-P., *Polymer Paint Colour Journal* **October 1995,** *185(4372),* pp. 7,10,12.

[19] Kressdorf, B.; Hoppe, L.; Lühmann, E.; Szablikowski, K. *EP 425,947* (May 8, 1991).

[20] Padget, J. C.; Pears, D. A. *EP 367,464* (May 9, 1990).

[21] Beck, E.; Keil, E.; Lokai, M. *Farbe und Lack* **March 1992,** *98(3),* pp. 165-170.

[22] Meixner, J.; Kremer, W., *EP 501,247* (September 2, 1992).

[23] Meixner, J.; Kremer, W., *Proceedings of XIX FATIPEC Conference;* Forschunginstitut für Pigmente und Lacke e.V.: Stuttgart, Germany, 1988; pp. 65-74.

[24] Meixner, J.; Pedain, J.; Fischer, W. *EP 381,862* (August 16, 1990).

[25] Beck, E. et al., *DE 4,031,732* (April 9, 1992).

[26] Garratt, P.G.; Klimesch, K. F. *Polymers Paint Colour Journal* **February 9, 1994** *184(4343),* pp. 30-32.

[27] Stenson, P. *Paint and Coatings Industry* **September 1992,** pp.22-29.

[28] Kressdorf, B. et al. *EP 574,775* (December 22, 1993).

[29] Haberle, K.; Weyland, P.; Eckert, G.; Renz, H. *EP 392,352* (October 17, 1990).

[30] Flakus, W. *EP 518,020* (December 16, 1992).

[31] Rosthauser, J. W.; Nachtkamp, K. In *Waterborne Polyurethanes;* Frisch, K. C.; Klempner, D., Eds.; Technomic: Westport, CT 1987; Vol. 10.

[32] Arnoldus, R. *Proceedings of RadTech Europe '89;* RadTech Europe: Fribourg, Switzerland, 1989; pp. 121-127.

[33] Zom, W. et al. *US 4,730,021* (March 8, 1988).

[34] Gould, N. P. *GB 2,270,917* (March 30, 1994).

[35] Baranowshi, T. R.; Bechara, I. *EP 541,289* (May 12, 1993).

[36] Uminski, M.; Saija, S. J. *Surface Coatings International* **1995,** *78(6),* pp. 244-249.

[37] Taylor, D. M. *Proceedings of RadTech Europe '93;* RadTech Europe: Fribourg, Switzerland, 1993; pp. 864-870.

[38] McGinniss, V. D. et al., *US 4,107,013* (August 15, 1978).

[39] Pons, D. A. *EP 330,246* (August 30, 1989).

[40] Stevens, E.; Lear, P., *EP 517,379* (December 9, 1992).

[41] Arnoldus, R.; Adolphs, R.; Zom, W.; Pollano, G. *Modern Paint and Coatings* **November 1991,** pp. 42-46.

[42] Warson, H. *Polymers Paint Colour Journal* **July 18, 1990,** *180(4265),* pp. 507-512.

[43] *Polymers Paint Colour Journal* **July 1995,** *185(4369),* p. 22.

[44] Wehman, E. et al. *Proceedings of XXII FATIPEC Conference;* Forschunginstitut für Pigmente und Lacke e.V.: Stuttgart, Germany, 1994; Vol. II, pp. 219-232.

[45] Pears, D. et al., *EP 442,653* (August 21, 1991).

[46] Pears, D; Heuts, M. P. J. *EP 442,652* (August 21, 1991).

[47] Erickson, E. R. et al., *US 4,107,013* (August 15, 1978).

[48] K. Moussa, K.; Decker, C. *Proceedings of Radtech '92 North America;* RadTech International North America, Northbrook IL, 1992; Vol. I, pp. 291-301.

[49] Arnoldus, R. *Proceedings of RadTech Europe '95;* RadTech Europe: Fribourg, Switzerland, 1995; pp. 469-473.

[50] Chang, C.-H. et al., In *Handbook of Coatings Additives, Vol. 2;* Calbo, L. J., Ed.; Marcel Dekker: New York, NY, 1992; pp. 1-50.

[51] Nahm, S. H., *Journal of Coatings Technology* **1991,** *63(798),* pp. 47-54.

[52] Catalina, F.; Peinado, C.; Corrales, T. *Rev. Plast. Mod.* **1992,** *63(431),* 561-569.

[53] Libassi, G.; Nicora, C. *EP 386,650* (September 12, 1990).

[54] Pietschmann, N. *J. Radiation Curing* **1994,** *21(4),* pp. 2-9.

[55] Decker, C.; Moussa, K. *J. Coatings Technology* **1990,** *62(786),* pp. 55-61.

[56] Tilley, M. G. et al. *Proceedings of Radtech '92 North America;* RadTech International North America: Northbrook IL, 1992; p. 48.

[57] Wang, X.; Gillham, J.K. *J. Coatings Technology* **April 1992,** *64(807),* pp. 37-45.

[58] Hoppe, L.; Lühmann, E., *Proceedings of RadTech Europe '91;* RadTech Europe: Fribourg, Switzerland, 1991; pp. 420-439.

[59] Lühmann, E.; Hoppe, L.; Szablikowski, K. *US 4,772,329* (September 20, 1988).

[60] Baumann, H. *I.-Lack* **April 1993,** pp. 144-147.

[61] Sutton, D. C. *EP 267,554* (May 18, 1988).

[62] Salem, M. S., Bate, N. *Polymers Paint Colour Journal* **June 14, 1989,** *179(4239),* pp. 400-402.

[63] Salem, M. S., Bate, N. *Polymers Paint Colour Journal* **March 14, 1990** *180(4256),* p. 150.

[64] Brosse, J.-C. et al. *Proceedings of RadTech Europe '95;* RadTech Europe: Fribourg, Switzerland, 1995; pp. 271-279.

[65] Muzyczko, T. M.; Thomas, D. C. *US 4,186,069* (January 29, 1980).

[66] Joerg, K.; Zertani, R. US 5,053,317 (October 1, 1991).

[67] Korvemaker, P. *Polymers Paint Colour Journal* **November 13, 1991,** *181(4295)* pp. 658-659.

[68] Kubota, T. et al. *EP 443,537* (August 28, 1991).

[69] Smith, G. A.; Rumack, D. T.; Frihart, C.R. *US 5,109,053* (April 28, 1992).

Chapter 8

The Application of Carbodiimide Chemistry to Coatings

J. W. Taylor[1] and D. R. Bassett[2]

[1]Waterborne Research Laboratory, Eastman Chemical Company,
P.O. Box 1955, Kingsport, TN 37662
[2]UCAR Emulsion Systems, Union Carbide Corporation,
410 Gregson Drive, Cary, NC 27511

The use of carbodiimide chemistry in coatings has been explored. Model studies show that in polar environments, at elevated temperatures and in the presence of amines, that carbodiimides react with acetic acid to form predominately N-acyl urea products. DSC results show that the N-acyl urea moiety is stable below 155 $^\circ$C. At ambient temperature, the half-life of 1,3-dicyclohexylcarbodiimide in the presence of an equal molar amount of acetic acid is two hours. Multifunctional carbodiimides were synthesized from multifunctional isocyanates, ureas and thioureas. Crosslinking studies showed that emulsions of multifunctional carbodiimides are excellent low-temperature crosslinkers for waterborne coatings. Blends of polycarbodiimide emulsions and waterborne carboxylic acid-containing particles dry to give films with improved tensile properties and excellent solvent resistance. Carboxylic acid-containing waterborne particles were modified using alkyl carbodiimide ethyl methacrylates to produce waterborne particles with polymerizable double bonds. Films prepared from these reactive particles and small amounts of t-butyl peroxybenzoate were shown to cure thermally giving films with high gel content. Carbodiimides were blocked with diethylamine and novolac resins to form guanidine and isourea moieties. The use of these carbodiimide blocking technologies is described for powder coatings and high resolution dual-tone photoresists.

The increasing knowledge of carbodiimide chemistry shows that it can be useful for crosslinking carboxylic acid-containing resins (1-8), and for imaging photoresists (9-12). Furthermore, biological studies on animals show that low-molecular weight polycarbodiimides can be prepared which have low toxicity and no mutagenicity (13). Although carbodiimides react with carboxylic acids to form anhydrides, in polar environments, at elevated temperatures, or in the presence of amines, carbodiimides react with carboxylic acids to form predominately N-acyl ureas. As illustrated in scheme 1, multifunctional carbodiimides are useful as crosslinkers for carboxylic acid-containing polymers under the appropriate cure conditions.

Scheme 1

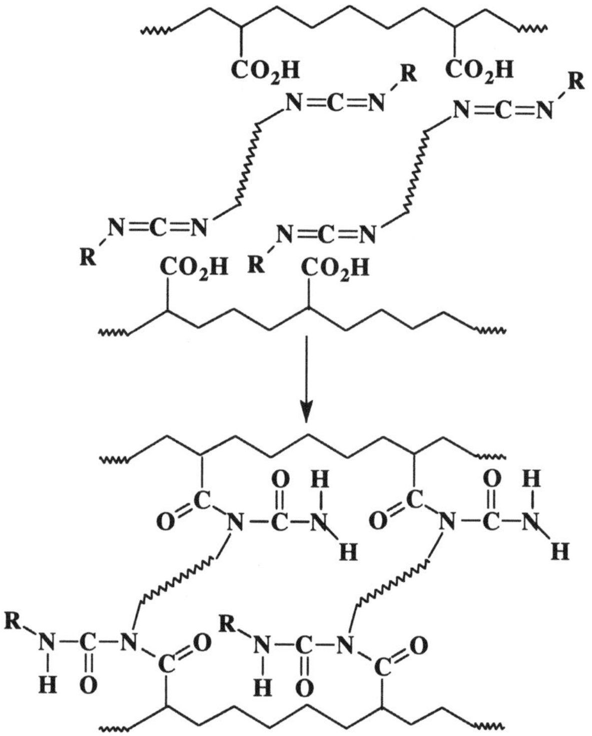

In the literature (14) it is known that carbodiimides react with carboxylic acids to form N-acyl ureas or anhydrides and ureas. These two pathways are illustrated in the scheme 2 below:

Scheme 2

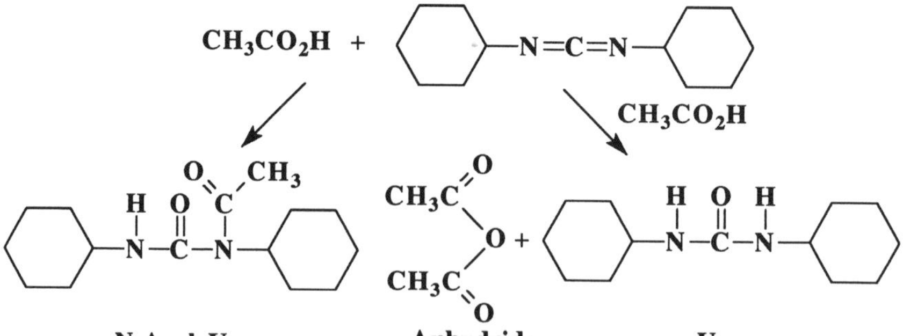

N-Acyl Urea Anhydride Urea

Since anhydrides rapidly react with water to reform carboxylic acids, the formation of anhydrides is highly undesirable for crosslinking carboxylic acid-containing waterborne polymers. Thus, one would predict that the mechanical properties and solvent resistance of cured coatings would deteriorate (a decrease in crosslink density) with time in a high humidity environment due to the hydrolysis of anhydride crosslinks.

Development of Carbodiimide Crosslinking Chemistry

During the development of carbodiimide crosslinking chemistry, the crosslinking reaction was modeled using 1,3-dicyclohexylcarbodiimide and acetic acid. It was observed that the ratio of N-acyl urea to acetic anhydride depends upon the environment of the reaction. Specifically, increasing the polarity (Table 1) or temperature of the reaction medium (Table 2) or adding bases to the reaction medium (Table 3) increases the ratio of N-acyl urea to acetic anhydride.

Table 1. Solvent Effect

Solvent[a]	Mole Percent Carbodiimide	Mole Percent Anhydride	Mole Percent N-Acyl Urea	Acyl Urea/ Anhydride
Carbon Tetrachloride	44	47	30	0.2[b]
Cyclohexane	59	24	17	0.7
Acetonitrile	20	34	46	1.4
Acetonitrile	32	31	37	1.2[b]
Tetrahydrofuran	3	8	89	11.1

a. Reaction conditions were at 30 °C for 45 h in tetrahydrofuran at 0.040 M for each reactant. b. DeTar, 25 °C, 0.04 M, 1966, from reference 14.

Table 2. Temperature Effect

Temperature (°C)	Mole Percent Carbodiimide	Mole Percent Anhydride	Mole Percent Acyl Urea	Acyl Urea/ Anhydride
0.0	36	18	46	2.6
30.0	4	8	88	11.1
48.0	11	4	85	21.3

Reaction conditions were for 44 h in tetrahydrofuran at 0.040 M for each reactant.

Table 3. Effect of Triethyl amine on Acyl Urea to Anhydride Ratio

Triethylamine Equivalence	Acyl Urea/Anhydride Molar Product Ratio[a]
0.0	10.0
0.5	17.6
1.0	41.6

Reaction conditions were 30 °C for 23 h in tetrahydrofuran at 0.045 M for each reactant.

Thermogravimetric analysis of the N-acyl urea product from the reaction between 1,3-dicyclohexylcarbodiimide and 2,2-dimethylpropionic acid shows that the N-acyl urea bond (bond between the nitrogen and carbonyl of the carboxylic acid) is not stable above 155 °C. These results suggest that coatings which require cure temperatures greater than 155 °C may give isocyanate decomposition products (15). To examine the reactivity of 1,3-dicyclohexylcarbodiimide with acetic acid, the reaction between the 1,3-dicyclohexylcarbodiimide and acetic acid in tetrahydrofuran was monitored by FT-IR spectroscopy by following the carbodiimide absorption at 2260 cm^{-1}. The data (Figure 1) show that approximately 50% of the 1,3-dicyclohexylcarbodiimide reacts in 2 h at ambient temperature (23 °C). The thermal stability of the N-acyl urea linkage, the rapid reaction between 1,3-dicyclohexylcarbodiimide and acetic acid at ambient temperature, and the ability to adjust the polarity, temperature, and basicity of the reaction medium (Tables 1, 2, and 3) to make the N-acyl urea derivative the major product of the reaction demonstrate that carbodiimide chemistry is useful chemistry for designing low-temperature crosslinkers for carboxylic acid-containing polymers.

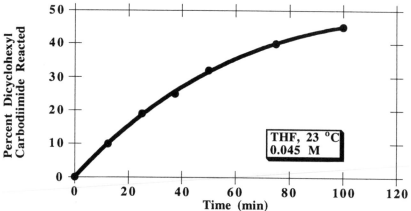

Figure 1. Reaction of acetic acid with 1,3-dicyclohexylcarbodiimide.

Preparation of Multifunctional Carbodiimides. Two types of multifunctional carbodiimides were prepared. The first type is a nonbranched polycarbodiimide prepared using a procedure described by Campbell (16, 17). Butyl isocyanate was reacted with the isophorone diisocyanate in amyl acetate using a phospholene oxide catalyst. The theoretical functionality of this multifunctional carbodiimide is four. The reaction is shown in scheme 3.

Scheme 3

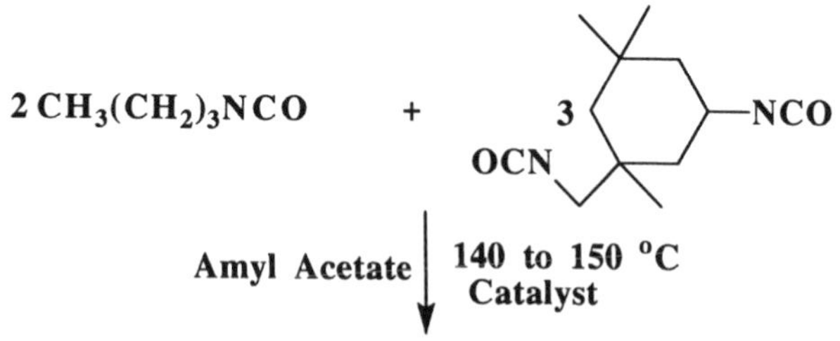

$2\ CH_3(CH_2)_3NCO$ +   3 [structure] —NCO

OCN

Amyl Acetate | 140 to 150 °C Catalyst

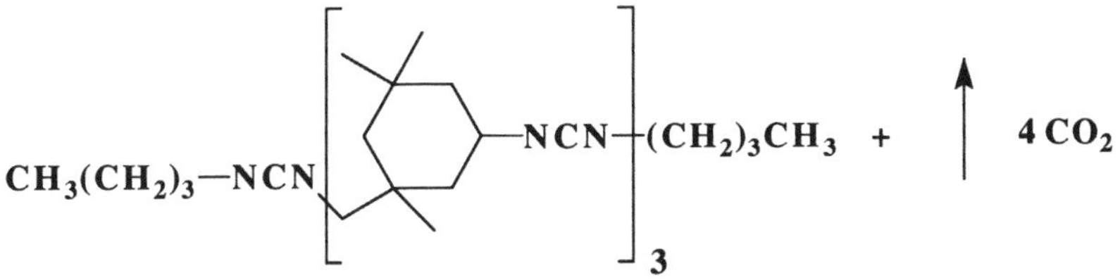

$$CH_3(CH_2)_3-NCN \left[\begin{array}{c} \\ \end{array} -NCN-(CH_2)_3CH_3 \right]_3 + \uparrow 4\,CO_2$$

1

Preparation of Monodispersed Trifunctional Carbodiimides. The second type of carbodiimide crosslinkers that were studied were trifunctional branched carbodiimides. The syntheses of the trifunctional carbodiimides were accomplished using two methods. In the first method, a trifunctional urea was first prepared. In a typical preparation of the trifunctional urea, a methylene chloride solution of the 4-aminomethyl-1,8-diaminooctane (50% by weight) was added gradually to a stirred methylene chloride solution of the alkyl isocyanate (11% by weight). During the addition of the triaminononane, the white trifunctional urea formed immediately. Yields of the trifunctional ureas varied between 78.1 and 99.8%. The preparation of the trifunctional carbodiimides was accomplished by dehydration of the corresponding trifunctional urea using bromotriphenyl phosphine bromide in the presence of triethylamine (18). The dehydration reaction was carried out in methylene chloride at temperatures between -5 and 5 °C. Although this route is not a feasible commercial route due to the high raw material cost of bromotriphenyl phosphine bromide, the process gives trifunctional carbodiimide crosslinkers with acceptable purity, stability, and color. The synthesis is shown scheme 4.

Scheme 4

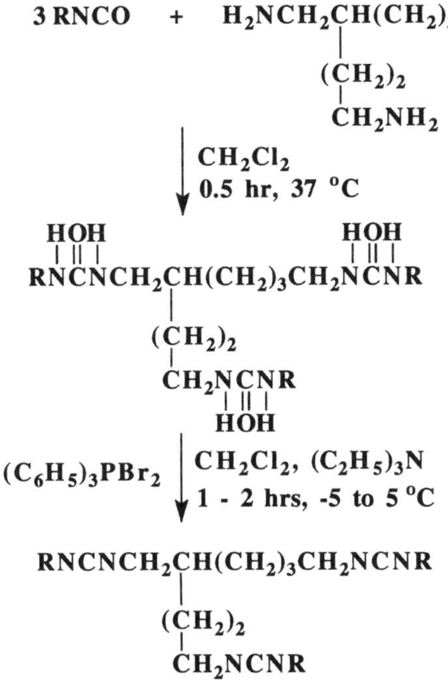

$$3\,RNCO + H_2NCH_2CH(CH_2)_3CH_2NH_2$$
$$|$$
$$(CH_2)_2$$
$$|$$
$$CH_2NH_2$$

$$\downarrow \quad \begin{array}{c} CH_2Cl_2 \\ 0.5\ hr,\ 37\ °C \end{array}$$

$$\overset{HOH}{\underset{|\ ||\ |}{RNCNCH_2CH(CH_2)_3CH_2NCNR}}$$
$$|$$
$$(CH_2)_2$$
$$|$$
$$\underset{|\ ||\ |}{CH_2NCNR}$$
$$HOH$$

$$(C_6H_5)_3PBr_2 \quad \downarrow \quad \begin{array}{c} CH_2Cl_2,\ (C_2H_5)_3N \\ 1 - 2\ hrs,\ -5\ to\ 5\ °C \end{array}$$

$$RNCNCH_2CH(CH_2)_3CH_2NCNR$$
$$|$$
$$(CH_2)_2$$
$$|$$
$$CH_2NCNR$$

The process produces yields from 21 to 80%. The percent of carbodiimide equivalence based on theory varies from 75 to 100%. The low viscosities obtained, as a result of the unsymmetrical nature of the trifunctional carbodiimides, should be noted (Table 4).

Table 4. Synthesis Results of Trifunctional Carbodiimides

R	Precursors Used for Preparation	Yields Percent[a]	NCN of Theory[b]	Brookfield Viscosity cps, LVT #1 60 rpm	Active Carbodiimide, %[e]
$(CH_3)_2CH-$	Urea	80	84	14	93
$(CH_3)_2CH-$	Thiourea	70	91	23	82
$CH_3(CH_2)_3-$	Urea	68	87	19	91
$(CH_3)_3C-$	Urea	62	75	160[c]	94
$(CH_3)_3C-$	Thiourea	52	81	-	52
C_6CH_5-	Urea	21	100	30[d]	84
$C_6H_{11}-$	Urea	43	90	152[d]	92
$C_6H_{11}-$	Thiourea	69	82	-	90

a. All yields are from an unoptimized process.
b. Determined by titration using the procedure of Ref. 19
c. Viscosity obtained at 30 rpm.
d. Viscosity obtained with a LVT #2 spindle.
e. Activity determined by drying sample at 120 °C for 1.5 h.

Preparation of the trifunctional carbodiimides from the second route was accomplished by the desulfurization of the corresponding trifunctional thiourea (18-20). In a typical preparation of the trifunctional thiourea, a methylene chloride solution of 4-aminomethyl-1,8-diaminooctane (50% by weight) was added to a stirred methylene chloride solution of the alkyl isothiocyanate (15% by weight). The reaction was easily followed by monitoring the isothiocyanate absorption using IR spectroscopy. After the reaction, the trifunctional thiourea was used as prepared or isolated. To prepare the carbodiimide the thiourea may be added as a solid to a dispersion of methylene chloride in a basic water solution of hypochlorite, or it may be added as a solution of the thiourea in methylene chloride to a stirred basic water solution of the hypochlorite. During the addition of thiourea, the temperature is kept between -5 and 8 °C. While chlorinated hydrocarbons are preferred as the solvent, the literature (18) suggests that other solvents (e.g. petroleum ether) are also suitable. Requirements for suitable solvents includes solvents which are immiscible with water to minimize hydrolysis of the carbodiimide and solvents that do not contain acidic hydrogen atoms that are reactive with the carbodiimide functionality. The synthesis is shown in scheme 5.

Scheme 5

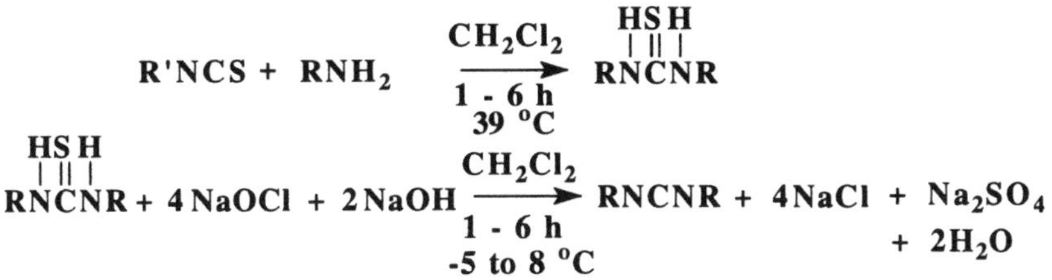

$$\text{R'NCS} + \text{RNH}_2 \xrightarrow[\substack{1 - 6\ h \\ 39\ ^\circ C}]{\text{CH}_2\text{Cl}_2} \overset{\text{HS H}}{\underset{\text{| || |}}{\text{RNCNR}}}$$

$$\overset{\text{HS H}}{\underset{\text{| || |}}{\text{RNCNR}}} + 4\,\text{NaOCl} + 2\,\text{NaOH} \xrightarrow[\substack{1 - 6\ h \\ -5\ to\ 8\ ^\circ C}]{\text{CH}_2\text{Cl}_2} \text{RNCNR} + 4\text{NaCl} + \text{Na}_2\text{SO}_4 + 2\text{H}_2\text{O}$$

The literature (20) suggests that the products of the desulfurization reaction are the carbodiimide, sodium sulfate, sodium chloride, and water with minimum amounts of elemental sulfur as a by-product; however, the laboratories at Union Carbide Corporation found that the above hypochlorite reaction produced major amounts of sulfur as a by-product. Commercialization of this process requires the complete conversion (oxidation) of sulfur to sulfate in order to obtain a product with acceptable purity; therefore, additional research is necessary to eliminate the by-product, sulfur. The above hypochlorite reaction has the advantage that theoretically, the reaction can be run in a continuous reactor (21, 22). As shown in Table 4, the yields (unoptimized) for the hypochlorite process vary from 52 to 70%. The production of sulfur may lower the yields by absorbing the trifunctional carbodiimide. Future work should include optimization of the commercial process via statistical methods to minimize the by-product, sulfur, while maximizing the yield.

Evaluations of Multifunctional Carbodiimides. The tetrafunctional "linear" polycarbodiimide (1) and trifunctional "star" carbodiimides prepared by the bromotriphenylphosphine bromide process (Table 4) were formulated in UCAR® Vehicle 462, a high-molecular weight styrene-acrylic water-borne polymer that contains two percent acrylic acid. Clear films were cast over Leneta paper and cured. A trifunctional aziridine, 2, was used as a control to gauge the efficiency of carbodiimide crosslinking chemistry. Although toxic, aziridinyl crosslinkers are extremely efficient crosslinkers for waterborne coatings prepared from carboxylic acid-containing waterborne polymers.

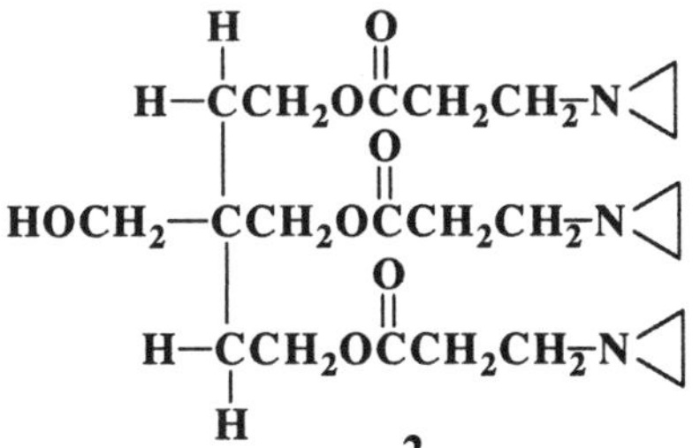

$$
\begin{array}{c}
\overset{H}{\underset{|}{\text{H}-\text{C}}}\text{CH}_2\text{O}\overset{O}{\overset{||}{\text{C}}}\text{CH}_2\text{CH}_2\text{N}\vartriangleleft \\
\overset{|}{\text{HOCH}_2}-\text{C}\text{CH}_2\text{O}\overset{O}{\overset{||}{\text{C}}}\text{CH}_2\text{CH}_2\text{N}\vartriangleleft \\
\overset{|}{\text{H}-\text{C}}\text{CH}_2\text{O}\overset{O}{\overset{||}{\text{C}}}\text{CH}_2\text{CH}_2\text{N}\vartriangleleft \\
\overset{|}{\text{H}} \qquad \mathbf{2}
\end{array}
$$

To measure the effectiveness of carbodiimide crosslinking, clear films prepared from UCAR® Vehicle 462 and the above carbodiimide crosslinkers were cured for 15 minutes at 127 °C (260 °F). FT-IR spectroscopy studies of the cured films show that more than 95% of the carbodiimide functionality reacts during cure. After cure, the films were evaluated for their crosslink density, tensile strength at break, elongation, and solvent resistance.

Since methyl ethyl ketone is used in the coatings industry to determine the solvent resistance of cured coatings, it was chosen as the solvent to study crosslinking efficiency. The crosslink density or molecular weight between crosslinks, M_c, of the cured films in Table 5, was determined by swelling the films in methyl ethyl ketone. The M_c was calculated using the equation shown below (23):

$$M_c = -\frac{V_m (1 - 2M_c/M) (V_2^{1/3} - 2V_2f^{-1})}{v \ln(1 - V_2) + V_2 + \chi V_2^2} \qquad (1)$$

where V_m is the molar volume of methyl ethyl ketone, 89.6 mL/mole; v is the specific volume of the polymer, 0.893 mL/g; f is the functionality of the junction points in the network, V_2 is the volume fraction of the polymer in the gel at equilibrium, M is the primary molecular weight as defined by Flory, and χ is Flory's interaction parameter, a parameter which characterizes the interaction energy per solvent molecule divided by the product of the Boltzmann constant and temperature (kT). Since the carboxylated polymeric particles of UCAR® Vehicle 462 are very high molecular weight, M in the term, $1 - M_c/M$, is very large relative to M_c; thus, this term reduces to 1.

To use Equation 1, Flory's χ parameter must be determined for methyl ethyl ketone and the polymer. It is known that Flory's χ parameter is related to the second virial coefficient A_2, a measure of the polymer-solvent interactions (23). The relationship is shown below:

$$\chi = 0.5 - \frac{A_2 M_1 p_2^2}{p_1} \qquad (2)$$

where M_1 is the molecular weight of methyl ethyl ketone, 72.12 g/mole; p_1 is the density of methyl ethyl ketone, 0.805 g/mL; and p_2 is the density of UCAR® Vehicle 462 resin. The second virial coefficient can be obtained from osmometry or light scattering measurements provided the polymer is soluble. However, UCAR® Vehicle 462 contains polymeric particles with a high gel content. A simple picture of the waterborne particles of UCAR® Vehicle 462 are that they are spherical polymer particles with a surface (depth undefined) high in styrene, butyl acrylate, and acrylic acid. The particles contain a multifunctional acrylate to control gel content. Although this description is an oversimplification, it will be a very useful model in discussing film formation and crosslinking mechanisms.

To obtain a soluble polymer with a similar solubility parameter, a model polymer was prepared by emulsion polymerization which contained methyl methacrylate, styrene, butyl acrylate and ethyl acrylate at 28, 20, 46, and 6% by weight, respectively. Dried films of this polymer were soluble in methyl ethyl ketone, producing a slightly hazy solution; however, light scattering experiments required a clear solution. To purify this polymer, dried films were dissolved in methyl ethyl ketone, (10% solids), then precipitated in a large volume of methanol (20 mL of methanol to 1 mL of polymer solution). The model polymer was then removed, dried and the procedure repeated. The model polymer was then redissolved in methyl ethyl ketone to produce a clear solution. Rayleigh light scattering results in methyl ethyl ketone give a second virial coefficient of 3.740×10^{-4} mL-mol/g^2 and a weight-averaged molecular weight of 4.93×10^5 g/ mol. Substitution of the second virial coefficient value into Equation 2 gives a Flory χ value of 0.458. In using the χ value the following assumptions are made:

Flory's χ parameter obtained from a solution of the model polymer dissolved in methyl ethyl ketone is equal to the χ parameter of the polymer from UCAR® Vehicle 462 swollen in methyl ethyl ketone,

Flory's χ parameter is a constant, and it is not a function of V_2, the volume fraction of the polymer in the gel,

small changes in the composition of the polymer from UCAR® Vehicle 462 by chemically incorporating a small level of crosslinker into the matrix of the polymer do not effect Flory's χ parameter, and

dilute solutions of the model polymer in methyl ethyl ketone used to obtain the second virial coefficient, A_2, are a continuous solution of noninteracting polymer chains. This assumption is true if the volume fraction of polymer in methyl ethyl ketone is adequate (greater than 0.05) or the polymer-solvent interaction is poor enough that the polymer molecules telescope one another freely without any net interaction. According to Flory, if the above assumption is invalid, A_2 as it relates to the Flory-Huggins theory (Equation 2) would be somewhat larger than the reported value of 3.740×10^{-4} mL-mol/g^2; hence, χ would be smaller.

It has been shown by others (25-27) that Flory's χ parameter is not always constant as predicted by theory, but that it often varies linearly with the volume fraction of polymer in the solvent. The M_c values calculated using Equation 1 are likely not absolute values. In particular, if the polymer or solvent possesses a significant dipole, χ may vary linearly over the total concentration often increasing for the type of polymer-solvent system under study. The increase may be as large as 0.3 units (23). The variability of χ poses no problems, however, for the crosslinking comparisons used in this study since for a given volume fraction of polymer in methyl ethyl ketone, one has a given χ value; hence, relative comparisons are possible between films crosslinked with different crosslinkers. The above statement is true only if the same polymer is used in the comparisons and the crosslinkers incorporated into the matrix of the polymer after cure do not effect the χ parameter. Equation 1 was used to determine the average molecular weight between crosslink points in a cured film at given levels of crosslinker. If all available carbodiimide or carboxyl groups have reacted, the M_c obtained is a measure of the crosslinking potential of the crosslinker.

To evaluate the crosslinkers in this study using Equation 1, crosslinked films were prepared from the following formulation:

Table 5. Formulation Used to Evaluate Multifunctional Carbodiimides

Formulation Components	Mass of Components in Formulation
UCAR® ® VEHICLE 462	100.00
Dimethylethanol amine (50% in water)	0.40
Butyl CELLOSOLVE	7.42
Water	7.42
Crosslinker Emulsion (27 % in water)	Variable

Emulsions of the polycarbodiimide crosslinkers were prepared using a previously published procedure (1). Emulsions of the trifunctional carbodiimides and **1** (27% active in water and amyl acetate) were added to the above partial formulation at 5 parts of dry carbodiimide crosslinker per 100 parts of dry polymer resin (phr). The

aziridine crosslinker, **2**, was added neat to the formulated latex. The pH of UCAR®
Vehicle 462 was 7.2 ± 0.2. The final pH of the formulations was 8.4 ± 0.3. A
formulation that was used as a control was also prepared as described in Table 5
except that no crosslinker was added to the formulation. Small aliquots of the final
formulations were placed in Teflon molds. After air drying for two or three days, the
films were cured for 15 minutes at 127 °C in a forced-air oven. All formulations
were used within 6 h of their preparation, and were allowed at least 30 minutes to
defoam. All carbodiimide emulsions were used within 2 days of their preparation.
The final dry thickness of the films was approximately 22 mils.

For Mc measurements, films were soaked in methyl ethyl ketone for at least
two days to obtain equilibrium, weighed, and the volume fraction of polymer in the
gel, V_2, calculated. For mechanical property measurements (23) the tensile strength
at break and elongation at break were determined using the films prepared as
described above. Stress is defined as

$$Stress = F / A_0 \qquad (3)$$

where F is the force on the sample, and A_0 is the initial cross sectional area of the
sample, and Elongation (or strain) is defined as

$$Elongation = L / (L - L_0) \qquad (4)$$

where L_0 is the initial length of the sample and L is the length of the sample at some
point in time. During the elongation process, at some point in time, the sample
breaks. This point in time will be referred to as elongation at break. The stress at this
point will be referred to as the tensile strength at break. The cured films were cut
with a "dogbone" die. Using a 10 pound load cell, the crosshead speed on the Instron
was set at one inch per minute.

The results are shown in Table 6. The control film (film containing no
crosslinker) has a M_c of 4.07×10^4 g/mol, a tensile strength at break of 820 psi, and
an elongation of 830%. The swollen gel from the control film results from the
heterogeneous nature of the waterborne particles from UCAR® Vehicle 462 and is
not indicative of cure.

Table 6. Ultimate Properties of Waterborne Coatings

Compound[a]	Mc (g/mole)	Tensile Strength (psi)	Elongation
1	4706	1420	430
2	2961	867	510
Isopropyl	4897	1425	450
cyclohexyl	3740	888	450
n-butyl	3050	866	485
t-butyl	3008	1190	510
phenyl	3088	1326	560

a. Level of crosslinker is 5 phr.

The data show that **1** and 1,3,6-tri(N-isopropyl-N'-methylene carbodiimide) hexane
are the least efficient crosslinkers in obtaining low M_c values at 5 phr of crosslinker
whereas **2**, 1,3,6-tri(N-t-butyl-N' methylene carbodiimide) hexane, 1,3,6-tri(N-n-
butyl-N'-methylene carbodiimide) hexane, and 1,3,6-tri(N-phenyl-N'-methylene
carbodiimide) hexane are the most efficient. Interestingly, 1,3,6 (N-cyclohexyl-N'-

methylene carbodiimide) hexane gives a M_C value between the two groups. Based on molecular weight, 1,3,6-tri(N isopropyl-N'-methylene carbodiimide) hexane should give films with the lowest M_C values; however, as shown in Table 6, experimentation did not support this assumption. It is postulated that the higher than expected M_C value of 1,3,6-tri(N-isopropyl-N'-methylene carbodiimide) hexane may result from a higher water solubility compared to the other carbodiimide crosslinkers because of its lower molecular weight; thus, more hydrolysis of the 1,3,6-tri(N-isopropyl-N' methylene carbodiimide) hexane may occur during cure.

As shown in Table 6, the carbodiimide crosslinkers with the best tensile properties at 5 phr of crosslinker are **1**, the isopropyl, t-butyl, and phenyl trifunctional carbodiimide derivatives. Figure 2 shows the stress-strain curves for cured films of UCAR® Vehicle 462 as a function of the level of 1,3,6-tri(N-isopropyl-N'-methylene carbodiimide) hexane.

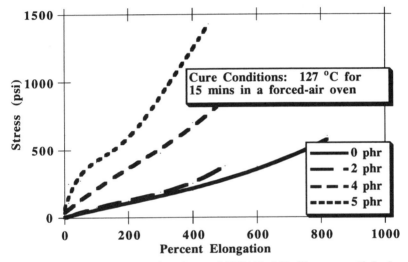

Figure 2. **Stress-strain plots of UCAR 462 films crosslinked with various levels of 1,3,6-tri(N-isopropyl-N'-methylene carbodiimide) hexane.**

The plots show that at a given elongation, the stress increases with increasing levels of crosslinker. A crosslinker level of 5 phr gives the greatest tensile strength at break, the highest modulus, and requires the most work (energy) to break the film.

Solvent Resistance Studies. The previous section dealt with the ultimate properties the crosslinkers impart to cured films   As previously noted, essentially all available reactive sites were shown by FT-IR spectroscopy to have reacted under the cure conditions employed; however, it is of interest to measure a property that the crosslinkers impart to cured films under a variety of cure conditions. A useful test is the resistance of crosslinked films to methyl ethyl ketone double rubs. This test indirectly measures the balance of film formation and final crosslink density. It is obvious that the proper balance of film formation and crosslink density is necessary for good resistance to methyl ethyl ketone double rubs.

All formulations were prepared as previously described. Wet films were air dried for 30 s then placed in a forced-air oven. During cure, the air velocity at the surface of the films was 400 ft /min ($\pm$ 50). The final dry film thickness was 1.2 mils on Leneta paper. During the course of this work, it was shown that the number of methyl ethyl ketone double rubs obtained on crosslinked films decreased as the thickness of the films decreased; therefore, smooth films of equivalent thickness were essential in obtaining reproducible results. Each methyl ethyl ketone double rub is an average of three measurements. The ratings reported in Table 7 were used to judge the solvent resistance of cured films. The results are shown in Table 6. As previously noted, the control film (films containing no crosslinker) has a Mc of 4.07 X 10^4 g/mol, a tensile strength at break of 820 psi, and an elongation of 830%.

Table 7. Ratings for Solvent Resistance

Methyl Ethyl Ketone Double Rubs	Rating
< 30	No Resistance
31-50	Poor
51-99	Fair
100-150	Good
151-250	Excellent
>250	Outstanding

Table 8 shows the methyl ethyl ketone double rubs obtained on crosslinked films which were cured under a variety of conditions at 2, 3, 4, and 5 phr of crosslinker. Examination of data in Table 8 shows that, in general, the 1,3,6-tri(N-n-butyl-N'-methylene carbodiimide) crosslinker gives films with the best solvent resistance at low cure temperatures whereas the t-butyl derivative and **1** give cured films with the least solvent resistance. Although the isopropyl, cyclohexyl, and phenyl derivatives give films with solvent resistance properties at low cure temperatures which on the average are less than the n-butyl derivative, all appear to impart better solvent resistance than **2**, the aziridine crosslinker. This data suggests that the carbodiimide crosslinking chemistry is competitive in cure speeds to aziridine crosslinking chemistry.

Film Formation Studies. To illustrate the effects of film formation versus crosslinking, the formulation as described in Table 5 was prepared using 3 phr of 1,3,6-tri(N-isopropyl-N'-methylene carbodiimide) hexane. Films were cast over Leneta paper and cured at 60 and 93 °C. The time of cure was adjusted to achieve essentially the same percent of crosslinking as measured by FT-IR spectroscopy. The results (Table 9) are shown below.

These results show the importance of achieving adequate film formation to obtain good resistance to methyl ethyl ketone double rubs. Better film formation is obtained at the higher cure temperature which results in better interfacial crosslinking and solvent resistance. This experiment illustrates why methyl ethyl ketone double rubs do not necessarily reflect the reactivity of the crosslinker.

Stability Studies. Carbodiimides are known to dimerize and trimerize at ambient temperature (28). Unfortunately, stability studies of the crosslinker show that only **1**, 1,3,6-tri(N-t-butyl-N'-methylene carbodiimide) hexane and possibly 1,3,6-tri(N-cyclohexyl-N'-methylene carbodiimide) hexane give crosslinkers with acceptable stability (Table 10).

Table 8. Solvent Resistance Studies of UCAR® Vehicle 462 Films Crosslinked with Carbodiimide Crosslinkers

2 phr

Cure Conditions					
Time (mins)	Temp. (°C)	1	2	isopropyl	n-butyl
15	60	43	37	40	95
5	85	45	56	75	103
5	93	---	---	---	131
15	127	300	300	157	300

Cure Conditions				
Time (mins)	Temp. (°C)	t-butyl	phenyl	cyclohexyl
15	60	34	55	68
5	85	42	67	70
5	93	88	60	97
15	127	300	300	300

3 phr

Cure Conditions					
Time (mins)	Temp. (°C)	1	2	isopropyl	n-butyl
15	60	94	50	93	140
5	85	86	73	140	300
5	93	112	225	---	300
15	127	300	285	300	300

Cure Conditions				
Time (mins)	Temp. (°C)	t-butyl	phenyl	cyclohexyl
15	60	42	82	89
5	85	54	70	300
5	93	90	121	300
15	127	300	300	300

4 phr

Cure Conditions					
Time (mins)	Temp. (°C)	1	2	isopropyl	n-butyl
15	60	55	48	86	94
5	85	73	80	164	300
5	93	106	300	300	300
15	127	300	300	300	300

Cure Conditions				
Time (mins)	Temp. (°C)	t-butyl	phenyl	cyclohexyl
15	60	42	79	125
5	85	72	126	235
5	93	74	282	300
15	127	300	300	300

5 phr

Cure Conditions					
Time (mins)	Temp. (°C)	1	2	isopropyl	n-butyl
15	60	79	55	94	87
5	85	90	188	108	150
5	93	97	---	158	300
15	127	300	300	300	300

Cure Conditions				
Time (mins)	Temp. (°C)	t-butyl	phenyl	cyclohexyl
15	60	54	115	67
5	85	59	252	131
5	93	73	300	213
15	127	300	300	300

All methyl ethyl ketone double rubs are an average of three experiments.

Table 9. Film Formation Studies

Cure Temperature (°C)	Percent NCN Reacted	M_c (g/mole)	Methyl Ethyl Ketone Double Rubs
60	80.0	9025	57
93	80.7	8964	300

Table 10. Time to Gel

Crosslinkers	Stability in Days (25 °C)
1[a]	> 3000
2	> 365
Isopropyl	200
cyclohexyl	300
n-butyl	48
t-butyl	>3000
phenyl[b]	> 151

a. Crosslinker 45% active in amyl acetate.
b. Crosslinker 83% active in hexane.

Since the trifunctional carbodiimide crosslinker, 1,3,6-tri(N-cyclohexyl-N' methylene carbodiimide) hexane imparted better resistance to methyl ethyl ketone double rubs than 1,3,6-tri(N-t-butyl-N'-methylene carbodiimide) hexane at low cure temperatures (Table 8), it was of interest to study the reactivity of these two crosslinkers in more detail. For quantitative measurements, films for ATR FT-IR spectroscopy were prepared from a blend of UCAR® Vehicle 462, carbodiimide emulsion, and 16 phr of butyl CELLOSOLVE. Films prepared from the model latex and the carbodiimide emulsions were used to develop calibration plots. All films were cast over Leneta paper then cured in a forced-air oven for 5 minutes at 93 °C. The dry thickness of the films was 0.80 mil ± 0.2. To correct for film contact on the ATR crystal, the absorbance for the carbodiimide group at 2132 cm^{-1} was normalized using the styrene bands at 3031 cm^{-1} and 1497 cm^{-1}. The FT-IR spectroscopy results are shown in Figure 3.

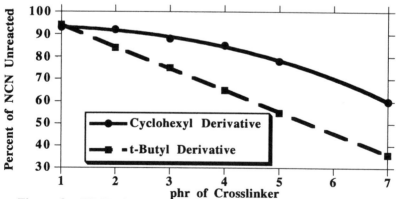

Figure 3. FT-IR ATR analysis of one mil films cured at 93 °C for five minutes.

The plots show that beyond 1 phr, 1,3,6-tri(N-cyclohexyl-N'-methylene carbodiimide) hexane is more efficient and hence more reactive than 1,3,6 tri(N-t-butyl methylene carbodiimide) hexane. The overall decrease in efficiency of the two crosslinkers with increasing level of crosslinker is expected. As the level of crosslinker is increased in the polymer, there are more carbodiimide groups competing for a limited supply of carboxylic acid groups. In fact, for the cyclohexyl and t-butyl derivative, beyond 5.1 phr of crosslinker (assuming no hydrolysis of the carbodiimide groups during cure) there is an excess of carbodiimide groups compared to carboxylic groups. Interestingly, the decrease in efficiency of 1,3,6-tri(N-t-butyl-N'-methylene carbodiimide) hexane was linear. It should be noted that titration of the carbodiimide functionality of 1,3,6-tri(N-cyclohexyl-N'-methylene carbodiimide) hexane and 1,3,6-tri(N-t-butyl-N'-methylene carbodiimide) hexane yielded an activity of 5.5 and 5.4 meq NCN/g of crosslinker, respectively. This fortuitous result means, for example, that 1 phr of either crosslinker adds essentially the same equivalents of carbodiimide to the matrix of the polymer; thus, any difference in efficiency is not due to concentration effects. The t-butyl moiety of 1,3,6-tri(N t-butyl-N'-methylene carbodiimide) hexane provides more steric hindrance than the cyclohexyl moiety of 1,3,6-tri(N-cyclohexyl-N'-methylene carbodiimide) hexane. The higher steric hindrance from the t-butyl moiety of 1,3,6-tri(N-t-butyl-N'-methylene carbodiimide) hexane gives excellent long term stability (Table 10) but slows the reaction between the carboxylic acid groups of the polymer and the carbodiimide moiety of the crosslinker. Curing films at high temperature and long times results in the complete reaction of both crosslinkers; so, no difference in reactivity is observed. As the cure temperature is lowered or the cure time shortened, the difference in reactivity is measurable.

The reactivity and efficiency of 1,3,6-tri(N-cyclohexyl-N'-methylene carbodiimide) hexane were compared to **1**. For this comparison, two formulations were prepared from UCAR® Vehicle 462. Each formulation contained 16 phr of butyl CELLOSOLVE as the filming aid and three phr of a selected crosslinker. Films were cast over Leneta paper then cured in a forced air oven at 104 °C (220 °F) for varying lengths of time to a dry film thickness of 1.2 mils ± 0.2. Samples were then cut and analyzed by ATR FT-IR spectroscopy. Figure 4 shows the plots of percent of carbodiimide reacted as a function of time for the two crosslinkers.

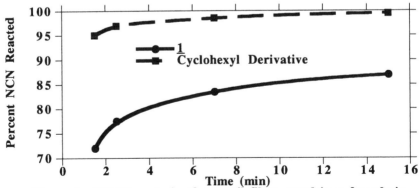

Figure 4. FT-IR analysis of one mil films cured in a forced-air oven at 104 °C.

As expected, most of the reaction occurs during the filming process. In addition, 1,3,6-tri(N-cyclohexyl-N'-methylene carbodiimide) hexane is shown to be more efficient at 3 than 1 phr. Both **1** and 1,3,6-tri(N-cyclohexyl-N'-methylene carbodiimide) hexane have essentially the same milliequivalence of carbodiimide per gram of dry crosslinker.

Elimination of Carbodiimide Emulsions: For the evaluations described in this work, **1** and the trifunctional carbodiimides were emulsified for easy dispersion into waterborne paint formulations. Industrially, the emulsification step is cumbersome, and increases the cost of carbodiimide-based crosslinkers. In addition, once crosslinkers are emulsified, their slow and unpreventable reaction with water to form multifunctional ureas limits their shelf life. For example, at ambient temperature (23 $^{\circ}$C) an emulsion of **1** held 90% of its functionality for six months, but an emulsion of 1,3,6-tri(N-isopropyl-N'-methylene carbodiimide) hexane lost 11% of its carbodiimide functionality within ten weeks. For the above chemical stability determinations, each crosslinker (dissolved in amyl acetate) was emulsified with identical surfactant compositions to a droplet size of 1.3 microns (27.5% active crosslinker). Since the hydrolysis of carbodiimides has a strong pH dependence, preparing emulsions of multifunctional carbodiimide crosslinkers at pHs between 8.5 and 9.5 is critical for long term survival of the carbodiimide functionality (29, 30). To eliminate hydrolysis, polycarbodiimides sold commercially are self-emulsifiable in the presence of waterborne particles (31).

Preparation of Reactive Latexes Using Carbodiimide Methacrylates

Water-dispersible polycarbodiimides can crosslink carboxylic acid-containing waterborne polymers under ambient conditions. A disadvantage of water-dispersible polycarbodiimides is that once dispersed in a waterborne formulation, they hydrolyze with time. The high level of surface-active agents introduced from the surface-active polycarbodiimide can result in crosslinked coatings which are more water sensitive than desirable. Groups which are reactive with carboxylic acid groups are by nature reactive with water. One solution to the hydrolysis of carbodiimides functionality is to model one of the oldest crosslinking technologies. Alkyds are polyunsaturated polymers which crosslink upon exposure to air. Unsaturation was first introduced into latex particles by Tillson (32) using allyl methacrylate as a monomer. Taylor, Collins, and Clark greatly improved Tillon's approach by developing a emulsion polymerization process where more than 95% of the allyl moieties from allyl methacrylate survives (33). These technologies, which introduced allyl moieties into waterborne particles, provide cure in the final coating by an oxidation process. McGinniss, Seidewand, and Robert (34) first introduced polymerizable functionality onto the surface of latex particles by reacting 1(2-aziridinyl)ethyl methacrylate with carboxylic acid-containing waterborne particles. When this chemistry was examined by nuclear magnetic resonance at Union Carbide Corporation laboratories, the results show that more than 90% of the theoretical level of methacrylate moieties survives the modification process (unpublished results). Later, similar modifications were accomplished by Mylonakis (35) by reacting glycidyl methacrylate with carboxylic acid-containing waterborne particles. Recently, Wolfersberger, Schindler, Beckley, and Novak at Rohm and Haas extended Mylonakis's work by reacting glycidyl methacrylate with carboxylic acid-containing waterborne particles prepared using a multistage process (36). Up to 80% of the glycidyl methacrylate was incorporated onto the surface of their waterborne particles. Nuclear magnetic resonance studies at Union Carbide Corporation laboratories show that 65% of the theoretical level of unsaturation available from glycidyl methacrylate survives the modification process (unpublished work). Nevertheless, Rohm and Haas has successfully commercialized reactive waterborne acrylic polymers prepared from the reaction between glycidyl methacrylate and carboxylic acid-containing acrylic particles. These polymers have found use as aqueous UV curable emulsions (8). More recently, Rohm and Haas has introduced unsaturation into waterborne acrylic particles by reacting ammonia with pendant acetoacetoxy moieties to produce enamine moieties which cure in the presence of peroxide forming materials (37, 38). Pears and Overbeek at ICI have improved on this approach by reacting acetoacetoxyethyl methacrylate with amine-containing polymers to produce waterborne particles with pendant enamine and methacrylate moieties (39).

Modification of Waterborne Particles. The laboratories at Union Carbide Corporation chose to explore the use of carbodiimide methacrylates for incorporating polymerizable unsaturation on the surface of waterborne particles. The preparation of carbodiimide methacrylates, and their use to modify latexes have been described in detail (7, 40). Carboxylic acid-containing latexes were reacted with emulsions of alkylcarbodiimide ethyl methacrylates to produce acrylic or vinyl ester-based latex particles with pendant methacrylate functionality. The reaction, to prepare latex particles which contain pendant methacrylate groups, is illustrated in scheme 6

Scheme 6

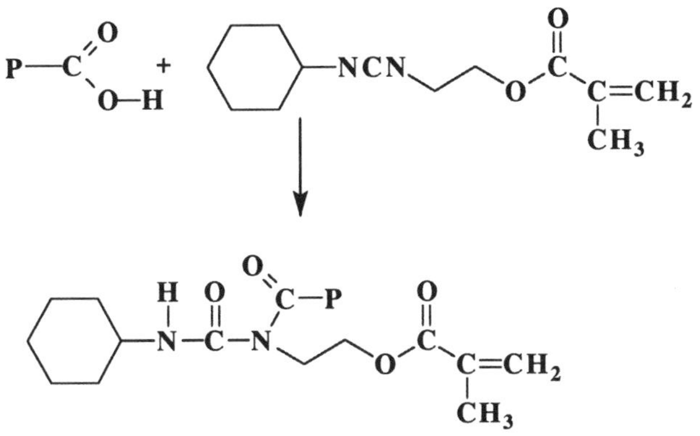

where P represents the surface of latex particles. Latexes were modified using the isopropyl, t-butyl, cyclohexyl, and phenyl derivatives of the alkyl (or aromatic) carbodiimide ethyl methacrylate. The reaction between the surface carboxylic acids and carbodiimide groups produces N-acyl urea groups which covalently link the methacrylate groups to the polymeric particles. In a typical latex modification, an emulsion of the alkylcarbodiimide ethyl methacrylate is stirred into a neutralized latex. The latex, which contains up to 10% methacrylic acid as part of its monomer composition, is neutralized to a pH of 8.5 with a volatile base such as ammonia or triethyl amine. The latex is heated at 80 °C until the surface modification is completed, and no carbodiimide absorption is visible by FT-IR spectroscopy. Analysis of the resulting latexes via nuclear magnetic resonance show that the latex particles contain more than 90% of the theoretical level of pendant methacrylate functionality.

To demonstrate the crosslinking from waterborne particles containing polymerizable unsaturation, the gel fraction of cured films (Figure 5) were obtained. A low-molecular weight waterborne polymer (Mn = 10,800 g/mol, Mw = 26,800 g/mol), which was prepared from a monomer composition containing 10%

methacrylic acid, was modified using various alkylcarbodiimide ethyl methacrylates. Films were cast from the modified waterborne polymers and cured thermally using t-butyl peroxybenzoate as the initiator. The gel fractions of the films were obtained by extraction with tetrahydrofuran. The results (Figure 5) show that modifications using the cyclohexyl or isopropyl derivatives give cured films with the highest gel fractions. A detailed account of the above work will be given in a future publication.

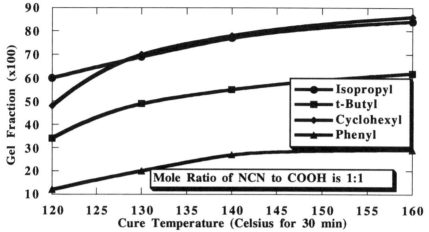

Figure 5. Gel fraction plots of cured films prepared from a latex modified with carbodiimide methacrylates.

Powder Coatings

The use of polycarbodiimides for powder coatings does not appear practical because carbodiimide crosslinkers are reactive at ambient temperatures in films prepared from carboxylic acid-containing polymers. Neutralization of the carboxylic acid moieties with amines improves stability, but does not prevent crosslinking within the film. To form a stable powder, one solution is to block the carbodiimide with a chemical agent (41, 42).

It is well known (28, 43) that guanidines are obtained by the addition reaction between aromatic carbodiimides and amines. Lyman and Sadri (44) have reported the physical properties of polyguanidines prepared from aromatic polycarbodiimides and primary amines. Iwakura et al (45-47) studied the addition reaction of p-phenylene bis(alkyl carbodiimides) with benzylamine, cyclohexylamine, morpholine, and hexamethylenediamine. They report pure bis(guanidines) could not be isolated from the reaction product of all-aliphatic bis(carbodiimides) and undisclosed amines; however, the formation of the guanidine moiety was detected by IR spectroscopy. Attempts at purification of the bis(guanidines) prepared from all-aliphatic bis(carbodiimides) and amines resulted in undefined decomposition products. The unidentified decomposition products from Iwakura's work suggest that guanidines may decompose to form carbodiimides and amines. The reaction would be analogous to many of the deblocking chemistries reported for blocked alkyl or aromatic isocyanates.

Preparation of a polyguanidine. To demonstrate the general feasibility of blocking and deblocking polycarbodiimides, a model all-aliphatic polyguanidine **4** was prepared, and its decomposition kinetics studied as a neat material, as a blend in poly(methyl methacrylate), and as a blend in a carboxylic acid-containing acrylic resin designed for powder coatings (41). As shown in scheme 7, the polycarbodiimides were reacted with diethylamine to give the desired polyguanidine.

Scheme 7

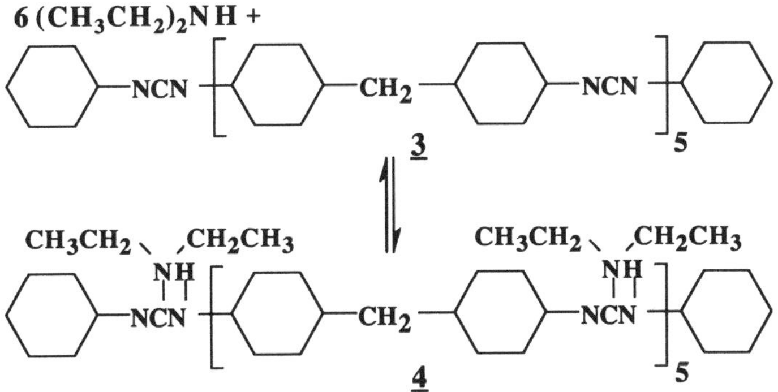

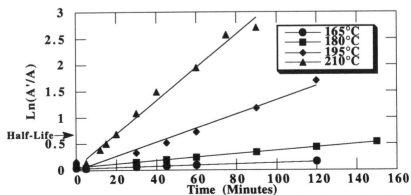

The model polyguanidine, **4** was chosen for kinetic studies because its decomposition product is analogous to 1,3-dicyclohexylcarbodiimide, a carbodiimide which as been extensively evaluated in its reaction with acetic acid (14). In addition, 1,3-dicyclohexylcarbodiimide does not readily dimerize or trimerize at temperatures below 225 °C.

Decomposition Studies of 4. The decomposition of **4** to **3** was confirmed by TGA, DSC, and FT-IR. The TGA result of **4** held at 215 °C shows that after 50 minutes the decomposition reaction is essentially complete. The polyguanidine lost 27.4% of its initial weight (theoretical loss, 25.3%). The DSC gives a small endothermic absorption at 54.6 °C and a large endothermic absorption at 240 °C.

The disappearance of the guanidine absorption at 1635 cm^{-1} was monitored by FT-IR spectroscopy at increasing times while holding the sample at a isothermal temperature.

As shown in Figure 6, the decomposition follows first order kinetics. The half-life of

Figure 6. The first order kinetic plots of a six-functional poly(guanidine) obtained by FT-IR spectroscopy.

4 at 180 °C is 178 minutes. The Arrhenius plot gives an activation energy of 31.4 Kcal/mol and an intercept (pre-exponential factor) of 8.7×10^{10} s^{-1}. When polyguanidines are blended into acrylic polymer a positive matrix effect is observed (Table 11) which results in an increase in the rate of decomposition of the polyguanidine. One can show that when **4** is blended into an acrylic polymer, then cured at 180 °C that more than 90% of the guanidine moieties of the polyguanidine decompose to carbodiimide moieties within 30 minutes. The evaluation of this technology in powder coatings has been discussed in detail (41).

Table 11. Half-Life Data at 180 oC for the decomposition of a polyguanidine

Compositions	Half-Life (minutes)
Polyguanidine	178
Polyguanidine in Poly(methyl methacrylate)	5
Polyguanidine in SCX-817C[a]	8

a. SCX-817C is a low molecular weight carboxylic-containing resin designed for powder coatings available from S. C. Johnson.

Dual-Tone Photoresists

Photoresists are photosensitive materials which change their solubility after exposure to light (48). Irradiation of photoresists through a mask (Step 4 of Figure 7) causes the irradiated area to become more soluble in basic aqueous solutions. Photoresists are typically novolac resins which have diazonapthaquinone (DANQ) sensitizers, **5**, attached to the polymer backbone or added to the resist formulation. DANQ sensitizers act as inhibitors to decrease the solubility of the photoresist in basic aqueous solutions. As shown in scheme 8 below, irradiation of photoresists causes the DANQ to form a carbene which then undergoes a Wolff rearrangement in the presence of water to form a base-soluble indenecarboxylic acid photoproduct, **6**:

Scheme 8

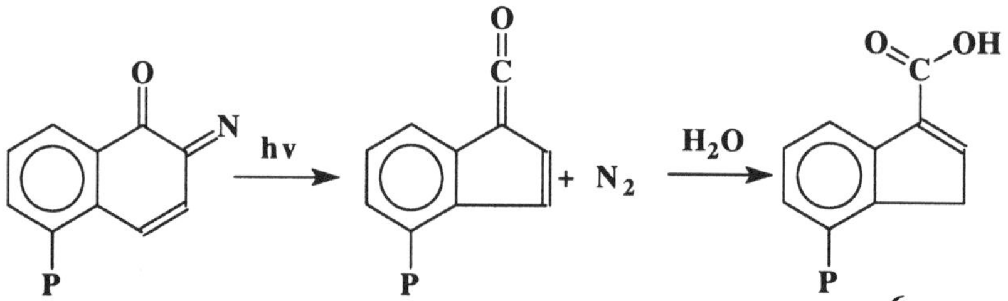

During development, preferential dissolution of the photoresist in the irradiated areas results in a positive image (step 5 of Figure 7, left side). To prepare negative images, the film is baked in step 5 of Figure 7 (right side) where reactive chemistry (to be defined) converts the carboxylic acid groups of **6** into a group which is not base soluble. Flood exposure followed by development then gives a negative image.

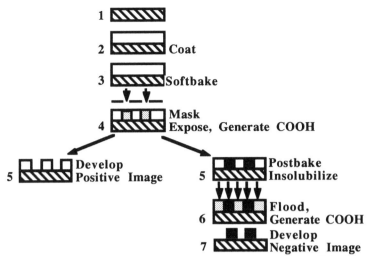

Figure 7. Process routes for processing negative and positive images

Dual-tone photoresists are defined as photoresists capable of producing positive or negative images. Negative images prepared from dual-tone photoresists can give improved resolution, depth of focus, and line-width control over underlying topography while minimizing standing wave effects when compared to positive images. High aspect ratios are possible, and the slope of the sidewalls can be controlled to give positive or negative profiles (49, 50). Image reversal for a photoresist resist designed for positive imaging was reported by Moritz and Paul (51). They found that the addition of one percent of monazoline to a positive working photoresist formulation allows negative images to be obtained by using the process illustrated in Figure 7. MacDonald obtained image reversal using imidazole as a additive to the photoresist formulation (52). Other approaches use bases such as triethanolamine (48-52). These approaches result in photoresist formulations with poor stability.

Carbodiimide Technology for Photoresists.   Union Carbide Corporation h a s explored the use of carbodiimide chemistry in photolithography (53). In 1987 their laboratory demonstrated that the addition of carbodiimides to positive working photoresist formulations results in dual-tone photoresist formulations (9), photoresist formulations which give positive or negative images (Figure 7) depending upon processing conditions. Novolac photoresist formulations containing polycarbodiimides give photoresist films which process to give positive or negative images; however, photoresist formulations containing polycarbodiimides gel within 24 h. On the other hand, photoresist formulations containing 1,3-dicyclohexylcarbodiimide are stable, and process to give positive or negative images. These interesting observations resulted in further exploration of the reaction between 1,3-dicyclohexylcarbodiimide and the acidic hydroxyl moieties of the novolac resin.

In 1963 Vowinkel (54) showed that long reaction times at 100 °C were required between 1,3-dicyclohexylcarbodiimide and phenol to prepare 1,3-dicyclohexyl-O-phenyl isourea; however, no kinetic data were given, and no kinetic data have been found in the literature regarding the reaction rate between carbodiimides and phenol. Phenol is intermediate in acidity between carboxylic acids and aliphatic alcohols; thus, if protonation of the carbodiimide group is the rate controlling step in the reaction, one would expect the reaction between carbodiimides and phenol to be faster than the reaction with the less acidic aliphatic alcohols but slower than with the more acidic carboxylic acids. Kinetic studies of 1,3-dicyclohexylcarbodiimide with phenol were carried out using FT-IR spectroscopy to resolve this issue. Figure 8 shows that the reaction between 1,3-dicyclohexylcarbodiimide and phenol is very slow at 23 °C. Assuming second order kinetics, the slope from the plot shown in Figure 8 gives a rate constant of 0.00337 $M^{-1}day^{-1}$ (half-life = 207 days) for this reaction in tetrahydrofuran.

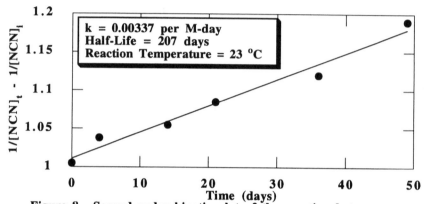

Figure 8. Second order kinetic plot of the reaction between dicyclhexylcarbodiimide and phenol.

Attempts at preparing the 1,3-dicyclohexyl-O-phenyl isourea from 1,3-dicyclohexylcarbodiimide and phenol at high temperatures resulted in an equilibrium

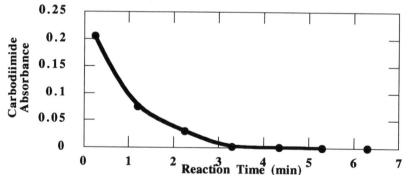

Figure 9. Infrared carbodiimide absorbance of dicyclohexyl-carbodiimide in a novolac formulation as a function of time.

between the reactants and the product. However, the addition of 1,3-dicyclohexylcarbodiimide to a novolac resin dissolved in propylene glycol methyl ether acetate (Figure 9) results in a rapid decrease of 1,3-dicyclohexylcarbodiimide in the resin solution at ambient temperature. FT-IR analysis shows that as the carbodiimide absorbance at 2120 cm^{-1} decreases, an absorption band at 1658 cm^{-1} increases. The absorption at 1658 cm^{-1} is attributed to the isourea moiety. Assuming second order kinetics, the rate constant is 1.02 M^{-1}day^{-1} (23 °C). We postulate that the higher rate constant in a novolac resin is due primarily to the more acidic novolac hydroxyl groups.   In support of this postulate, Hawtrey showed that 1,3-dicyclohexylcarbodiimide reacts with 2,4,6-trinitrophenol in ethyl acetate to form the isourea within 2.5 h at ambient temperature (55).

It was of interest to investigate the chemistry of novolac-bound isoureas. As a result of the solution equilibrium reaction between 1,3-dicyclohexylcarbodiimide and phenol at elevated temperatures, it was postulated that novolac-bound isoureas would fragment at elevated temperatures to form novolac hydroxyl goups and 1,3-dicyclohexyl carbodiimide.   In the irradiated photoresist 1,3-dicyclohexylcarbodiimide rapidly reacts during baking with the indenecarboxylic acid derivatives to deactivate the acid by formation of the N-acyl ureas (step 5 of Figure 7, right side). To support our argument a modified novolac resin, which contained isoureas and no DANQ, was prepared by reacting 10 g of 1,3-dicyclohexylcarbodiimide with 100 g of novolac resin. A film was spun over a sodium chloride disc then dried at 90 °C to a film thickness of 1.5 microns. While the novolac film was heated at 100, 110, 120, and 130 °C, FT-IR spectra were taken at 6 minute intervals.   The results (Figure 10) show that free 1,3-dicyclohexylcarbodiimide begins to appear at 100 °C (2% of theoretical), and that the amount of free 1,3-dicyclohexylcarbodiimide in the resin increases with increasing temperature. At each temperature the concentration of 1,3-dicyclohexylcarbodiimide reaches an apparent equilibrium. After 110 °C, the temperature was initially raised to 124 °C then rapidly reduced to 120 °C.   The level of free 1,3-dicyclohexylcarbodiimide immediately falls when the temperature is lowered from 124 to 120 °C to a constant value.

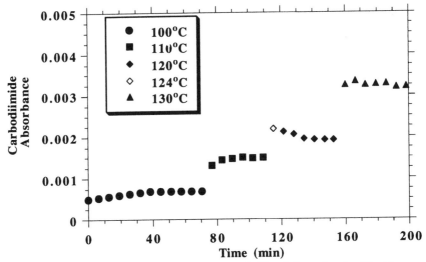

Figure 10. Equilibrium studies of dicyclohexylcarbodiimide in a novolac resin.

The equilibrium reaction is illustrated in the scheme 9 below:

Scheme 9

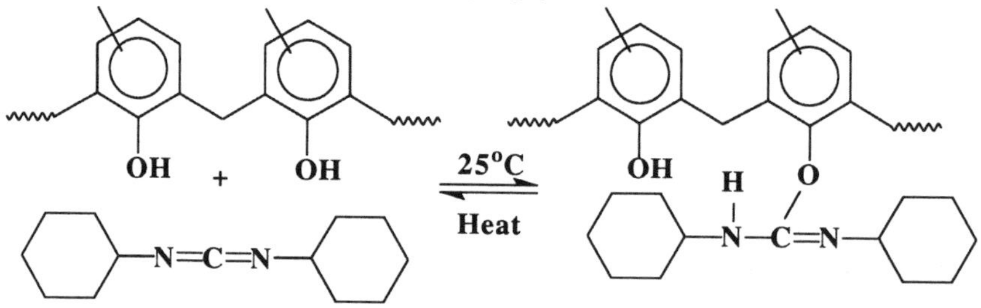

The formation of free 1,3-dicyclohexylcarbodiimide in a polar environment at elevated temperatures suggests that during postbake, 1,3-dicyclohexylcarbodiimide deactivates the indenecarboxylic acid groups in the irradiated areas by formation of N-acyl ureas. The proposed model is shown in scheme 10 below:

Scheme 10

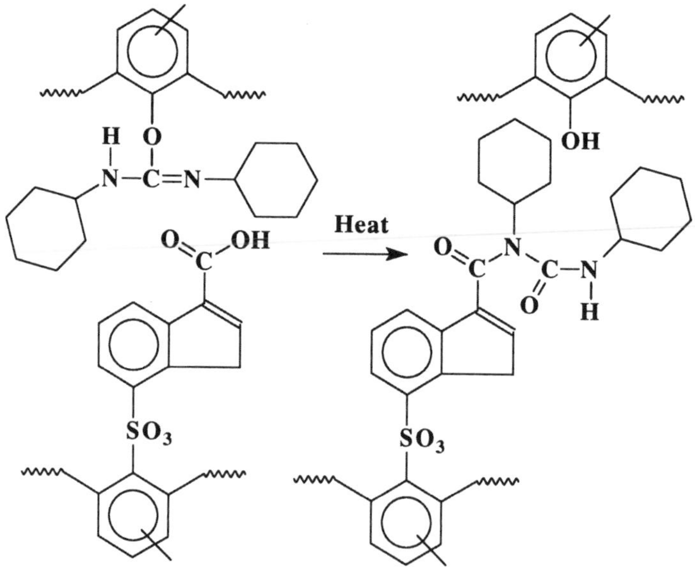

The data in Figures 9 and 10 support the argument that during postbake 1,3-dicyclohexylcarbodiimide deactivates the indenecarboxylic acid groups in the irradiated areas by formation of N-acyl ureas; thus, this allows the photoresist to be processed to give a negative image. The micrographs in Figure 11 show positive (left side) and negative images (right side) obtained from an i-line dual-tone photoresist. During development the film loss from unexposed areas is less than 5% for the positive or negative image. The positive image was produced using the process described on the left side of Figure 7 whereas the negative image was produced using the process described on the right side of Figure 7.

DualTone Resist
Positive Tone 0.5 μm Line/Space

DualTone Resist
Negative Tone 0.5 μm Line/Space

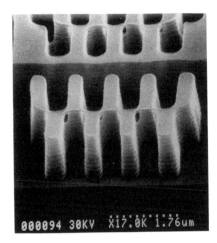
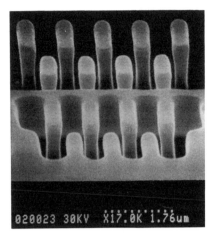

Figure 11. Left side is a positive tone image (0.5 micron line/space). Right side is a negative tone image (0.5 micron line/space).

Future of Carbodiimides in Coatings

Since 1984, carbodiimide technology for coatings has grown continuously (56). Carbodiimide chemistry has found use as a crosslinking technology in waterborne coatings designed for wood, automotive, hardboard primer, general metals, and as curing agents for coatings over plastics (56-62). The reactivity of polycarbodiimides compares to that of aziridine-based crosslinkers. Carbodiimides can be blocked with amines and phenolic-based polymers. Today, however, when they are compared to other crosslinkers such as polyisocyanates, carbodiimide crosslinking technology remains expensive. The expense of carbodiimide chemistry has limited its expansion into more coating markets. Contributions to improve the commercial synthesis of nonsymmetrical carbodiimide chemistry are needed. In particular, the commercial preparation of carbodiimide methacrylates represent a challenge for today's organic chemist. Polycarbodiimides based on aromatic isocyanates are less expensive, but slowly polymerize with time by dimerization and trimerization reactions to higher molecular weight materials. Perhaps, more sterically hindered aromatic isocyanates will eliminate the stability of aromatic polycarbodiimides (63).

References

1. Taylor, J. W. U.S. Patent 4 820 863, 1989.
2. Taylor, J. W. U.S. Patent 5 047 588, 1991.
3. Taylor, J. W. U.S. Patent 5 081 173, 1992.
4. Taylor, J. W. U.S. Patent 5 108 653, 1992.
5. Taylor, J. W.; Collins, M. J.; Bassett D. R. Proc. Am. Chem. Soc., Div. PMSE, **1992**, 67, 335.

6. Collins, M. J.; Taylor, J. W.; Bassett, D. R. Proc. Am. Chem. Soc., Div. PMSE, **1992**, 67, 338.

7. Taylor, J. W.; Collins, M. J.; Bassett, D. R. U.S. Patent 5 371 148, 1994.

8. Trentini, M. C.; Gerosa, P.; Carlson, V. Eur. Coat. J., **1995**, 5, 362

9. Taylor, J. W.; Bassett, D. R. U.S. Patent 5 087 547, 1992.

10. Taylor, J. W.; Jiang Y.; Bassett, D. R. Proc. Am. Chem. Soc., Div. PMSE, **1991**, 64, 50.

11. Taylor, J. W.; Bassett, D. R. Proc. SPIE- International Opt. Soc. Eng., Adv. Resist Tech. Processing **1989**, VII, 1262, 538

12. Taylor, J. W.; Bassett, D. R. U.S. Patent 5 240 811, 1993.

13. Bushy Run Research Center, Acute Toxicity and Primary Irritancy Studies of an All-aliphatic Polycarbodiimide, Unpublished Results.

14. DeTar, D. F.; Silverstein, R J. Am. Chem. Soc., **1966**, 88, 1020

16. Campbell, T. W.; Smeltz, K. J. Org Chem,. **1963**, 28, 2069.

17. Campbell, T. W.; Monagle, J. J. J. Am. Chem. Soc., **1962**, 84, 1493.

18. Palomo, C.; Mestres, R. Synthesis, **1981**, 373.

19. Zarembo, J. E.; Watt, M. M. Microchem. J., **1962**, Ser. 2, 591

20. Schmidt, E.; Seefelder M. Ann., **1951**, 871, 83.

21. Schmidt, E.; Reichenhall, B.; Schnegg R. U.S. Patent 2 656 383 1953.

22. Schmidt, E.; Zaller, F.; Moosmuller, F.; Kammerl, E. Ann., **1954**, 585, 230.

23. Flory, P. J. Principles of Polymer Chemistry; Cornell Unversity Press: London, 1953; Chapter 12.

24. Collins, E.; Barnes, J.; Billmeyer Jr., F. Experiments in Polymer Science; John Wiley and Sons: New York, 1973; p 130.

25. Bawn C; Freeman R.; Kamaliddin A.,Trans, Faraday, J. Chem. Soc., **1950**, 46, 677.

26. Gee, G. J. Chem. Soc., **1947**, 280.

27. Gee, G.; Orr, W. Trans. Faraday Soc., **1946**, 42, 507.

28. Williams, A.; Ibrahim, I. T. Chem. Rev., **1981**, 81, 589.

29. Ibrahim, I. T.; William, A. J. Am. Chem. Soc., **1978**, 100, 7420.

30. Williams, A.; Ibrahim, I. T. J. Am. Chem. Soc., **1981**, 103, 7090.

31. Taylor, J. W. US Patent 5 117 059, 1992

32. Tillson, H. C. US Patent 3 219 610, 1965.

33. Taylor, J. W.; Collins, M. J.; Clark, M. D. US Patent 5 539 073, 1996.

34. McGinniss, V. D.; Seidewand, J. R.; Robert, J. US Patent 4 107 013, 1978.

35. Mylonakis, S. G. US Patent 4 244 850, 1981.

36. Wolfersberger, M. H.; Schinder, F. J.; Beckely, R. S.; Novak, R. W. US Patent 5 306 744, 1994.

37. Bors, D. A. Eur. Patent Appl. 492 847 A2, 1991.

38. Bors, D. A.; Lavoie, A. C.; Emmons, W. D. US Patent 5 484 849, 1996.

39. Pears, D. A.; Overbeek, G. C. European Patent Application, 442 653 A2, 1991.

40. Taylor, J. W.; Collins, M. J.; Bassett, D. R. Proc. Am. Chem. Soc., Div. PMSE, **1995**, 73, 102.

41. Taylor, J. W.; Collins, M. J.; Bassett D. R. J. Coat. Tech., **1995**, 67, 43.

42. Wick, Z. W.; Jones, F. N.; Pappas, S. P. Organic Coatings: Science and Technology, John Wiley and Sons: New York, 1992; Vol 1, Chapter 3.

43. Khorana, H. G. Chem. Rev., **1953**, 53, 145.

44. Lyman, D. J.; Sadri, N. Makromol. Chem., **1963**, 67, 1.

45 Iwakura,Y.; Tsuzuki, R.; Noguchi, K. Makromol. Chem., **1966**, 98, 21.

46. Iwakura, Y.; Noguchi, K. Bull. Chem. Soc. Jpn, **1967**, 40, 2383.

47. Iwakura, Y.; Noguchi K. J. Poly. Sci., **1969**, Part A1, 801.

48. Moreau, W. M., "Semiconductor Lithograph: Principles, Practices, and Materials," Plenum Press, New York, 1988.

49. Alling, E.; Stauffer, C. Solid State Tech., **1988**, 37.
50. Marriot, V.; Garza, C. M.; Spak, M. SPIE Proc. Advances in Resist Technology and Processing IV, **1987**, 771, 221.
51. Moritz, H.; Paul, G. U.S. Patent 4 104 070, 1978.
52. MacDondald, S. A.; Miller, R. D.; Willson, C. G.; Feinberg, G. M.; Gleason, R. T.; Halveson, R. M.; MacIntyre, M. W.; Mostsiff, W. T.; Proc. Microelectron. Seminar, INTERFACE, **1982**, 144.
53. Taylor, J. W.; Brown, T. L.; Bassett, D. R. Proc. SPIE- International Opt. Soc. Eng., Adv. Resist Tech. Processing, **1990**, VII, 538.
54. Vowinkel, E. Chem. Ber., **1963**, 96, 1702.
55. Hawtrey, A. O. Tetrahedron Letters, **1966**, 6103.
56. Watson, S. L.; Humphreys, G. R. U.S. Patent 4 487 964, 1984.
57. Brown, W. T. Surf. Coat. Int., 1996, 78(6), 238.
58. Brown, W. T. Proc. Water-Borne, High Solids Powder Coat. Symp., 1994, 21st (Pt.1), 40.
59. Brown, W. T.; Day, J. C. Eur. Patent Appl. 62852 A2, 1994.
60. Tye, A. J.; Beck, G.; Mormile, P. J. U.S. Patent 5 357 021, 1994.
61. Mallon, C. B.; Chu, H. U.S. Patent 5 008 363, 1991.
62. Sundararamon, P.; Claar, J. A.; Kanie, C. M. U.S. Patent 5 105 010, 1992.
63. Imashiro, Y.; Takahashi, I. U.S. Patent 5 360 933, 1994.

Chapter 9

Synthesis and Coating Properties of Novel Waterborne Polyurethane Dispersions

Valentino J. Tramontano, Michael E. Thomas, and Robert D. Coughlin

Specialty Chemicals, King Industries, Inc., Science Road,
Norwalk, CT 06852

A series of novel waterborne polyurethane dispersions were prepared by non-isocyanate chemistry and their properties in cured coating formulations were studied. The combination of free film surface analysis by infrared spectroscopy, solvent swelling studies and stress-strain analysis reveal more homogeneous network formation in films of these polyurethane dispersions, as compared to polyurethane dispersions prepared via isocyanate processes. The novel polyurethane dispersions are solvent-free and were formulated into coatings which are low in volatile organic content, and fit the need for water-based resins that conform to current government regulations. It was also discovered that one of the polyurethane dispersions possesses unique wetting characteristics, and functions as an efficient dispersing resin for a variety of inorganic and organic pigments.

Waterborne polyurethane dispersions (PUR) find use in many industrial coating applications. Certain applications involve the use of waterborne polyurethanes as uncrosslinked thermoplastic coatings. In many other areas where solvent resistance and improved coating performance properties are essential, it is possible to achieve crosslinking with a polyisocyanate, an amino-formaldehyde crosslinker or other crosslinking agent (1-5). Coatings cured with amino-formaldehyde (melamine) resins find widespread use, and are advantageous because they are stable in a one-package system. Typical commercial solvent-borne polyurethane coatings are relatively high in solvent content and do not qualify in low volatile organic content (VOC) applications. There is currently a great need for low or near zero VOC coating systems due to current government regulations. For this reason, there is an increased interest in developing waterborne resin technology that meets new environmental standards (6).

Most conventional PUR dispersions are high molecular weight ionic polymers which are commonly prepared by one of two processes: the polyurethane is polymerized in solvent then dispersed in water, or an isocyanate terminated pre-polymer is prepared in the melt or in an aprotic solvent, and chain extended with a diamine in the water phase in the presence of a neutralizing tertiary amine (*7,8*). Urethane dispersions are also prepared through the attachment of polyoxyethylene side chains that enable the polymer to be dispersed in water utilizing a non-ionic mechanism (*9*).   We have prepared novel low MW hydroxy/carboxy functional polyurethane dispersions through the use of non-isocyanate chemistry (*10,11,12*). When crosslinked with a melamine resin these PUR dispersions produce films of more uniform structure and crosslink density.   Due to the higher reactivity of the hydroxyl groups with the melamine resin, there is a substantial reduction of the melamine self-condensation reaction.   Extensive melamine self-condensation reaction is associated with a reduction in the performance properties of the resultant coating (*13*).   The present study demonstrates the film properties and network formation of these novel low MW dispersions cured with a melamine resin.

One of the emerging alternative technologies is the use of a water-dispersible polyisocyanate as a crosslinker for hydroxy functional water-based resins (*14-17*). The concept involves the modification of a polyisocyanate crosslinker with water soluble moieties, and dispersing the crosslinker in a water-based resin formulation, while attempting to minimize the amount of time the crosslinker is in direct contact with water. One of the objectives is to reduce the amount of side reaction of isocyanate with water, which generates carbon dioxide, and can cause imperfections as the formulation dries to a finished coating (*18*). One of the four novel polyurethane dispersions developed in our series is designed for crosslinking with a water dispersible polyisocyanate.

EXPERIMENTAL
Materials
The four hydroxy/carboxy functional PUR dispersions (XM-2311, XM-2312, XM-4310 and XM-4316) were prepared by a non-isocyanate process.   Physical characteristics are given in Table 1 and the synthetic scheme is shown in Figure 1. The polymers were synthesized by first reacting the diester with the polyol in the presence of an organometallic catalyst at 200° - 220°C in vacuo.   Methanol is the byproduct of the trans-esterification reaction.   Next, a hydroxy-functional urethane diol was added, and propylene glycol was removed in vacuo at 180°C. The hydroxy-functional urethane diol was previously prepared by a non-isocyanate process utilizing the reaction between a cyclic carbonate and a diamine.   The resin was then carboxy-functionalized and dispersed in water with the aid of a neutralizing tertiary amine (Figure 1).   Number average molecular weights for the PUR dispersions were in the 3000 - 4000 g/mol range.

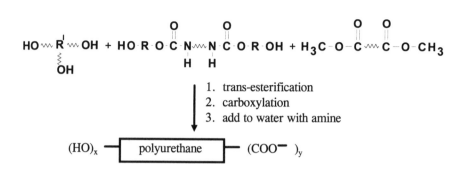

(R, R' and ᴧᴧᴧᴧ = aliphatic groups)

Figure 1.   Synthesis of waterborne anionic polyurethane dispersion by a non-isocyanate process.

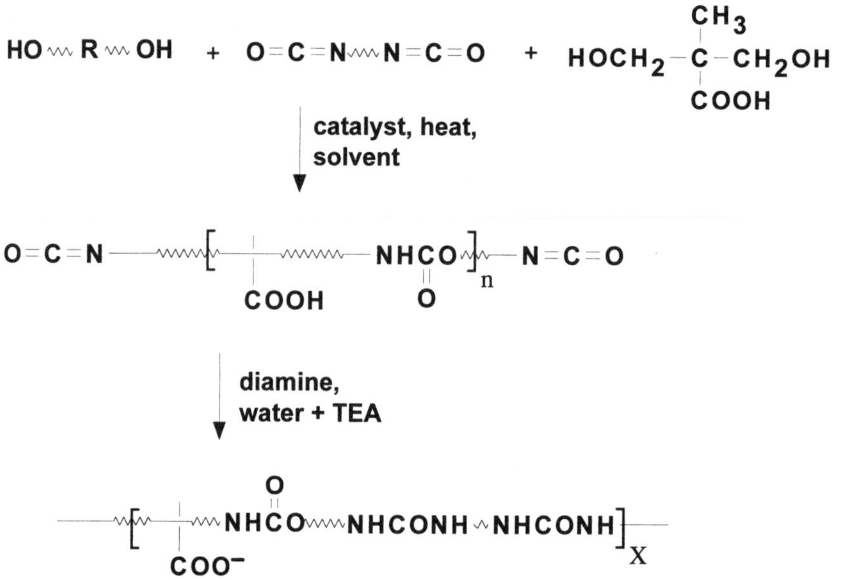

HO ⱯⱯ R ⱯⱯ OH   +   O=C=N⟋⟋N=C=O   +   HOCH$_2$–C(CH$_3$)(COOH)–CH$_2$OH

catalyst, heat,
solvent

O=C=N ——⟋⟋[⟋⟋ — NHCO]$_n$— N=C=O
            | ‖
          COOH   O

diamine,
water + TEA

                     O
                     ‖
——⟋[⟋⟋NHCO⟋⟋⟋NHCONH⟋NHCONH]$_X$
      |
     COO$^-$

Figure 2.   Synthesis of conventional waterborne polyurethane dispersion by the isocyanate prepolymer process.

The XM-2311 PUR was also prepared in an oligomeric form (1,500 g/mol) for use as a pigment dispersing resin. Comparisons of a melamine-cured PUR were made to a PUR (designated XP-7) prepared by a conventional isocyanate process. Two polyurethane dispersions, designated XP-7 and XP-4, were prepared by the conventional prepolymer isocyanate process given in Figure 2.   As seen in the reaction scheme, this process actually produces a polyurethane-urea polymer, although such polymers are typically abbreviated as polyurethanes.   The chain extension reaction of the isocyanate terminated polyurethane with the diamine forms the urea moiety.

The melamine resin (crosslinker) used in this study was a commercially available version of hexakis(methoxymethyl)melamine (HMMM), which has a degree of polymerization of about 1.5, an average molecular weight of 554 and an average theoretical functionality of 8.3.   The waterborne (WB) acrylic dispersion used for formulating was Acrysol WS-68 from Rohm and Haas, a hydroxy/carboxy functional resin.   A water-dispersible polyisocyanate from Bayer Corporation (Bayhydur XP-7007, a modified aliphatic isocyanate trimer) was used for crosslinking XM-4316 in one study, and another PUR prepared via isocyanate chemistry (designated XP-4) was used for further comparison to XM-4316. Pigments were obtained from commercial suppliers such as Bayer, Cabot, DuPont and BASF.

Coating Formulations

The PUR dispersions were formulated with 10 to 35% HMMM on total resin solids, and catalyzed with 1% of amine-blocked p-toluenesulfonic acid. Coatings were cast on Bonderite 1000 iron-phosphated cold-rolled steel panels at a 25 micron (1.0 mil) dry film thickness, and were cured for 20 minutes at 120°C and at 150°C. Pigmented films were cured at 150°C for 15 minutes.   Coatings prepared with the water-dispersible isocyanate were cured at ambient temperature for two weeks (21°C $\pm 2°$), and also cured at 80°C for 30 minutes.   Coated panels were tested for dry film thickness (DFT), pencil and Knoop hardness (KHN), Konig hardness, cross-hatch adhesion, direct and reverse impact resistance, MEK (methyl ethyl ketone) double rubs, and gloss (60° and 20°) in accordance with ASTM procedures.

Surface analysis was performed by using Fourier Transform Infrared Analysis Attenuated Total Reflectance Spectroscopy (FT-IR ATR).   An IBM Model IR44 instrument was used with a Spectra Tech zinc selenide reflectance attachment. Stress-strain analysis of coating free films was performed on an Instron Model 1011 materials tester. Particle size measurements were made with a Leeds and Northrup Microtrac Model UPA 150 instrument.   Cured coatings were tested for abrasion resistance on a Taber Abraser Model 5130 using CS-17 abrasive wheels in accordance with ASTM D4060. All abrasion data were normalized to wear index values based on 1000 cycles. Gloss measurements were obtained with a Byk-Gardner Multi-Gloss meter. Molecular weight values were obtained by size exclusion chromatography on a Waters 510 system with a Waters 410 refractive index detector, and values are

Table 1. Physical characteristics of waterborne polyurethane dispersions

PROPERTY	XM-2311	XM-2312	XM-4310	XM-4316	XP-7	XP-4
Method of synthesis*	NI	NI	NI	NI	I	I
% Nonvolatile, 60 min., 110°C	43.8	36.0	40.9	42.0	36.0	40
Brookfield Viscosity, cP, 25°C	1720	700	506	2300	650	300
Acid Number (on solids)	60	30	45	45	7.5	32
Hydroxyl Number (on solids)	33	70	65	75	0	56
Number Avg. MW	1500	3100	3100	3000	30000	3750
pH of dispersion	8.4	8.3	8.4	7.4	8.1	7.5
Particle Size, mean diameter (nm)	11	61	15	24	66	23
Solvent Content, %	0	0	0	0	5	10
Neutralizing Amine	TEA & DIOPA	DIOPA	DMEA	NMM	TEA	TEA

* I = isocyanate process; NI = non-isocyanate process

Table 2. Coating properties of PUR dispersions cured with HMMM

PROPERTIES	XM-4310	XM-4316	XM-4310	XM-4316
			(WITH 10% XM-2311)	
Knoop Hardness (KHN$_{25}$)	12.1	16.2	10.0	13.1
Konig Hardness (seconds)	128	145	109	130
Direct Impact Resistance (in*lb)	120	120	160	160
Tensile Strength (psi)	3000	6500	-----	-----
Percent Elongation	15	4	-----	-----

relative to polystyrene standards.   The system was operated at 30°C using tetrahydrofuran as the eluent at a flow rate of 1.0 ml/min.

RESULTS AND DISCUSSION

Polyurethane Dispersion Properties and Resin Design

The basic physical properties of all six dispersions are given in Table 1.   All but the XP-7 and XP-4 dispersions were prepared using the non-isocyanate synthetic process. These two dispersions were prepared by the conventional diisocyanate prepolymer process.   Note that these two dispersions also contain solvent, which is 1-Methyl-2-pyrrolidinone (commonly named N-methyl-pyrrolidone).   The solvent is present during the isocyanate prepolymer synthesis step for viscosity reduction.   The presence of the solvent increases the VOC of the final coating formulation.   The polyurethane dispersions prepared by the non-isocyanate process (XM type) do not require any solvent during synthesis, and thus yield coatings with much lower VOC levels.

Different neutralizing amines were chosen to prepare the anionic dispersions depending on the end-use application and on the nature of the resin backbone. Triethylamine (TEA) has traditionally been used for conventional PUR's due to its high basicity and high volatility at ambient temperature.   We have found that N-methylmorpholine (NMM) yields a dispersion with superior shelf stability and rather unique colloidal behavior.   A mixed amine approach of diisopropanolamine (DIOPA) and TEA was used for XM-2311 because of an enhancement in colloidal stability. Dimethylethanolamine (DMEA) was useful in certain instances, but is not desirable for use with water dispersible isocyanates due to its reactive hydroxyl functionality.

The XM type PUR dispersions have particle size values that are typical for waterborne polyurethanes.   Amongst the XM PUR's, the largest particle size was obtained with the resin with the lowest level of carboxyl functionality.   Such is expected due to the larger hydrophobic particle center associated with less carboxyl groups on the exterior of the particle.

In varying the type of comonomer components in the transesterification reaction, different final film properties were obtained.   One variable that was employed was the balance between diesters and polyols of straight-chain alkyl type and cycloaliphatic type.   The introduction of a cycloaliphatic diester or diol provided a convenient means of increasing the glass transition temperature (T_g) of the polymers. The increase in polymer T_g resulted in an increase in the final cured film hardness. The XM polymer T_g increases in the order of XM-2311 < XM-2312 < XM-4310 < XM-4316.   Polymer XM-4316 has the highest level of cycloaliphatic segment, and only XM-2311 has no cycloaliphatic segments.   XM-2312 and XM-4310 have the same base resin composition, and only differ in acid number (amount of carboxyl functionality).   Table 2 gives some of the coating properties when the dispersions are cured with 25% HMMM and 1% blocked acid catalyst (both on total resin solids) at 150°C.   The introduction of 10% XM-2311 results in a softening of the films and also yields a coating with greater impact resistance (flexibility).

Coating Properties

The film property results of the XM-2312 / HMMM system (Table 3) clearly indicate that the optimum level of HMMM crosslinker is 25 to 30%. At this level there is no loss in impact resistance while the film retains acceptable elongation. The optimum film cured at the 25 to 30% HMMM level has an excellent balance of hardness and flexibility, which is usually difficult to achieve considering the tendency for the melamine resin to self condense (crosslink) and form brittle films. These findings provide evidence for a more uniformly crosslinked network. At the 35% HMMM level there is a drastic loss in film elongation along with a reduction in direct impact resistance. At this point, the theoretical amount of HMMM required is exceeded leading to self HMMM reaction, which leads to embrittlement of the film. Excellent compatibility of the resin and the HMMM crosslinker is evidenced in the high gloss values for the clear-coat films.

Table 3. Coating properties of XM-2312 PUR cured with HMMM at 150°C

% HMMM	KNOOP HARD.	DIRECT IMPACT (IN*LB)	MEK 2X RUBS	60° GLOSS %	MAX. TENSILE (PSI)	% ELONG.
10	2.7	>160	35	>90	----	----
15	3.0	>160	120	>90	1000	32
20	5.0	>160	>200	>90	2350	34
25	9.4	>160	>200	>90	2800	14
30	15.0	>160	>200	>90	3550	13
35	18.0	140	>200	>90	3450	2

In a second coating system, XM-4310 was evaluated as a modifier for an acrylic resin. The XM-4310 resin imparted significant improvements to the performance properties of a white thermoset acrylic coating, as shown in Table 4. The XM-4310 was used at a 16 weight percent level on total resin solids, with an HMMM content of 20 weight percent. The coating was compared to the acrylic with the same HMMM level. The HMMM level was adjusted (lower than standard 25 - 30%) in this case because the acrylic requires a lower HMMM level. The white pigment (DuPont R-900) was initially dispersed in the acrylic resin, then let-back with the acrylic or PUR/acrylic. Both formulations were catalyzed with only 0.2 weight percent of blocked p-toluenesulfonic acid, and the final pigment to binder ratio was 0.63 to 1.0. The formulations required less catalyst than the clear-coat PUR/HMMM formulations. The final solids of both formulations was 42.5%.

Improvements were found in flexibility, chemical resistance, and in salt spray resistance. Increased abrasion resistance of the film occurred upon introduction of the XM-4310. It is clear that the introduction of urethane character into an acrylic coating provides benefits in abrasion resistance, which will enhance the long term durability of the coating. The fact that the MEK resistance increased upon introduction of the XM-4310 provides evidence for the formation of a more highly crosslinked network. The XM-4310 has the available primary hydroxy functionality for curing with the HMMM. The acrylic polymer has a much higher molecular weight than the XM-4310 polyurethane. Thus, the molecular diffusion of acrylic polymer chains is slower, affecting the availability of the hydroxyl reactive sites.

Test panels were prepared for weathering under a constant salt-spray environment for 250 hours in accordance with ASTM D117. At the conclusion of the test the acrylic coating was completely rusted whereas the coating modified with the XM-4310 PUR had only 3 mm of rust creep. Therefore, it is shown that by modifying the acrylic coating with XM-4310 that a number of coating performance advantages are realized. The VOC for this system may be further reduced since some of the cosolvents present are required for the acrylic component. We have found that the use of the XM type dispersions for modifying acrylics enhances film formation (film coalescence) of the coatings, thus requiring less cosolvent to accomplish this function, leading to a favorable reduction in VOC.

Table 4. Coating properties of acrylic and PUR/acrylic cured with HMMM at 150°C (white pigmented)

Mechanical & Appearance Properties	Acrylic / HMMM	Acrylic / XM-4310 / HMMM
Pencil Hardness	4H	4H
Knoop Hardness	19.7	15.3
Direct Impact (inch*lb)	60	120
Reverse Impact (inch*lb)	< 5	60
Gloss, 60°, %	94	92
Gloss, 20°, %	79	75
MEK Double Rubs	155	>200
Taber Abrasion, Wear Index	620	412
Salt Spray, Creep in mm (250 Hours exposure)	99% rusted	3 millimeters

A series of coatings were prepared where various resins were melamine cured on test panels, and tested for abrasion resistance. The resins were used alone; not as modifying resins for other systems. The Taber abrasion wear index data is given in Table 5. The lower the wear index number, the more abrasion resistant the coating, and testing was performed as indicated in ASTM D4060. The XM-4310/HMMM coating is the most abrasion resistant coating tested when compared to other melamine cured waterborne polyurethane, acrylic, and polyester coatings.

Table 5. Taber abrasion data for various waterborne resins cured with HMMM at 150°C

WB resin cured with HMMM	Taber Abrasion Wear Index
XM-4310 PUR	87
conventional PUR (XP-4 type)	335
acrylic	620
polyester	646

Waterborne Polyurethane Dispersions As Pigment Dispersing Resins
The starting point in any water-based coating application is the initial dispersing of the pigment in water. In designing a polyurethane dispersion that functions as an efficient pigment wetting resin, certain design features were implemented. A high degree of branching is present, along with hydroxyl functionality and increased levels of urethane and carboxyl functionality. The presence of all three of these moieties in the same molecule, along with the proper resin molecular weight distribution, allow the groups to work synergistically to "wet" the pigment's surface. XM-2311 dispersed a wide range of organic and inorganic pigments, and readily dispersed pigments that are traditionally considered difficult to disperse in water (phthalo blue, perylene red).
Base pastes were readily prepared by ball-mill grinding the pigment in XM-2311 and deionized water. Such pastes were stable for up to one year when stored at ambient temperature.
     One of the key attributes of XM-2311 is that it functions as the sole pigment dispersing vehicle, not requiring the use of any additional dispersing resin. Table 6 illustrates the formulas used to disperse a number of pigments with XM-2311. Note that the phthalo green pigment loading is very high, at a level of 3 to 1 pigment to binder (P/B) ratio. An effort was made to obtain the highest loading level for this pigment as a way to illustrate the fact that XM-2311 is efficient in wetting pigment. Many of the other formula P/B values can likely be increased as well, and only the green-pigmented formula has been optimized thus far.

Table 6. Low molecular weight PUR used to disperse various pigments

Material	Carbon Black	Phthalo Blue	Perylene Red	Phthalo Green	TiO$_2$ White
XM-2311	36.22	48.27	48.27	37.05	28.73
Pigment	15.00	21.00	21.00	48.75	68.13
Water	45.11	30.73	30.73	14.2	3.14
DMEA	3.67	-----	-----	-----	-----
TOTAL	100.00	100.00	100.00	100.00	100.00
Resin Solids, WT %	15	21	21	16	12.3
P/B	1/1	1/1	1/1	3/1	5.5/1

The TiO_2 (white) pigment required only 10-15 minutes of dispersing time on a Cowles disperser. The dispersing ability of the XM-2311 PUR was compared to other commercially available acrylic and alkyd resins, as well as a commonly used surfactant system. It was determined that the other commercial systems required longer grind times in dispersing the TiO_2 pigment A shorter grind time is desirable from an industrial standpoint due to the lower cost associated with the more efficient process. The carbon black pigment required some pre-neutralization with DMEA. The grind pastes generated high-performance coatings when formulated and cured with XM-4310 and HMMM. The grind pastes were also successfully formulated into a number of other commercially available resin vehicles. The XM-2311 grind pastes were compatible with waterborne acrylic, alkyd, polyester, and polyurethane resins, indicating a "universality" of the XM-2311 PUR and its broad use in water-based systems. A coating formulation was prepared whereby XM-2311 was used to grind the phthalo blue pigment, and was combined with the XM-4310 PUR and melamine. The formulation was cast on B-1000 steel panels, and cured at 150°C for 20 minutes. The properties of the coating are given in Table 7, which exhibit an excellent combination of flexibility, chemical resistance and high gloss.

Table 7. Coating properties of phthalo-blue pigmented XM-2311 / XM-4310 system cured at 150°C with HMMM

PROPERTIES	CURED COATING
Pencil Hardness	H-2H
Knoop Hardness	4.1
Direct Impact (in*lb)	>160
Reverse Impact (in*lb)	>160
Gloss, 60 degree, %	91
Gloss. 20 degree, %	78
MEK Double Rubs	>200

Curing Polyurethane Dispersions with Water-Dispersible Isocyanates

XM-4316 proved to be readily crosslinkable at ambient temperature with a water-dispersible polyisocyanate. One of the inherent drawbacks of this new technology is the formation of carbon dioxide due to the side reaction of isocyanate with water. When an isocyanate reacts with water, the products are a urea linkage (via an amine intermediate) and carbon dioxide. The carbon dioxide formation is problematic in that it causes imperfections in the coating during cure, such as blistering and pin-hole formation. An excess of isocyanate relative to polymer hydroxyl functionality is recommended due to this side reaction. It is quite common to prepare such water-based formulations with a 2.0/1.0 isocyanate to hydroxy-polymer mole ratio.

One way to minimize such effects is to prepare the PUR with a neutralizing amine that does <u>not</u> function as an efficient catalyst for the isocyanate/water reaction. The use of N-methylmorpholine (NMM) as the neutralizing amine for the XM-4316

PUR proved to be beneficial in this respect. NMM is a weaker base than most other tertiary amines, and is not a strong catalyst for the side reaction of isocyanate and water (19,20,21). The use of triethylamine in this case is not favored, due to its strong basicity and ability to catalyze the isocyanate/water side reaction. Triethylamine is the most commonly used amine for neutralizing commercial PUR dispersions. The pK_a values for triethylamine and N-methylmorpholine are 10.75 and 7.38, respectively, at 25°C. Triethylamine has a rate constant 7.4 times greater than N-methylmorpholine in catalyzing the reaction between phenyl isocyanate and n-butanol. Formulation characteristics and clear-coat film properties are given in Table 8 for the XM-4316 PUR cured with a water-dispersible polyisocyanate.

Table 8. Properties of PUR cured with water-dispersible polyisocyanate

PROPERTY	AMBIENT CURE	80°C, 30 MIN
NCO / OH Ratio	1.7 / 1.0	1.7 / 1.0
Theoretical VOC (gram/liter)	89	89
pH	7.5	7.5
Dry Time, Set to touch (hours)	0.25	----
surface dry (hours)	2.25	----
through-dry (hours)	5.75	----
film thickness (microns)	25	25
Konig Hardness (sec)	41	70
MEK Double Rubs	90	160
Direct Impact Resistance (in*lb)	> 160	> 160
Gloss, 60°, 20°, %	> 90, > 90	> 90, > 90

The high gloss values indicate that there is no problem of entrained carbon dioxide in the coating system. The film has acceptable hardness and solvent resistance upon ambient curing, and these values are much improved upon a low-temperature bake cycle. Ambient cured panels were tested two weeks after initial film casting. The improvement in MEK resistance for the low-bake coating indicates the formation of a more highly crosslinked network. Since the low-bake cycle allows for faster water release from the film, there is less time for the side reaction of water and isocyanate to take place. Therefore, the isocyanate/PUR reaction is favored, leading to optimum crosslink formation. After mixing the XM-4316 PUR and the water-dispersible polyisocyanate the formulation has a pot-life of 4 to 6 hours, without undergoing any foaming from carbon dioxide formation.

In comparison, when formulating the XP-4 PUR into a similar formulation, the use of cosolvents was required to obtain acceptable film formation and coating gloss. As a result an almost three-fold increase in VOC level resulted (255 grams/liter), thus favoring the formulation with the XM-4316 PUR for VOC reduction.

Stress-Strain Analysis

Figures 3 and 4 show the stress-strain behavior of the free films of the XM-2312 and the conventional PUR (designated XP-7) cured with the HMMM crosslinker at 120°C. The extension ratio is defined as the ratio of the extended sample length to the original sample length.   The conventional PUR (XP-7) behaves in a two-phase fashion, undergoing an initial rapid increase in stress accompanied by very little elongation, indicating behavior of a more highly crosslinked phase. Next, there is a steady elongation of the sample.

In considering all of the other experimental evidence thus far, it is likely that the rapid increase in tensile strength is related to the HMMM self-crosslinked network, followed by the elongation of the uncrosslinked polyurethane phase.   The 30,000 g/mol PUR can exhibit such a distinct elongation profile because the resin itself is thermoplastic.   Conversely, the XM-2312 film undergoes a more uniform change in its stress-strain relationship, providing evidence for a more homogeneous network.   Such behavior is more closely related to the ideal (theoretical) stress-strain curve given in Figure 5, which is derived from the classical relationship (*11*) relating retractive stress of an elastomer during deformation (elongation).

As seen in Figure 3, once the level of HMMM is exceeded (see 35% curve) beyond the amount needed for proper crosslinking, the same effect becomes noticeable in the XM-2312 film.   Although less HMMM is required for curing the XP-7 resin, it is clear that even at the 10% level (Figure 4) the two phase stress-strain behavior is observed.   Theoretical HMMM resin levels are determined from the total available hydroxyl and carboxyl functionality in the polymers.   The theoretical molecular weight per crosslink point is 561 g/mol for XM-2312 and 7480 g/mol for the XP-7 polymer.   The stress-strain data provide evidence of the formation of a more uniform and homogeneous network in the XM-2312 cured film.   Network homogeneity translates to improved mechanical and performance properties of the resultant coating.

In our earlier work (*22*), we demonstrated the same stress-strain behavior in cured films of the XP-4 polymer.   In this instance, the water-dispersible isocyanate was used as the crosslinking agent.   The XP-4 polymer is prepared by an isocyanate process, and is close to the molecular weight of the XM-2312 and XM-4316 polymers (see Table 1), thus eliminating PUR molecular weight as a variable.   Similar two-phase behavior was demonstrated by the cured XP-4 polymer, whereas the XM-4316 system exhibits a stress-strain curve more closely resembling an ideal system.

Free Film Solvent Swelling Experiments

Swelling experiments of the HMMM crosslinked free films was performed using methylene chloride as the swelling solvent.   The Flory-Rehner equation (Equation 1) was employed to calculate the crosslink density of the films (*23*).

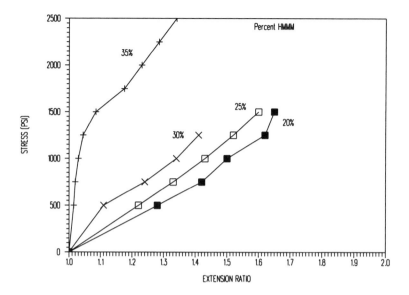

Figure 3.   Stress-strain curves for XM-2312 PUR cured with various levels of HMMM.

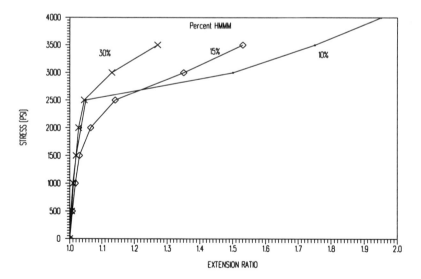

Figure 4. Stress-strain curves for XP-7 PUR cured with various levels of HMMM.

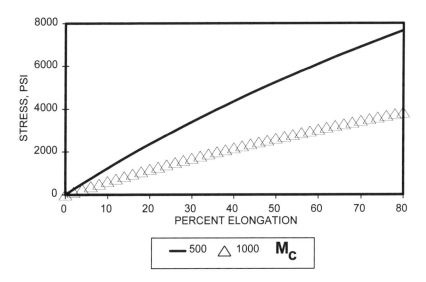

Figure 5. Theoretical stress-strain curves at different molecular weight per crosslink values for an ideal network.

$$-[\ln(1-\upsilon_2) + \upsilon_2 + \chi_1(\upsilon_2)^2] = V_1 n[(\upsilon_2)^{1/3} - \upsilon_2/2] \qquad \text{Equation 1.}$$

The value υ_2 is the volume fraction of polymer in the swollen mass, V_1 is the molar volume of the swelling solvent, χ_1 is the Flory-Huggins polymer-solvent dimensionless interaction term, and n is the number of active chain segments per unit volume. The quantity n equals ρ/M_c, where ρ is the density and M_c is the molecular weight between crosslinks. Published data for the interaction term (or parameter) in chlorinated solvents indicates a value in the range of 0.4 - 0.5 (*24*). In this study, an interaction parameter of 0.5 was used. The value of the molar volume of methylene chloride is 64.22 ml/mol. A previously reported method was used for the solvent swelling study where the dimensions of the film were accurately measured before and after swelling (*11*).

Figures 6 and 7 illustrate the data in graph form. Three different curves are shown in each figure. One curve is for the measured (actual) crosslink density values, one is for the calculated HMMM/polymer values, and one is for the values calculated for the scenario where the HMMM cures with the polymer and with itself (remaining curing by self-condensation of HMMM). This last scenario occurs if there is insufficient hydroxyl functionality present to cure with the HMMM (*25*). The XM-2312 film has measured crosslink density values that are close to the calculated values. However, the XP-7 film has measured crosslink density values that are closer to the values calculated for self-condensed HMMM. The data clearly show that the XM-2312 film has more ideal behavior and that the XP-7 film has a two-phase composition, with one phase composed of mostly HMMM self-crosslinked material. The presence of such material is known to lead to brittle coatings and reduced coating performance.

The isocyanate prepolymer process allows for the chain extension of an isocyanate terminated prepolymer with a diamine, such that any hydroxyl end groups are present at approximately one per 30,000 molecular weight unit. Therefore, the XP-7 polymer has little or no pendant hydroxy functionality by virtue of its method of synthesis (see Table 1), and thus can not obtain a lower M_c without the occurrence of the melamine self-condensation reaction. As indicated earlier, the carboxyl groups may participate in crosslinking with HMMM, but this process occurs at a much slower rate and is less likely to occur.

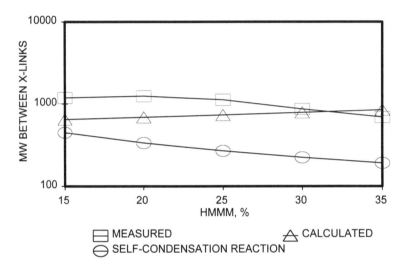

Figure 6. The XM-2312 / HMMM free film crosslink density data.

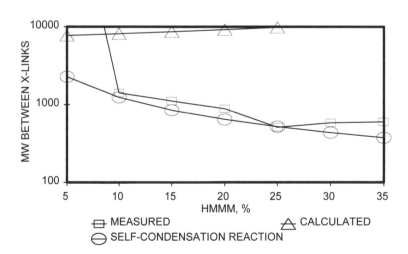

Figure 7. The XP-7 / HMMM free film crosslink density data.

FT-IR ATR Analysis of Free Films

Surface analysis of the cured coating free films by FT-IR ATR proved to be very useful in comparing the relative concentrations of HMMM at the film-air interface, which is an indirect measure of film homogeneity and network formation.   The spectra are given in Figure 8. A comparison was made between the XM-2312 and the XP-7 films cured with 30% HMMM at 150°C.

The IR band of interest is the out-of-plane deformation of the HMMM triazine ring which occurs at 815 cm^{-1}(26). All of the spectra were baseline corrected so that they may be compared. The individual polymers (XM-2312 and XP-7) did not have any absorptions in the region of interest, thus eliminating any interference problems. The depth of penetration of the IR beam is given by Equation 2, where λ_o is the wavelength of the infrared radiation, n_1 is the refractive index of the ATR crystal,

$$d_p = \lambda_o \ / \ 2 \ \pi \ n_1 (\sin^2 \theta \ - \ n_{21}^2)^{1/2} \hspace{3cm} \text{Equation 2.}$$

θ is the angle of incidence and n_{21} is the ratio of refractive indices of the sample and the ATR crystal (27). Using Equation 2, the depth of penetration at 815 cm^{-1} is 3.5 microns at the film-air interface. The films prepared for this study were 25 microns in total thickness.

The surface spectra indicate a significant increase in the HMMM absorption for the XP-7 film, indicating a higher HMMM content on the film surface, and a less homogeneous film. Our studies indicate a level of HMMM twice that on the surface than is theoretically present in bulk for the XP-7 coating. Conversely, the XM-2312 coating had the expected level of HMMM at the surface. Unreacted melamine resin tends to migrate to the film-air interface and self react if there are a lack of hydroxyl functional groups available for crosslinking (28). Such a self-crosslinked melamine network functions only as a temporary barrier. Once abraded or degraded under weathering conditions the exposed coating has reduced performance characteristics. These results provide further evidence that the XM-2312 type resin yields coatings that have a more homogeneous composition, which ultimately leads to coatings with improved performance and weathering resistant properties.

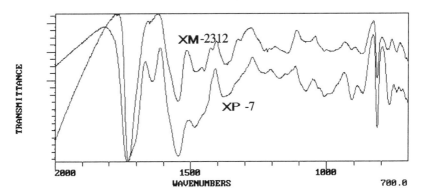

Figure 8. FT-IR ATR spectra of XM-2312 and XP-7 films cured with 30 weight percent HMMM at 150°C.

Summary

Low VOC coatings with superior performance properties are obtained when formulating with the waterborne polyurethane dispersions, XM-2311, XM-2312, XM-4310 and XM-4316. The polyurethanes are prepared by non-isocyanate chemistry which eliminates the exposure hazards of working with isocyanate raw materials. The XM-2311 PUR functions as a highly efficient pigment dispersing resin by virtue of the presence of hydroxyl, carboxyl and urethane functionality. The proper molecular weight and degree of branching incorporated into the resin design also enables the resin to effectively "wet" pigment surfaces, and not undergo any mechanical (shear) degradation. Other acrylic type dispersing agents are subject to shear degradation during the dispersion process due to the higher molecular weight values. Also, there is no need for the use of solvents or surfactants in dispersing the pigments with XM-2311. These additives are often necessary in using other waterborne resins to disperse pigments. XM-2311 is able to disperse a wide variety of pigments, both inorganic and organic types, which is unique among conventional dispersing resins.

The XM-2312 PUR coating cured with melamine has a VOC level of 46 grams/liter (0.38 lb/gallon), and the two-component formulation of the XM-4316 PUR with the water-dispersible isocyanate has a level of 89 grams/liter (0.74 lb/gallon). These VOC levels are well below the current government regulation limits and yield coatings which are considered environmentally friendly. Recent advances in the water-dispersible isocyanate technology have made it possible for us to prepare formulations at VOC levels of less than 50 grams/liter with the XM-4316 PUR.

The use of the XM-4310 PUR as a modifying resin in an acrylic/melamine coating imparts performance advantages in flexibility, abrasion resistance and in salt-spray weathering resistance. The XM-4310 PUR also has the highest abrasion resistance when compared to other waterborne resin types cured with the same melamine coating system.

The combination of stress-strain analysis, solvent swelling studies, and surface FTIR ATR provides evidence for more uniform network formation in the coatings prepared from the XM type polyurethanes. It is clear that the formation of a more uniform network leads to coatings with improved performance properties.

Acknowledgments

The authors would like to acknowledge Mr. Michael Emmet and Mr. Anthony Prokopowicz for their assistance in testing the coatings. VJT thanks Prof. Ben Bangerter at the Yale University Instrumentation Center for providing access to their facility, and for assistance in obtaining NMR spectra.

Abbreviations

WB	waterborne
PUR	waterborne polyurethane dispersion
VOC	volatile organic content
HMMM	hexakis(methoxymethyl)melamine crosslinker
melamine	hexakis(methoxymethyl)melamine crosslinker
DMEA	dimethylethanolamine
TEA	triethylamine

NMM	N-methylmorpholine
DIOPA	diisopropanolamine
P/B	pigment to binder ratio
M_c	molecular weight per crosslink
XM	waterborne polyurethane dispersion prepared by non-isocyanate synthetic route
XP	waterborne polyurethane dispersion prepared by conventional isocyanate prepolymer synthetic route
cP	centipoise, a unit of viscosity
P/B	pigment to binder ratio
MEK	methyl ethyl ketone
cosolvent	coalescing solvent (solvent used to achieve film coalescence)
MW	molecular weight
ASTM	American Society for Testing and Materials

Literature Cited

1. Chang, W.H. and Hartman, M.E. U.S. Patent 3 912 790, 1975.
2. DiDomenico, E. U.S. Patent 4 451 622, 1984.
3. Altschuler, L. and Simms, J. U.S. Patent 4 632 964, 1986.
4. Tirpak, R.E. and Markusch, P.H., *J. Coatings Technology.*, **1986**, Vol. 58, No. 738, 49-54.
5. Potter, T.A. and Williams, J.L., *J. Coatings Technology*, **1987**, Vol. 59, No. 749, 63-72.
6. Product Report, *Chem. Eng. News*, September 25, 1995, 40-44.
7. Dietrich, D., *Die Ang. Makromol. Chem.*, **1981**, *98*, 133-165.
8. Kim, B.K. and Lee, J.C., *J. Polym. Sci., Polym. Chem. Ed.*, **1996**, Vol. 34, 1095-1104.
9. Coogan, R.G. and Boghossian, R.V. U.S. Patent 5 043 381, 1991.
10. Tramontano, V.J. and Blank, W.J, *Preprints of Waterborne, Higher-Solids and Powder Coatings Symposium, FSCT*, 1994, New Orleans, pages 77-100.
11. Tramontano, V.J. and Blank, W.J., *J. Coatings Technology*, **1995,** Vol. 67, No. 848, 89-99.
12. Blank, W.J. and Tramontano, V.J., *Prog. Organic Coatings*, **1996**, Vol. 27, 1-15.
13. Reference 11.
14. Jacobs, P.B. and Potter, T.A. U.S. Patent 5 200 489, 1993.
15. Hart, R.E. U.S. Patent 5 508 340, 1996.
16. Hart, R.E. U.S. Patent 5 472 634, 1995.
17. Blum, H. *et al,* U.S. Patent 5 331 039, 1994
18. Bittner, A. and Ziegler, P., *Preprints of Third Nurnberg Congress-New Technologies for Coatings and Inks*, 1995, Paper No. 38.
19. Schwetlick, K., Noak, R. and Stebner, F., *J. Chem. Soc., Perkin Trans.*, **1994**, Vol. 2, 599-608.
20. Wong, S. and Frisch, K.C., *J. Polym Sci., Polym. Chem. Ed.*, **1986**, Vol. 24, 2867-75.

21. Sojecki, R., *Acta Polymerica*, **1990**, Vol. 41, No. 6, 315-318.
22. Tramontano, V.J. and Blank, W.J., *Preprints of Waterborne, Higher-Solids and Powder Coatings Symposium, FSCT*, 1995, New Orleans, pages 245-253.
23. Sperling, L.H., *Introduction to Physical Polymer Science*, Wiley-Interscience, New York, 1986, page 343.
24. Wolf, B.A. In *Polymer Handbook*, Brandup, J. and Immergut, E.H. Ed., 2nd Ed., Wiley -Interscience, New York, 1975.
25. Yamamoto, T. *et al*, *J. Coatings Technology.*, **1988**, Vol. 60, 762, 51-59.
26. Vandeberg, J.T. *et al*, *An Infrared Spectroscopy Atlas for the Coatings Industry*, FSCT, Philadelphia, PA, 1980, page 35.
27. Urban, M.W., *Vibrational Spectroscopy of Molecules and Macromolecules on Surfaces*, Wiley-Interscience, New York, 1993, page 82.
28. Hirayama, T. and Urban, M.W., *Prog. Organic Coatings*, **1992**, Vol. 20, 81-96.

Chapter 10

Making Paint from Alkyd Emulsions

A. Hofland

DSM Resins bv, P.O. Box 615, 8000 AP Zwolle, Netherlands

Every producer and user of paints is familiar with binders that are dissolved in organic solvents. When changing from those 'conventional' systems towards water based *alkyd emulsions* and/or physically drying *acrylic dispersions,* certain difficulties concerning paint-formulation, application, open-time, and film formation can arise. The properties of coatings strongly depend on film formation. It turns out that in this respect, paints based on binder dispersions are more critical than solvent-based paints. This is because coalescence of the binder particles is a critical and property-determining step in the case of dispersions. Film formation, and hence protective properties, are mainly governed by *viscosity* and *miscibility* of the resins that constitute the binder particles. By means of a comparison of these two parameters for acrylic and alkyd resins, the differences in film formation between acrylic dispersions and alkyd emulsions are highlighted. It is shown that during film formation an alkyd can spread completely because it inverts from an o/w emulsion to a w/o emulsion. The particles that form a physically drying acrylic dispersion, on the contrary, only coalesce to a certain degree during film formation, even when coalescing agents are used. Complete spreading of the alkyd is very favourable in obtaining high gloss, barrier properties, adhesion/penetration etc. Favourable properties of both acrylic dispersions and alkyd emulsions can be combined to obtain solvent-free paints with good properties.

The System

It is hard to imagine that there are more complex systems than solvent based paints, yet there are: water based paints. In water based paints, one not only combines the non-miscible components, such as binder and pigment, but also suspends them in a medium they are not compatible with: water. On all relevant interfaces, surface active agents should be present. This is already the situation for a simple, model-like paint, without additives like anti-skin, anti-foam and anti-wrinkling agents (the latter two of which are also very surface active), leaving water based paints little less than small miracles. Using the proper combination of binder and pigment as well as the right (and not too many) additives, it is possible.

One precondition is that the surfactant will have to stay on the interface it was meant for. This sounds logical, but examples of surfactant migration from water / resin interface to water / pigment interface are numerous. These alterations invariably result in flocculation of at least one of the two dispersed phases. In this context, a possible cosolvent can also be regarded as a surfactant. For example, butyl diglycol has an affinity (hydrophilic/lipophilic balance, HLB) towards water / resin interfaces, almost equal to some widely used nonionic surfactants. Unfortunately, it does not have the accompanying stabilizing properties. Although the effect of such a replacement (surfactant --> solvent) is not great, it will influence the time of flocculation / coagulation of the binder, resulting in poor gloss properties.

To ensure that during the paint preparation no problems arise, it is wise to be acquainted with some effects that introduce instability:

*The temperature can change the stabilizer's affinity towards the surface (temperature instability). For most emulsions this limit is 60 °C. This problem presents itself predominantly with nonionic surfactant stabilized emulsions.

* Addition of metal salts will change the charge stabilization (minus repels minus), and so will lowing the pH below 3. Small amounts of salts like the driers will not present a problem. This is called charge instability and will present itself with ionically stabilized systems. The higher the valency of the cations, the more the electric double layer will be compressed. This will dramatically reduce electrostatic stabilization.

Most commercial alkyd emulsions are stabilized in a mixed fashion: both non-ionically and by charge. This means that as long as either the temperature criterium or the salt criterium are met, there is no problem.

* Applying high shear to an emulsion may mechanically remove the stabilizer from the surface. This means that grinding pigment in the emulsion should be avoided. Gear pump stability differs per product, and care should always be taken with grinding alkyd emulsions.

* Finally, during freezing of an emulsion, the remaining impurities like calcium and magnesium salts in the water are concentrated. Because of the pure nature of the ice crystals, these impurities are driven out. This means that cosolvent destabilization or charge destabilization can occur (freeze-thaw stability), unless this is prevented by cosolvents that cause freezing point depression.

The Film

The process of film formation of water based paints is governed by three physico-chemical principles:

1. Two liquid phases having a large difference in viscosity
are difficult to mix.
2. The range of miscibility of polymers becomes narrower when
the molecular weight is increasing (*1*).
3. Structural similarity favours miscibility (*2*).

Film Formation: Alkyd Emulsion vs. Acrylic Dispersion

The difference in behaviour of physically drying acrylic dispersions compared to alkyd emulsions can be explained by the properties of the two systems (Table I) together with the guiding principles discussed above.

Table I. Important Parameters of the Two Binder Systems

	Acrylic Dispersion	*Alkyd Emulsion*
M_w (g/Mol)	$10^5 - 10^6$	2000 - 8000
binder-viscosity	'high'	'low'
T_g (°C)	$-37 \Rightarrow +104$	$-90 \Rightarrow -30$

As soon as an acrylic dispersion has been applied, water begins to evaporate from the film. At room temperature, the rate determining factor in the evaporation is the humidity of the air above the paint film. As soon as the polymer particles begin to touch each other, partial coalescence takes place. At the same time water is transported from the lower part of the film to the upper part.

The driving force for both the sintering and the water flow is capillary pressure caused by the high surface tension of the water together with the curvature of the interfacial boundaries (*3,4*). This capillary flow proceeds very fast compared to the evaporation of water at the film-air interface. Hence, there is only a small gradient in the concentration of water in the film in the vertical direction. This is the reason, fortunately, that complete coalescence in the upper part of the film does not take place in this first stage of drying.

At a certain moment, the polymer particles reach a distance from each other where the Critical Interparticle Distance (CID) is attained in a vertical direction in the film (*5*). From this moment on, deformation (coalescence) of the polymer particles can take place. The driving force for this process, called wet sintering, is again the capillary pressure. Of course, the deformation is opposed by the viscosity of the polymer. It should be noted that the shape in which the water between the particles is forced is unfavourable from an enthalpy point of view:

because of its high surface tension, water is trying to adopt a form in which the surface-to-volume ratio is as small as possible. The high viscosity of the polymer at the interface is preventing this. When all the water has evaporated out of the film, the well known polyhedric structure of the polymeric particles remains. From this moment on autohesion has to take place: interfacial diffusion of polymeric chains occurs to obtain a continuous and homogeneous film. The interdiffusion process is only driven by cohesive energy between the diffusing polymeric chains. Capillary forces do not play a role any more.

In describing polymeric particles that are dispersed in water, two phases are very often distinguished (6). It is energetically favourable if hydrophilic parts of the resins are in contact with the surrounding water. Hydrophobic material, on the other hand, does not want to be in contact with the water. These preferences are the reason why a dispersed polymeric particle can be described as a hydrophilic shell surrounding the hydrophobic core. It is known that the core-shell structure, apart from other properties, has a noticeable influence on the autodiffusion process. The formation of a completely homogeneous film is not expected for two reasons. The first reason has a thermodynamic origin. Because of the high molecular weight of the acrylic resin, there is only a limited miscibility of the acrylic chains (and the emulsifier). This is especially so if the core-shell structure of the particles is taken into account: the solubility parameter of core and shell will be different. There is also a kinetic reason. The mobility of the components in the particles (including the emulsifier) will be very low because of the high viscosity of the acrylic resin. This means that the interfaces will not disappear easily. Note that even for polybutylacrylate dispersions, having a T_g far below 0°C, the high viscosity of the acrylic still prevents (fast) interdiffusion.

For alkyd emulsions, the film forming process is in some respects comparable to the film formation of acrylic dispersions. There are however, important differences. In the first stage of film formation, water evaporates and is transported upwards because of capillary pressure. The gradient in humidity in the air above the film will again be the rate determining step in stage 1. At a certain moment, the alkyd droplets are starting to touch each other and partial interfacial diffusion of alkyd chains will take place. Until this moment the film formation proceeds in the same way as in the case of an acrylic dispersions.

When a vertical slice of the film has a certain composition, a process takes place that is very much wanted by coating technologists: the oil-in-water emulsion inverts through a metastable homogeneous state into a water-in-oil emulsion! This process can be seen very easily by applying a varnish of alkyd emulsion on a glass panel. After some time four different stages of this process are visible: The central zone consists of an alkyd-in-water emulsion and has a white colour. Surrounding it is the grey water-in-alkyd phase. Between those two areas, a clear and transparent zone can be seen. The transparency of the film is telling us that in the film there are no inhomogeneities with dimensions comparable to the wavelength of visible light (or larger). The material in this clear zone is in the metastable transition state as the film inverts from an alkyd-in-water emulsion into a water-in-alkyd emulsion (7). In this metastable phase, the composition of the film is such that water and oil are miscible in a metastable sense. This composition is called Critical Volume Composition. As soon as more water evaporates the composition of the material makes a water-in-alkyd emulsion more favourable from an energetical point of view. After this inversion has taken place, water continues to evaporate from the film. Transport of water to the surface takes place by means of diffusion instead of capillary flow. After some time the film becomes clear because the water-droplets are so small that they are not visible anymore, i.e., smaller than half the wavelength of visible light.

The last phase that the alkyd goes through is the cross-linking with oxygen from the air. The big difference between the film formation of an acrylic dispersion and that of an alkyd emulsion is that in the case of an alkyd emulsion the interfaces between the binder droplets disappear only within a few seconds of time. Of course, new interfaces appear as water droplets are being formed a short time later, but the main point is that alkyd chains can merge on a molecular scale and acrylic can not.

Basically, film formation in a water based paint is based upon the one phenomenon we definitely try to avoid during storage: flocculation. The alkyd resin particles will have to have a reason to start flocculating at the appropriate time. As has been explained before, this reason lies in inversion, brought about by the disappearance of the continuous phase: water. The quality of the resulting film is dependent upon how the following two problems are solved: How to get rid of the water that might be enclosed and how to give good surface wetting of the pigments and extenders, which are possibly suffering from the same problems?

Since the alkyd resin can flow freely, in contrast to the pigment, it is essential that flocculation of the resin precedes that of the pigment. Both problems are solved automatically if the resin particles can be deformed into hexagons, i.e., with alkyd emulsions. In hexagons, capillary forces are greater than in closed packed spheres, see Figure 1.

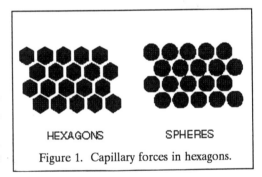

Figure 1. Capillary forces in hexagons.

These forces aid in flocculation and force the water out at the same time. Hard spheres like acrylic or styrene/acrylic particles (i.e., when no coalescing aid is used) and/or pigment particles do not have these capillary flocculation aids, and hence will remain available for covering by the resin. This inhibits film formation and raises the apparent pigment volume concentration (PVC). For this reason especially in blends of alkyd and acrylic resins, the normal calculation rules for PVC are by definition not valid.

What happens next is completely different from what happens in a conventional alkyd paint. During film formation in a conventional paint, the polarity of the continuous phase does not change at all, going from solvent to alkyd. During film formation in a water based alkyd paint, the continuous phase changes from highly polar (water) to virtually non-polar (resin). The behavior of the pigment particle will have to be adjusted likewise, mostly within several minutes. This means that its surface will have to exhibit a somewhat "schizophrenic" behavior, which can be accomplished either by two surface active agents, or by one with more than two affinities. One could introduce the expressions "alkydophilic" and "pigmentophilic", next to the term hydrophilic. Such a product is the adduct of polyethylene glycol and an acid functionality to linseed oil. There are numerous ways of producing such an product. An example is DSM Resins' Urad DD518. This additive is shown in Figure 2.

Due to its oily nature (DD518 is based on linseed oil) this product is suitable for use with long oil alkyds predominantly. Its uses include pigment wetting and dispersing agents, aid in incorporating cobalt soaps, as well as newtonic rheology modifier, to be used in combination with associative polyurethane thickeners.

For use with shorter oil alkyd emulsions, where the yellowing of the oily part can be a problem, BYK Chemie (Germany) has developed a similar compound without using oil: Disperbyk 190.

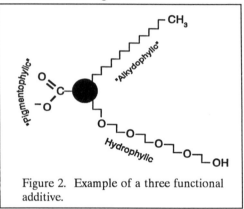

Figure 2. Example of a three functional additive.

Dispersing Pigments

In fact, three processes precede a proper pigment dispersion: grinding the particles, wetting them with continuous phase, and covering them with stabilizer to prevent flocculation. This is the case for solvent borne as well as for waterborne paints.

Grinding the Particles. In fact, there is very little to be said about this. The same equipment can be used for water based paints, provided that it is stainless steel. Cowles-type mills can be used for pigment pastes that already incorporate the rheology modifiers that have proven in the past to be essential for water based paints. The structure should be the right balance between short and creamy (created by cellulose-type thickeners) and long like in a traditional alkyd (created by some associative thickeners like Acrysol 1020). In fact, a long structure can be obtained by any additive that dissolves in the water phase and has a relatively low molecular weight.

Since the particle size and its distribution are far more important in water based paints than in their solvent borne equivalents, pearl mills are recommended for making paints with alkyd emulsions. Also, temperature is under greater control due to the short duration of the process. This is essential since many of the wetting and dispersing agents are in fact nonionic surfactants that have a cloud point, above which they lose their affinity towards water.

Wetting the Surface. During the grinding process, pigment / pigment as well as pigment / air interfaces will have to be replaced by the pigment / water interface. Following the laws governing total surface tension, the (normally high) surface tension of the water phase will have to be lowered. The optimal grind additives should have moderate water solubility, a high molecular weight, and a high affinity toward the pigment. In this way it can be assured that there will be almost no agent in the water phase, yielding low foaming systems even with anionic stabilization mechanisms. *If foam problems occur, the selection of alternative wetting and dispersing agents is to be preferred above the use of antifoam additives.*

When antifoaming agents are used, the types that base their activity on incompatibility (Foamex 810, 815, 825 from Tego) or the ones based on acetylenic diols (Surfynol range by Air Products) are preferred. Silicone based antifoams like Surfynol DF58 require, by definition, an anti-cratering agent. Proper ones tested in alkyd emulsion based paints are Byk 348 and Surfynol 420.

Stabilizing the Particles. Stabilization of the "primary" particles is normally obtained in two ways. First, the surface will have to be covered with material giving protection against flocculation upon collision. Secondly, to reduce these collisions the continuous phase is thickened by the amount of nonionic material that did not adsorb onto the particle. For these two processes, the same material can be used. Again, a properly chosen, multi-functional additive can perform both tasks.

Some Formulations

Using the knowledge gathered thus far, we can have a closer look at some very basic formulations, shown below (Table II).

Glossy exterior coating. As is the case with solvent based alkyds, the preferred type is an alkyd with a high level of fatty acid modification, to make it hydrophobic. In this case, the fatty acid modification is based on soybean oil. The water is then added to the wetting and dispersing agent, and after inversion, an opalescent solution of the DD518 AZ in water is obtained. Adding amines or alkaline material like Tamol 731 will make the solution clear, but this will increase the water solubility and hence create foam problems. The use of heat will simplify this procedure. Since most proper grinding additives are stabilized in a combined nonionic/anionic fashion, tap water is typically not a problem. Overheating can be, though, since most grinding agents have cloud points.

To this solution, the pigment is added. As common with practical formulations, part of the pigment can be replaced by carbonates, silicates and/or other extenders for cost reasons. Grinding is best done in a pearl mill, but adjusting the structure of the mill base (i.e., removing water, adding a rheology modifier from the letdown, or using a dissolver disc) is another alternative. To lower the surface tension and to make brush-cleaning easier, a nonionic surfactant should be added. After adding a suitable anti-foam agent (Foamex 810), the millbase can be ground. To this grind, the letdown, a clear varnish, is added. The most critical point in this clear varnish is the drier. Past drying problems have originated in hydrolysis of the binder. This is often catalyzed by the pigment surface, on which it is adsorbed. Cobalt hydroxide, the drier in alkyds that promotes free-radical crosslinking, then remains inactive on the pigment surface. It can not be denied that hydrolysis of the binder also takes place to some extent, but this can be prevented by the right choice of ingredients.

Replacing phthalic anhydride by isophtalic acid is a well-known and very effective example. Nevertheless, it can be shown that fatty acids (the cross linking "anchors") are split off the alkyd. Only three fatty acids per alkyd molecule are needed to ensure a three-dimensional network. It can be calculated that there are approximately six fatty acids per alkyd chain, so hydrolysis of only 5 % of the fatty acids, as has been found to be the maximum, does not present a problem. *The best way of avoiding hydrolysis of both binder and drier is to adjust the pH to 7.*

Table II. Composition of a Glossy Exterior Paint

Component	Weight %	Supplier
Urad DD518	0.8	DSM Resins
Tamol 731, 25 % in water	0.6	Rohm & Haas
Demineralized water	14.9	
TiPure 706	29.2	DuPont
Borchigen DFN	0.4	Borchers
Tego Foamex 810	0.1	Goldschmidt
Uradil AZ516 Z-60	51.5	DSM Resins
Cobalt Hydrocure II	0.6	OM Group
Exskin 2	0.2	Hüls
Acrysol RM 1020	0.6	Rohm & Haas
Acrysol RM 8 (10 % in water)	0.6	Rohm & Haas
Surfynol 420	0.5	Air Products

The loss of "drying" in clear varnishes is much lower than that observed in pigmented white paints. This suggests that the pigment can be held partly responsible for this phenomenon. Upon addition of more drier soaps to the pigmented coatings, the drying problem is significantly reduced; this is a clear sign that the pigment inactivates the drier. Indeed, cobalt hydroxide can be found on the surface of these non-drying paints. When a pigment with the wrong surface and surface area is used, the drier can be deactivated very quickly. This is shown in

Figure 3, where the effect of pigment on the Beck-Koller drying stages (tack-free) and (touch-dry) is shown.

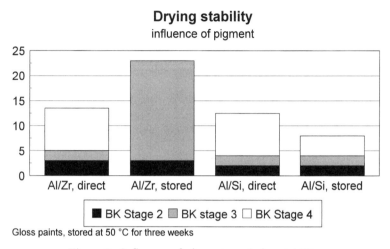

Figure 3. Influence of pigment on drying stability.

Note Above: Not only is the surface area different, but there is also a change in the loading amounts. The pigment leading to loss of drying (small surface area) is coated with aluminium- and zirconium oxides, whereas the other one is coated with aluminum and silicium. At present, the general understanding (Sachtleben, Kronos) is that the presence of over 0.5 % of silicium oxide on the surface will greatly improve the drying stability.

From a performance point of view, there is not much difference with its solvent borne counterpart. Gloss values of 70 and 95 units at 20° and 60° angles, respectively are the rule rather than the exception. Also, in terms of water and water vapor permeability, the alkyd emulsion closely resembles the solvent borne alkyd. This permeability has been measured in several ways:

Method 1. A 20 g beach ball was coated with 100 μm of paint, after which it was submerged in water. The weight gain after 100 hours was measured in grams.

Method 2. A comb-shaped electrode was coated with 100 μm of clearcoat, after which the dielectric constant was measured at several frequencies. From this, a volume fraction of water could be calculated.

Method 3. A detached film was placed over a cup filled with water at known relative humidity (RH). The film was placed with the paint / substrate interface down. The weight loss of the cup was measured in grams / m^2 / 24 hours.

Method 4. A wooden panel, as is commonly used for exterior wood protection in Scandinavia, was flushed with water for 2 days. After this period, the panel was placed at standard conditions, i.e. temperature 23°C and relative humidity of 54 %. The moisture weight exceeding 20 % of the panel (generally assumed to be the rot limit) was integrated over the total time, wetting as well as drying. In this way, insight could be gained regarding the wetting as well as drying out behavior (8). The 20% limit was chosen since this is generally agreed upon as the wood rot limit. The results are summarized in the next table (Table III).

The aesthetic performance of some alkyd emulsion based paints, applied to housing projects throughout the Netherlands, has been monitored. After three years, mainly on spruce, the following results have been derived from 1060 observations on a total of 29 systems (Table IV).

Glossy Interior Coating. Again, the procedure is fairly straightforward: The pigment paste is added to a clear varnish. The only difference is in the oil length of the alkyd. As might be expected for an interior paint, the alkyd used is only 40 % in oil and chain stopped with benzoic acid. To ensure a good compatibility with the alkyd, the wetting/dispersing agent has been replaced by Disperbyk 190, which will also improves the resistance to dark yellowing properties. This and other properties are listed on the next page (Table V and Table VI):

Table VI clearly stipulates the typical difference between a paint based on an alkyd emulsion and one based on a polymer dispersion, as have been discussed in the section on film formation.

Other Typical Application Differences.

Finally, there are some other intrinsic characteristics of alkyd emulsions.

Rheological behavior. An alkyd consists of an impressive "cocktail" of compounds, ranging from low molecular weight materials like the mono-ester of a polyalcohol with a fatty acid to almost gelatinous polycondensates with molecular weights of over 10.000. The colloidal states range from dissolved, low MW, highly polar materials, to truly over emulsified alkyds, to suspended gel-like materials. This has two main effects on the alkyd emulsion. First, due to the large spread in particle size, the solids content that can be obtained is very high; 70 % solids is no exception, especially with the higher oil length alkyds. Secondly, the low MW material that is water soluble gives the system a more Newtonian behavior than a true dispersion would. The dissolved material is actually operating as a thickener.

Open time. A second effect of the dissolved material is that it tends to "keep water in". This means that the inversion process is delayed and without typical open time additives, like propylene glycol, 15 minute time delays can be reached. Due to the high molecular weight and the fact that freezing point depression works on a molecular rather than a weight basis, this dissolved material does not yield adequate protection towards freezing. If a high freeze-thaw stability is required, one must add solvent.

Table III. Moisture Balance of Some Paint Systems

System	Method 1	Method 2	Method 3	Method 4
Alkyd in solvent	0.4	0.8	20	15-30
Alkyd emulsion	0.6	1.9	40	20-40
Polymer dispersion	3.7	8.6	70-80	8-11
Unit	grams	vol %	$g/m^2/24$ h	day%

Table IV. The COT (Dutch Center for Surface Technology) Durability Project

System ☞	Polymer dispersion	Alkyd emulsion	Alkyd in solvent	High solids alkyd
Number of painted substrates	540	120	240	160
Number of paint systems involved	16	3	6	4
Average gloss level in units at 20 ° angle	40	37	50	33
% of cracked panels	2.8	3.3	2.9	2.5
% of structural damage	1.3	0.8	2.1	3.1
% of wood rot	1.1	0.8	2.5	1.9
Number of painted substrates, disapproved by panel on:				
chalking	10	8	11	35
crack in paint	39	30	23	21
blisters	5	5	4	5

Table V. A Formulation For a Glossy Interior Coating

Component	Weight %	Supplier
Demineralized water	11.4	
Acrysol RM8, 15 % in water	2.8	Rohm & Haas
Byk 024	0.1	Byk Chemie
Disperbyk 190	1.3	Byk Chemie
TiPure 706	24.7	Rohm & Haas
Uradil AZ554 Z-50	53	DSM Resins BV
Acrysol RM8, 15 % in water	5.5	Rohm & Haas
Co Hydrocure II	0.5	OM Group
Exskin 2	0.3	Hüls
Byk 348	0.4	Byk Chemie

Table VI. Properties of a Glossy Interior Coating Based on Alkyd Emulsion

Properties, applied 100 μm wet

Beck-Koller drying recorder, hh:mm:		
Phase 1 (open time)	00:15	
Phase 2 (dust dry)	00:45	
Phase 3 (touch dry)	01:30	
Phase 4 (through hard)	15:30	
König hardness (seconds):		
24 hours	35	
4 days	40	
1 week	60	
2 weeks	70	
4 weeks	80	
Gloss (20 ° angle), after:		
24 hours	97	
4 weeks	94	
Yellow-index, after:		
24 hours	-1.3	
4 weeks in the dark	-0.7	

Yellowing. An alkyd remains an alkyd, even when it is put in water. This means that there will always be some yellowing, since the yellowing can be seen as a "derailment" of the drying process. The only alkyds that do not yellow (and smell) are the ones that do not dry. It is therefore strongly recommended to compare the alkyd with a solvent borne alkyd instead of with an acrylic waterborne system.

When alkyd emulsions are used, special care should be taken with respect to the presence of amines. Although some suppliers (a.o. DSM Resins) do not use amines to neutralize the emulsions, amines and ammonia can "sneak in" the paint formulation, even after application. Many problems have occurred when the trimming of a room was done with an alkyd emulsion and the walls and ceiling were treated with a latex, containing high amounts of ammonia. The two applications should not be done at the same time.

Hardness development. The hardness development of alkyd emulsions lags behind that of a solvent borne alkyd. The main reason for this is the plasticizing action of the hydrophilic part of the emulsifier, mostly polyethylene oxide. This part, especially when there is some hydration, has a very low T_g. For this reason, when an alkyd emulsion is compared with a solvent borne alkyd, it should be compared with one that is 10 % higher in oil length to compensate for this plasticizer.

Acknowledgment.

The Help of Jochum Beetsma, DSM Resins, in studying the film physicochemical aspects of film formation, is gratefully acknowledged.

Literature Cited.
1. Coleman, C.M.; Graf, J. *Painter, P.C.: Specific Interactions and the Miscibility of Polymer Blends;* Technomic Publishing Co. Inc.: Lancaster-Basel, 1991.
2. Van Krevelen, D.W. *Properties of Polymers;* lsevier Sc. Publ. Comp: Amsterdam, 1990.
3. Lyklema, J. *Fundamentals of Interface and Colloid Science;* Academic Press Inc.: San Diego, CA, 1991; Vol. 1.
4. Patton, T.C. *A Rheological Approach to Coating and Ink Technology;* John Wiley & Sons: New York, 1979.
5. Van Tent, A. *Turbidity Study of the Process of Film Formation of Polymer Particles in Drying Thin Films of Acrylic Latices;* Thesis, University of Delft, The Netherlands.
6. Van der Kolk, C.E.M.; Kruijt, R.; De Rouville, E.A. *JOCCA.* **1993** vol. 7, 280.
7. Becher, P. *Encyclopedia of Emulsion Technology;* Marcel Dekker Inc.: New York, 1983, Vol. 1.
8. Hjort, Stefan. Department of Building Materials, Chalmers University of Gothenburg, Sweden. Presented at the Asia-Pacific Conference in Coatings Technology, Singapore, April, 1995.

Chapter 11

Application of Electrosterically Stabilized Latex in Waterborne Coatings

D. D. Huang, S. Nandy[1], and E. J. Thorgerson

Thin Film Coating Technology, Film Imaging Research Division,
Polaroid Corporation, Waltham, MA 02254

The stability/aggregation behavior of carboxylated latex particles used in water-borne coatings has been investigated. Three forces - van der Waal's attraction, electrostatic repulsion, and steric repulsion - are the principal factors affecting latex stability. The electrosteric repulsion is a combination of electrostatic and steric potentials - which result from ionization of carboxylic acid on the surface of the latex. These effects make the latex particles more resilient in high ionic strength and high shear environment. An experimental investigation has been conducted using colloidal latex solution as a carrier layer in a three-layer coating to simulate photographic coating operation. The results demonstrate that the coating defects associated with formation of aggregates of carboxylated-latex correlate well with the swelling characteristic of these latex particles as a function of pH. Latex particles that do not swell, when pH adjusted from original 4.8 to final 7.2, show a tendency to form coating streaks and develop build up of latex aggregates on the surface of the coating applicator.

In photographic industry, coating is a prime technology used in the manufacturing of negative films. In instant color films, polymer colloids or latexes are one of the critical ingredients other than light sensitive emulsions and three primary color dyes needed for finished photographic products (1). During coating operations, a polymer latex solution layer is used as a carrier layer adjacent to a highly conductive dye layer in a multilayer coating structure as shown in Figure 1. On top of the dye layer different silver emulsion layers and other photographic functional materials are also coated simultaneously. In this method, coating fluids are metered through individual channels, exit out of slotted

[1]Current address: Moltech Corporation, Tucson, AZ 85747

heads, and flow down on an inclined plane (2). These fluid layers having very little intermixing are deposited on a fast moving web to form the final color film. It is easy to realize that the latex carrier layer is the most important layer for determining fluid stability because the bottom layer experiences the highest shear rate in the multilayer coating system (3).

From the coating process viewpoint, the properties of the latex systems crucial to this application are governed by the colloidal stability. It has been evident that the stability of the latex systems can be related to the force balance between spherical particles. In general, forces involved in the particle-particle interactions as particles approach each other can be attractive as well as repulsive in nature. The principal force affecting aggregation is the van der Waals attractive force between particles, whereas stabilizing force opposing aggregation is a result of the repulsive forces from surface charge and steric interaction between particles. In a latex dispersion system, it is the net effects of attractive and repulsive forces between the particles which govern their behavior and determine if the latex system is stable or will aggregate. There are many situations where van der Waals attraction is balanced only by electrostatic repulsion. However, in coating application stability is needed at high solid content and high electrolyte concentration, as well as at extreme conditions of temperature and flow. Under such circumstances, the repulsive force from surface charge alone is not sufficient to protect the latex particles. An additional force, called steric force, arising from the polymer interaction is required for stabilizing the system.

The evidence of adsorption of polymers or nonionic surfactants on particle surfaces for steric stabilization effect has been widely reported in the literature (4,5). In order to achieve stability, one part of the adsorbed polymer layer has to be anchored firmly onto the particle surface and the other extended into the medium. In many cases, the molecule used in this situation has both hydrophilic and hydrophobic moieties and the hydrophilic ends are extended into the aqueous phase. PEO/PPO and PEO/PMMA block copolymers are frequently used for this purpose (4). However, steric stabilization based on an adsorption mechanism sometimes can result in adverse effects. For example, depletion flocculation often occurs when polymers desorb from the particle surface. This condition can be avoided either by grafting or copolymerizing hydrophilic molecules on the outer surface of the colloid particle.

Carboxylic functional groups are often used in small amounts during polymerization to provide a specific property needed in a particular application (4). One of the major reasons is that the inclusion of acid groups will enhance not only the electrostatic effects but also the steric interactions of the latex particles (6). Currently, this combination of electrostatic and steric forces is called "electrosteric effects." This concept has been recently demonstrated in the case of aqueous polyurethane latex particles - which are electrosterically stabilized in high pH environment by copolymerization of carboxylic acid chains on the surface (7). The purpose of this paper is to demonstrate that the coating performance as determined by levels of streaks, using latex containing the carboxylic acids, can be explained by the "electrosteric effects" of the latex particles (8).

Experimental

Carboxylated latexes prepared by emulsion polymerization with butyl acrylate, methyl methacrylate, and methacrylic acid monomers are used in this investigation. The concentration of the latex solution used in coating experiments has been 34 w/w%. Particle sizes are measured by dynamic light scattering technique, zeta potential by ESA technique, flow properties by rheometry, and streaking propensity of latex coatings on clear polyester bases by image analysis. Each of these experimental techniques is now described as follows:

Latex Swell measurement by Dynamic Light Scattering   Dynamic   light scattering (also known as photon correlation spectroscopy of quasi-elastic light scattering) involves measurement of the time dependence of laser light scattering for a sample solution. The fluctuations of the scattered intensity are correlated by computer to yield a correlation function, which in turn is analyzed to give diffusion coefficients of the scattering bodies. Diffusion coefficients can be used to calculate the particle's hydrodynamic radius by the Stokes-Einstein equation. The light scattering experiments reported in this study were done on both Nicoli HN15-90 and Brookhaven BI-9000 systems.

Zeta Potential Measurement by Matec ESA System   The zeta potentials of the latex lots have been measured as function of amounts of 0.18N alkali added to titrate a fixed volume of these latex fluids using a Matec ESA 8000 system. The Matec ESA (Electrokinetic Sonic Amplitude) technique is based on an acoustic wave which is generated when a high frequency alternating electric field is applied to a pair of inert electrodes in a suspension of charged particles. The amplitude of the wave is dependent on the magnitude of the charges displaced per particle, the particle concentration and the amplitude of the electric field. In the unit used, the ESA is detected and converted to a voltage by a piezoelectric transducer. The zeta potential can be calculated from the magnitude of the voltage generated in the system.

Rheological Measurement   The shear stress vs shear rate experiments were performed by using Haake Rotovisco RV20-M10 Viscometer with a high shear sensor (HS1). This measuring system is best suitable for high shear and stability tests; thus, avoiding the usual change of measuring heads to get a complete flow curve. The thixotropic curves were generated by ramping the shear rate from 0 to 18,050 sec^{-1} in 5 minutes and then decelerating back to a complete stop in the same period of time.

Coating Experiments   A three-layer slide coating model was used to check the streaking propensity of latex materials. This model consisted of three layers - a latex layer as a carrier layer at a coverage of 2600 mg/m^2, a dye layer at a coverage of 900 mg/m^2, and a gel layer on the top at 2000 mg/m^2 - all coated on a clear polyester base. This model is represented in Figure 3. A cyan dye layer has been used in this case because the physical quality of the coating (i.e., the level of streaks) is easily visible in transmitted light through the clear polyester base.

Streak Level Evaluation   The level of streaks produced when different latex lots are coated on clear base was analyzed using gray level intensity measurements by transmitted light. The image was first captured by a CCD camera. The original image is passed through a lowpass filter to obtain a second image. The second image is subtracted from the first so that the "difference" image contains all the high-frequency information. The difference image is then averaged in the streak direction. The contrast for gray level close to the mean is next increased by density transformation (using lookup tables). The results are then sampled perpendicular to the streak direction, and the two extreme ends of the image are eliminated so that these would not affect the image processing quality. Finally, the standard deviation of the gray level histogram is calculated - which is a measure of the number and degree of streaks in the scanning area.

Results and Discussion

Colloid Stability   It is known that the stability of the polymer colloids can be related to the force balance between spherical particles (9,10). Here we consider three types of fundamental force which result in attraction and repulsion between a pair of colloidal particles. These are van der Waals attraction, electrostatic repulsion, and steric interaction. The evaluation of colloid stability is commonly made by calculating the potential energies with respect to distance between the particles. Figure 2 illustrates the interactions between two particles when approaching each other. The following section discusses the attractive and repulsive potentials resulted from these forces:

London-van der Waals Potential   It has been recognized that the van der Waals attraction becomes significant compared to the thermal energy, kT, of two particles at separations of the order of one-tenth of the diameter or less. The attractive potential thus, has an effective range of the order of 10 nm for our typical latex of 100 nm in diameter. In general, coagulation of such particles would occur when the net attraction at the minimum separation is several kT. For the particle size of 100 nm, the distance between the surfaces of particles should be larger than 4 nm to prevent the latex from aggregation. From the point of view of an individual particle, the thickness of the protective layer should be larger than about 2 nm in order to have a stable dispersion.

Electrostatic Potential   In an emulsifier-containing system, both chemically-bound surface charges from the sulfate groups of the initiator, and ionic groups from the adsorbed surfactant on the latex particle surface, help to stabilize the latex. For polymer latex particles containing acid groups, such as acrylic acid and methacrylic acid, the interpretation of their electrokinetic potentials presents uncertainties about the exact nature of the surface region of such particles during neutralization (11). Nevertheless,

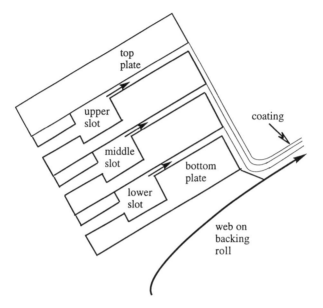

Figure 1. Schematic diagram of slide coating

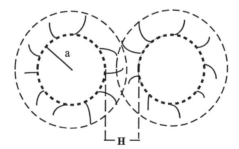

Figure 2. Two interacting particles with the radius a, carrying both electrostatic and steric layers approach each other separating by the distance H.

one of the most important quantities to emerge from the electrostatic charge effect is the double layer thickness which acts as an invisible shield for the latex particles.

The potential energy of repulsion from the charged surface may extend to appreciable distances toward the bulk solution, but its range is compressed by increasing the electrolyte content of the system. At low ionic strengths, the magnitude of the electrostatic potential is either comparable to or greater than the effective range of the van der Waals attraction, which provides a significant energy barrier against flocculation.

Steric Potential   As discussed earlier, polymer latex containing carboxylic acids exhibits a profound phenomenon of changing the surface charge effects through ionization process. Also, the hydrodynamic radius of latex particle increases the with increase in pH due to the ionization of carboxyl groups, breaking of hydrogen bonds and consequent adsorption of water molecules. The important variables which control the swellability of the carboxylated latices are the acid content, the hydrophilicity of the comonomers, the glass transition temperature and the molecular weight of the polymer.

An important characteristic of a carboxylated latex is the distribution of the carboxylic acid groups in the polymer particles. A high proportion of carboxylic acid groups at or near the surface of the latex particle significantly enhance the colloid stability of the latex. By contrast, acid groups which have become "buried" in the latex particles are less effective, at least until the particles are totally dissolved.

The hydrodynamic thickness of the extended polymer chains around the latex particles resulting from neutralization of carboxylic groups provides a critical steric force for colloidal stability. This is analogous to the steric stabilization of the adsorbed polymer on latex particles. However, the chemically combined carboxylate anions are more effective in providing the steric stabilization than the same number of carboxylate anions anchored to the polymer surface by adsorption.

It is important to note that the steric potential is radically different from electrostatic potential. The steric interaction usually is "shorter distance" due to its physical reach from the polymer chains; however, once the particles move into the domain of chain-chain interaction, the steric force increases at a dramatic rate. With an increase in length of polymer chain on the latex particle surface, the total potential energy becomes significantly repulsive in nature which prevents the latex particle from coagulation.

Multilayer Coatings and Osmotic Pressure Difference In a multilayer coating system, the bottom layer is the most important layer for determining the coatability limits because it is subjected to the highest shear rates occurring next to the coating head and the web. As indicated, a latex solution is used as a carrier layer in this application. Hence not only its flow behavior but also the interaction with the adjacent layer becomes an important factor in determining coating stability. The interface between the latex layer and the adjacent dye layer in the coating system used in this study may be considered as a semipermeable membrane. In this case, the water has a higher free energy in the latex layer and will flow from the latex layer to the dye layer. On the other hand, electrolytes will flow from the dye layer into the latex layer (See Figure 3). Water will continue to flow into the dye layer-until the pressure in the dye layer rises enough to hold back this flow. When the flow ceases, the system is in equilibrium. The osmotic pressure difference gives a measure of equilibrium concentration of water in each layer (12). One way to reduce the mass transfer of water from the latex layer to the dye layer (which would result in coagulation of the latex particles) would be to increase the electrolyte level in the latex layer, thus reducing the driving force for water and electrolyte transfer between these two adjacent layers. Reducing the water transfer from the latex layer would prevent coagulation of the latex particles on the applicator lip, and improve the physical quality of the coating (i.e., reduce the level of streaks). In instant film coatings, a highly conductive dye layer is usually coated adjacent to a less conductive latex layer. Therefore, an amount of 500 ppm of Na_2SO_4 was added to the latex solution during pH adjustment to bring up the conductivity of this layer.

Particle-Size Measurements   In order to determine the reproducibility of particle size measurement by dynamic-light scattering technique, particle sizes of two different lots were measured thirty times. The results are shown in Table I. The extreme tightness in the particle size measurement experiments is shown by the small value of the standard deviation. This shows the reproducibility of the measurements. From this an estimate is made of sample-size, so as to obtain particle size of latex lots with 95% confidence. It is determined that the particle size of each latex lot can be determined with 95% confidence by measuring it four times to achieve standard deviation of 0.6 nm.

Swell of Carboxylated Latex Particles   The swell in each latex lot (from pH 4.8 to 7.2) is presented in Table II. The particle sizes of different latex lots are measured at pH 4.8 and at pH 7.2. From Table II and Figure 4, it is observed that latex lots show different swelling characteristics as a function of pH. Lots 6 and 9 do not swell between pH 4.8 and 7.2, however other lots demonstrate various degrees of swell from 1 to 7 nm. It is postulated that at a low pH, the hydrogen of the carboxylic acid group may form a hydrogen bond to the nearest available electronegative atom. As hydroxyl concentration increases, carboxylic acid groups become ionized, breaking the hydrogen bonds as well as repelling each other. As this occurs, the molecules are allowed to relax, stretching outward from the anchor and increasing the hydrodynamic radius. The reason why different latex lots show variations in swelling behavior may be due to the difference in distribution of carboxylic acid groups on or near the surface of latex particles. These "hairy" molecules on the surface provide the steric protection for the latex particle.

Zeta Potential Measurements   Table III show the zeta potential data for different latex lots at pH 7.2. The zeta potential value varies between -11.0 mV and -14.0 mV for the latex lots studied. Figures 5 a and b show zeta potential of latex lots 2 and 10 as functions of pH of the latex solution. From Figures 5 a and b, it is observed that the zeta potential of latex 10 is initially slightly more than that of lot 2 (-14.5 mV as

THREE LAYER MODEL COATING SYSTEM

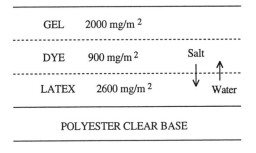

Figure 3. Interlayer diffusion due to osmotic pressure difference

Table I. Test for Reproducibility of Latex Particle Size Determination by Dynamic Light Scattering Measurements

Latex Lot Number	pH	Mean Diameter nm	Standard Deviation nm
A	4.8	98.5	0.6
B	4.8	100.6	0.6

Table II. Swelling of Latex Particle from pH 4.8 to 7.2

Latex Lot Number	Diameter, nm at pH=4.8	Diameter, nm at pH=7.2	Swell, nm
1	106	110	4
2	90	91	1
3	86	88	2
4	104	108	4
5	100	105	5
6	90	89	-1
7	94	101	7
8	97	101	4
9	86	86	0
10	105	108	3
11	100	102	2
12	97	100	3

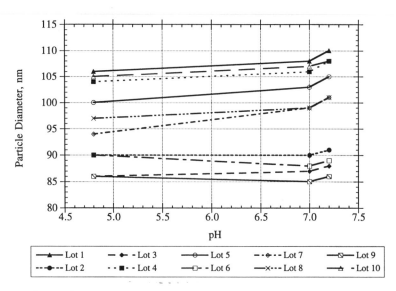

Figure 4. Particle diameter of different latex lots vs pH

Table III. Swell Number, Zeta Potential and Streak Standard Deviation
of Different Latex Lots

Latex Lot Number	Swell, nm	Zeta Potential at pH=7.2, mV	Streak Standard Deviation
1	4	-11.7	3.2
2	1	-12.7	5.7
3	2	-14.0	5.2
4	4	-13.1	3.1
5	5	-12.0	2.8
6	-1	-12.3	8.8
7	7	-11.8	1.9
8	4	-12.1	3.5
9	0	-13.7	6.1
10	3	-13.1	3.6

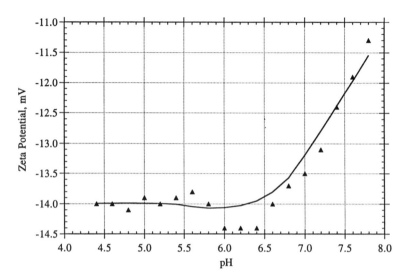

Figure 5(a). Zeta potential versus pH of latex [lot 2] solution

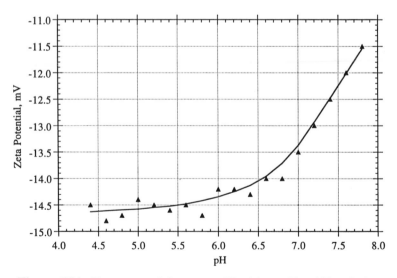

Figure 5(b). Zeta potential versus pH of latex [lot 10] solution

compared to -14 mV). This may indicate that there is initially slightly better electrostatic stabilization for lot 10. But with the addition of the alkali, the zeta potentials of both lots decrease in magnitude at the same rate, showing that electrostatic protection from double layer decreases at the same rate for both lots. The reduction in zeta potential may also be attributed to the desorption of the surfactants as acid groups convert into salts. The zeta potentials of other latex lots show similar characteristics as functions of pH of the latex solution. It is known from particle size measurements that lot 10 swells by 3 nm, where as lot 2 shows no swell. Although zeta potentials of various latex lots are similar, it is the additional swelling characteristic that prevents certain latex lots from coagulation during high shear coating as discussed in model calculations. Thus, electrosteric effects from both electrostatic and steric interactions between particles provides the stability for the colloidal latex systems.

Rheological Properties   In many cases, much of the diversity of the rheology of latex systems can be measured by simple shear flow experiments. For example, a thixotropic loop provides valuable information regarding the stability of the latex system under high shear conditions. During experiments a high magnitude fluctuation of the viscosity reading has been observed. Hence, the rheogram is shown in terms of shear stress as a function of shear rate. Figure 6a is a plot of shear stress versus shear rate for the carboxylated latex system at pH 4. Similar plots are shown in Figure 6 b, c, and d for the latex system at pH 5, 6, and 7 respectively. When the rheogram for a latex lot shows a high degree of chatter especially shown in Figure 8a, it has been observed that the latex solution has turned into severe coagulant after being subjected to high shear. This same phenomenon has been noticed for every lot of latex under investigation. In many applications, coagulation occurs under high shear conditions and particle collisions are predominantly caused by fluid flow (orthokinetic coagulation) rather than by Brownian diffusion (perikinetic coagulation) (13). This is why the latex solution can be stored for a long period of time at low pH, although it is not rheologically stable in the high shear environment. The thixotropic loop experiments also demonstrate that the shear instability occurs during the ramp up of shear and the magnitude of chatter reduces as pH increases from 5 to 7 shown in Figure 6 b, c, and d. One interesting finding is that except at pH 4, the ramp down curves show the unperturbed flow behavior suggesting the break down of the aggregate after very high shear. Nevertheless, the curves of the shear stress versus shear rate at pH 7 are the smoothest profiles without the presence of a large amount of chatter. This indicates that when latex swells at pH 7, the latex system is more stable and does not coagulate in the high shear conditions.

Coating and Imaging Analysis   The particle sizes of different latex lots are measured at pH 4.8 and at pH 7.2. The difference in particle size at these two pHs is characterized as the swell. The swell (from pH 4.8 to 7.2) of latex lots used in the first set of model-system coating tests, and the corresponding standard deviations in a gray-level histogram are shown in Table III. The standard deviation of a gray-level histogram is obtained from the coatings on clear base using the technique described earlier. This

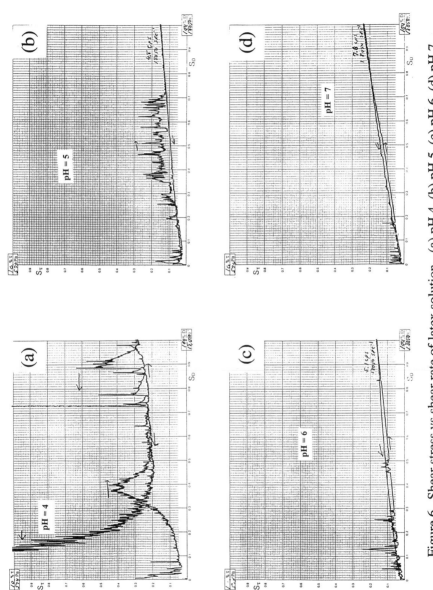

Figure 6.   Shear stress *vs* shear rate of latex solution.   (a) pH 4, (b) pH 5, (c) pH 6, (d) pH 7

standard deviation is an indication of the streak characteristics of the coating; i.e., higher the standard deviation higher is the level of streaks.

Results from a series of coating tests are shown in Figure 7. Figure 7 is a plot of the relationship between standard deviation in gray level intensity versus swell in the latex particle (from 4.8 to 7.2). It is observed that with increased swelling of the latex particles, there is a decrease in the standard deviation of gray level. This signifies that latex lots that swell at least 2 nm (14) (when pH adjusted from 4.8 to 7.2) or more, produce relatively lower level of streaks, while latex particles that do not swell produce coatings with higher levels of streaks. These results support the hypothesis that expansion of carboxylic acid chains on the latex surface produced by neutralization (15) provides the necessary steric stabilization of latex particles, thus preventing coagulation and producing lower levels of streaks in the coating.

Conclusion

The performance of latex solution used as a carrier layer in the multilayer water-borne coating system depends on colloidal stability of latex particles in high shear and high electrolyte environment. In this study a latex layer was coated adjacent to a highly conductive dye layer with a gel layer on the top to simulate actual photographic coating situations. It is recognized that the stability/aggregation of polymer colloids in aqueous solutions is a consequence of the force balance between the particles. In this system, van der Waals potential is the primary force for attraction, whereas electrostatic and steric potentials are the counterbalancing forces for separating the particles. Based on London dispersion theory, the van der Waals attractive potential is essentially proportional to the Hamaker constant, hence there is little opportunity to alter its influence on stability since the Hamaker constant originates from the nature of the materials in the system. As a result, most investigations have been focused on the possibility of modifying electrostatic and steric potentials. DLVO theory, which provides a quantitative description of the interactions between charged particles, consists of two important features for the electrostatic stabilization. The first is that the higher the potential at the surface of a particle and throughout the double layer, the larger the repulsion between the particles. The second feature is that as the concentration of an indifferent electrolyte is decreased, the distance from the surface at which repulsion significantly drops becomes greater. This concept has been applied in industry for many years; however, in some situations, such as when high electrolyte concentration is needed for balancing the osmotic pressure and high shear flow required for high-speed coating, the electrostatic repulsion potential may not be sufficient to prevent particle agglomeration. The modern approach is to enhance the steric potential either by the copolymerization of a hydrophilic polymer or by the adsorption of water soluble polymers on the surface of the particles. The steric barrier is the result of both an enthalpic and an entropic effect during interpenetration and compression of the polymeric chains. The advantage of steric stabilization is that the polymer chain-chain interactions are less sensitive to ionic species in the system compared to electrostatic interactions. The potential of a steric layer is usually short in range; however, it has a higher magnitude when it is brought into play. If the attraction potential beyond the

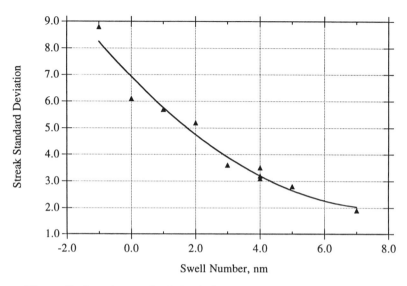

Figure 7. Streak standard deviation vs swell of latex particles

region of steric repulsion is of significant magnitude, coagulation of such particles will occur, despite the presence of an assumed infinitely strong steric layer. Results from the current experiments suggest that latex lots which swell more from pH 4.8 to 7.2 coat with lower levels of streaks. It can be concluded from this study that swelling characteristics of the latex particles is an important parameter which can provide the qualitative information about the streaking propensity of these latex lots. It is postulated that carboxylic acid chains on the surface of the latex particles swell when these latex solutions are pH adjusted from 4.8 to 7.2. The growth in the carboxylic acid chains provides the necessary steric stabilization to latex particles, which in addition to electrostatic protection, prevent the particles from coagulation during high-shear coating operations.

Acknowledgments

We are grateful to Dr. S. Herchen for his support on this investigation. Thanks are also due to Dr. B. Holland, M. Navarro, and pilot coating members for their experimental assistance. The support of literature search by Dr. D. Rickter is gratefully acknowledged.

Literature Cited

1. Walworth, V. K.; Mervis, S. H. In *Imaging Processes and Materials: Neblette's 8th Edition*; Sturge, J.; Walworth, V.; Shepp. A., Ed.; Van Nostrand Reinhold: New York, NY, 1989; pp 181-225.
2. Chen, K. S. A. PhD Thesis, University of Minnesota, 1992.

3. *Modern Coating and Drying Technology*; Cohen, E.; Gutoff, E., Eds.; VCH: New York, NY, 1992.

3. Costello, B. A.; Luckham, P. F.; Tadros, Th. F. *J. Colloid Interface Sci.* 1992, 152, pp. 237

4. Bassett, D. R.; Hoy, K. L. In *Polymer Colloids II*, Fitch, R. M., Ed., Plenum Press: New York, NY, 1980, pp 1-25.

5. Sato, T.; Ruch, R. *Stabilization of Colloidal Dispersions by Polymer Adsorption*; Marcel Dekker, New York, NY, 1983

6. Heijman, S.G. and Stein, N. H. *Chem. Eng. Sci.* 1993, 48, pp 313.

7. Satguru, R.; McMahon, J.; Padget, J.C.; Coogan, R.G. *J. Coatings Tech.* 1994, 66, pp 47.

8. Einarson, M.; Berg, J. C. *J. Colloid Interface Sci.* 1993, 155, 165.

9. Goodwin, J. W. In *Water-Borne Coatings*; Glass, J. E. Ed.; Chapter 3, 1996.

10. Buscall, R. In *Science and Technology of Polymer Colloids*; Poehlein, G. W.; Ottewill, R. H.; Goodwin, J. W., Ed.; NATO ASI Series E 68; Martinus Nijhoff: Hague, Netherlands, 1983, Vol. 2; pp 279-313.

11. Rosen, L. A.; Saville, D. A. *J. of Colloid Interface Sci.* 1992, 149, pp 542-552.

12. Cussler, E. L. *Diffusion*, Cambridge University Press: Cambridge, England, 1984, pp 378-386.

13. van de Ven, T. G. M.; Manson, S. G. *J. Colloid Interface Sci.* 1976, 57, pp 505.

14. Fowkes, F.M.; Pugh, R. J. In *Polymer Adsorption and Dispersion Stability*, Goodard E.D.; Bincent, B. Ed., ACS Symp. Ser. 240, Washington, DC, 1984, pp 331-354.

15. Yokoyama, A.; Srinivasan, K.R.; Fogler, H. S. *J. of Colloid Interface Sci.* 1988, 126, pp 141-149.

Chapter 12

Spectroscopic Studies of Surfactant Mobility and Stratification in Films from Homopolymer Latex Blends

Amy P. Chu, Lara K. Tebelius, and Marek W. Urban[1]

Department of Polymers and Coatings, North Dakota State University, Fargo, ND 58105

Behavior of sodium dioctylsulfosuccinate (SDOSS) surfactant and stratification processes in poly(butyl acrylate)/polystyrene (p-BA/p-Sty) latex blends were investigated using attenuated total reflectance (ATR) Fourier transform infrared (FT-IR) spectroscopy. These studies indicate that at the early stages of coalescence, SDOSS surfactant exudes to the film-substrate (F-S) interface and its $SO_3^-Na^+$ hydrophilic heads are oriented preferentially parallel to the film surface. Although the p-Sty phase shows no stratification at either interface, p-Sty rings assume preferentially perpendicular orientations at both interfaces. However, at extended coalescence times, surfactant molecules migrate to the film-air (F-A) interface, and maintain their parallel orientation. At the same time, the p-Sty phase forms a stratified layer at approximately 1.4 µm from the F-A interface. At this depth, the p-Sty rings change their orientation to become preferentially parallel to the surface.

Interactions between latex components and surfactants, along with other factors that influence surfactant mobility, represent critical issues for latex film formation because their magnitude may ultimately influence macroscopic film characteristics. For that reason, our recent studies[1-13] focused on the behavior of sodium dioctylsulfosuccinate (SDOSS) surfactant molecules in latex matrices. The primary focus of these studies was the examination of surface and interfacial structures formed in ethyl acrylate/methacrylic acid (EA/MAA, 4 % w/w) and styrene/n-butyl acrylate (Sty/n-BA) latex copolymer interfaces. It became apparent that the mobility and structural changes of SDOSS after coalescence near the film-air (F-A) and film-substrate (F-S) interfaces may alter numerous physical and chemical film properties.

[1]Corresponding author

We extended these studies to examine the behavior of a 50/50 mixture of separately homopolymerized polystyrene (p-Sty) and poly(butyl acrylate) (p-BA) blended latex dispersions, mixed in equal volume ratios prior to coalescence.[14] Although the primary interest was to understand SDOSS distribution across the film thickness, these studies opened up another avenue concerned with the stratification processes of individual latex components that may occur during or after coalescence. It appears that the presence of hard and soft latex particles does influence the distribution and mobility of individual components within the films,[14] and the phase separation is not a uniform phenomenon across the film thickness, but it occurs in the direction normal to the film surface. Furthermore, the presence of core/shell type of particles may influence the direction of surfactant migration.[15]

In this study, we will expand the scope of previous findings and alter the latex composition by using homopolymerized polystyrene (p-Sty) and poly(ethyl acrylate) (p-EA) blended latex suspensions which, upon mixing, will be allowed to coalesce. This approach will allow us to investigate surfactant-polymer interactions, and how latex composition will affect the distribution, mobility, and orientation of SDOSS. The 1046 and 1056 cm^{-1} bands due to the splitting of the 1050 cm^{-1} band, resulting from the S-O stretching of $SO_3^-Na^+$ hydrophilic ends on SDOSS will be used to trace these groups near the F-A and F-S interfaces. Similar to the previous studies, we will utilize attenuated total reflectance Fourier transform infrared (ATR FT-IR) spectroscopy along with the Q-ATR algorithm[16] to perform surface depth profiling experiments. Monitoring the intensity changes of the 700 cm^{-1} band will allow us to follow the behavior of p-Sty within the p-EA/p-Sty blended latex films.

EXPERIMENTAL

LATEX PREPARATION

Ethyl acrylate (EA) and styrene (Sty) monomers were individually polymerized using a previously reported semi-continuous emulsion polymerization process.[3,13] After synthesis of separate batches of poly(ethyl acrylate) (p-EA) and polystyrene (p-Sty), the homopolymers were mixed in a 50/50 w/w % ratio, stirred, and allowed to store for 3 days prior to coalescence. Such latex blends were deposited on a polytetra-fluoroethylene (PTFE) substrate to achieve a film thickness ranging from 100 to 150 μm, and allowed to coalesce for 24 to 72 hours under 40% relative humidity.

SPECTROSCOPIC ANALYSIS

ATR FT-IR spectra were recorded on a Mattson Sirus 100 spectrometer equipped with a variable angle rectangular ATR attachment (Spectra Tech) with a KRS-5 crystal. Typically, 200 coadded sample scans were acquired at a resolution of 4 cm^{-1}, and ratioed against the same number of scans of a single beam spectrum of an empty ATR cell. Polarization experiments were accomplished using a Specac 12000 IR polarizer. All spectra were corrected for optical distortions using recently developed algorithms allowing simultaneous corrections for optical effects of strong and weak bonds.[16,17] All spectra were normalized to the band at 853 cm^{-1} (not shown) which is due to the C-C skeletal modes of copolymer main chain.[18, 19]

RESULTS AND DISCUSSION

Previous studies[1-14] have established that the presence of the 1046 and 1056 cm^{-1} bands in the ATR FT-IR spectra recorded from the film-air (F-A) and film-substrate (F-S) interfaces is attributed to the splitting of the S-O stretching band at 1050 cm^{-1} of the SO$_3$⁻Na$^+$ hydrophilic groups of sodium dioctylsulfosuccinate (SDOSS). The splitting results from the formation of the following molecular entities: the association of SO$_3^-$ Na$^+$ entities with H$_2$O and acid groups. The hydrophilic end of SDOSS and its associations are illustrated below:

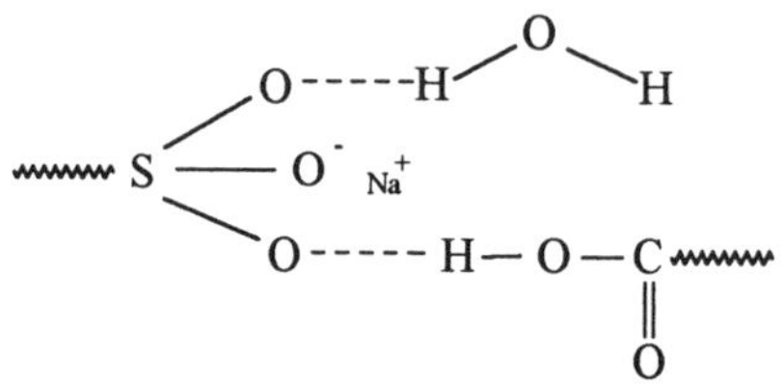

We also have determined that, among other factors, the simultaneous presence of hard and soft particles in a latex suspension produces stratification processes in latex films which may occur during and after coalescence.[14] In this study, we will analyze the exudation of SDOSS surfactant to the interfaces of p-EA/p-Sty blended latex films, with the focus on mobility and orientation of the surfactant molecules as a function of coalescence times, and how stratification processes of individual latex particles may influence surfactant concentration at the F-A and F-S interfaces.

To set the stage, let us examine a series of ATR FT-IR spectra obtained from the interfaces of a 50/50 p-EA/p-Sty blended latex films, which were allowed to coalesce for 24 hours under 40% relative humidity. Figure 1, traces A and C, show the transverse magnetic (TM) polarized spectra recorded from the F-A and F-S interfaces, respectively. The most pronounced spectral changes are detected at the F-S interface (traces C and D), where the presence of the 1046 cm^{-1} surfactant band in TM and TE polarizations is detected. A comparison of the TE (Trace D) and TM (Trace C) polarized spectra indicates that the band enhancement in the TE polarized spectrum results from orientation of the surfactant molecules. In this experimental setup, the TE wave is defined as having its electric vector parallel to the crystal plane (or perpendicular to the plane of incidence), and the TM wave has its electric vector perpendicular to the crystal plane. A pictorial definition of the TE and TM polarizations, and the orientation at the electric vector of the incident beam with respect to the film surface at the ATR crystal, are shown in Figure 2.[8,9,13,14,19] The enhanced intensity of the 1046 cm^{-1} band in the TE polarization indicates that the SO$_3^-$ Na$^+$ groups of SDOSS are preferentially oriented parallel to the F-S interface. These

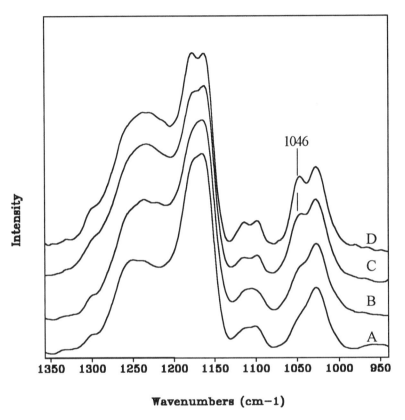

Figure 1. ATR FT-IR spectra of a 50/50 p-EA/p-Sty blended latex in the 1350-950 cm^{-1} region after 24 hours of coalescence. A) F-A, TM polarization; B) F-A, TE polarization; C) F-S, TM polarization; D) F-S, TE polarization.

results support our previous findings[1] which indicated that the initial surfactant migration to the F-S interface is driven to alleviate interfacial tension between a substrate and latex suspension. These results also confirm preferentially parallel orientation of the sulfonate groups at the F-S interface.[13]

For comparison purposes, let us now examine ATR FT-IR spectra recorded at the film interfaces after 72 hours of coalescence under 40% relative humidity. In contrast to the data presented in Figure 1, Figure 3 shows that the appearance of the 1046 and 1056 cm^{-1} bands is detected at the F-A interface (traces A and B). Along with the surfactant bands at 1046 and 1056 cm^{-1}, several other bands are enhanced in the spectra recorded from the F-A interface using TM and TE polarizations.

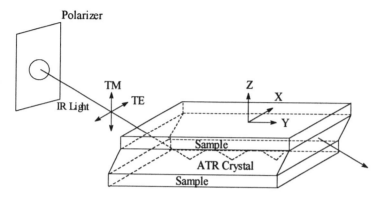

Figure 2. Schematic diagram of an ATR FT-IR experimental setup.

These bands are detected at 1207, 1233, 1261, and 1288 cm^{-1}. As was previously determined,[13,19] the 1207 and 1261 cm^{-1} bands enhanced in the TE polarization result from the splitting of the 1216 cm^{-1} band which is attributed to the asymmetric S-O stretching modes of the sulfonate groups of SDOSS. The 1233 and 1288 cm^{-1} bands, enhanced in the TM polarization, are due to the splitting of the 1241 cm^{-1} band, and are attributed to the C-O stretching of the surfactant backbone.

This information, along with the band intensity changes under different polarization conditions, indicates that when water is present in the system, hydrophilic $SO_3^-Na^+$ groups are oriented preferentially parallel near the F-A interface, whereas hydrophobic $(CH_2)_n$ surfactant backbone exhibits preferentially perpendicular orientation to the F-A interface which minimizes the "cross section" of the SDOSS molecules to allow water molecules to more freely migrate to the surface and diffuse out of the film.

Even though these conclusions agree with the previously reported results on p-BA/p-Sty latex films,[13] there are several differences between two latex systems which will inherently influence dynamics of the surfactant migration. Besides composition, the primary difference between p-BA/p-Sty and p-EA/p-Sty latexes is the time required for $SO_3^-Na^+$ heads to orient preferentially parallel to the F-A interface, and for the hydrocarbon tails to orient perpendicular to this interface. While in p-EA/p-Sty the hydrocarbon tails follow orientation of the sulfonate heads during early stages of coalescence, it takes about 15 days for the hydrocarbon tails to follow sulfonate orientation in p-BA/p-Sty. This transient effect of SDOSS tails results from the fact that the p-BA/p-Sty latex system was composed of a 5:1 ratio of p-BA/p-Sty,[10] whereas the p-EA/p-Sty latex is a 50/50 mixture. Furthermore, having excessive amounts of p-BA, which has a lower glass transition temperature than the p-EA component, provides a significantly higher degree of free volume, thus permitting a random orientation of the hydrocarbon tails until water is removed from the latex. Therefore, hydrocarbon tails take longer time to position themselves against the water flux.

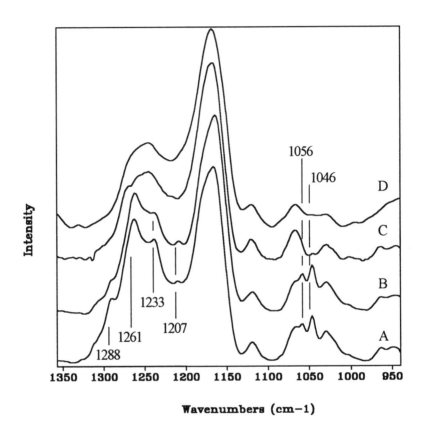

Figure 3. ATR FT-IR spectra of a 50/50 p-EA/p-Sty blended latex in the 1350-950 cm⁻¹ region after 72 hours of coalescence.   A) F-A, TM polarization; B) F-A, TE polarization; C) F-S, TM polarization; D) F-S, TE polarization.

Let us focus on molecular level changes as a function of depth from the F-A and F-S interfaces. Figure 4, Traces A through E, illustrate spectra recorded after 72 hours of coalescence using TE polarized light, and obtained at angles of incidence between 60° and 40°. By using this range of incidence angles, we are able to vary the depth of penetration of light into the film surface from 1.3 μm to 2.3 μm.[19,20] Thus, molecular level information from different depths within the film can be obtained; specifically, the distribution of surfactant molecules near the F-A interface after 72 hours of coalescence. Following the results shown in Figure 4, Trace A, which exhibits a spectrum recorded from the shallowest depths, the surfactant bands of 1046 and 1056 cm^{-1}, resulting from the association of $SO_3^-Na^+$ entities with H_2O and acid groups, respectively, exhibit the highest intensities. As the depth of penetration increases (Figure 4, traces B through E), the band intensities decrease. These observations indicate that the highest concentration of SDOSS molecules is detected at the F-A interface, and decreases at greater depths. As water-soluble SDOSS surfactant molecules migrate with the water flux toward the F-A interface, the surfactant stays at this interface, while water diffuses out of the film. In contrast, the studies on poly(ethyl acrylate)/methacrylic acid (EA/MAA) latexes[8] determined that when coalescing particles came into contact, surfactants are displaced into the aqueous phase, where they become free to migrate with the water front moving toward the F-A interface. Once at the interface, surfactants become trapped within partially coalesced particles, while water continues to diffuse out of the film. In the case of p-BA/p-Sty,[14] the presence of the surfactant bands at 1046 and 1056 cm^{-1} at various depths from the F-A interface was detected, with the highest band intensities at about 1.3 μm. At greater depths, a significant decrease of intensities was observed, indicating that the largest distribution of surfactants in the p-BA/p-Sty suspension exists at the F-A interface. As shown above, the same trends are detected for p-EA/p-Sty latex suspensions.

The next question is if surfactant distribution in 50/50 p-EA/p-Sty blended latex films is affected by stratification processes which may occur during or after coalescence. As we recall, stratification processes have been detected for hard p-Sty and soft p-BA particles.[14] In this case, monitoring intensity changes of the 700 cm^{-1} band resulting from the aromatic out-of-plane C-H normal deformation modes of styrene,[19,20] distribution of the p-Sty latex component can be monitored. For the p-EA/p-Sty latex system, we will take advantage of these findings and follow the p-Sty phase by monitoring the same band. Figure 5 shows ATR FT-IR spectra obtained from the interfaces recorded in both polarizations after 24 hours of coalescence. It appears that the 700 cm^{-1} band intensity is stronger in the spectra recorded from the F-A interface (traces B and D). This observation indicates that, initially, there is a greater concentration of p-Sty component at the F-A interface. Again, this observation agrees with the previous findings on p-BA/p-Sty latex films,[14] where hydrophobic p-Sty phase stratified near the F-A interface allowing the p-EA phase exposure to humidity. Furthermore, the 700 cm^{-1} band is enhanced in the TM polarization at the F-A and F-S interfaces (traces C and D), indicating that the styrene rings of p-Sty are oriented preferentially perpendicular to the F-A interface. Their perpendicular orientation results from the fact that when residual water molecules diffuse out of the film, hydrophobic styrene units facilitate this process.

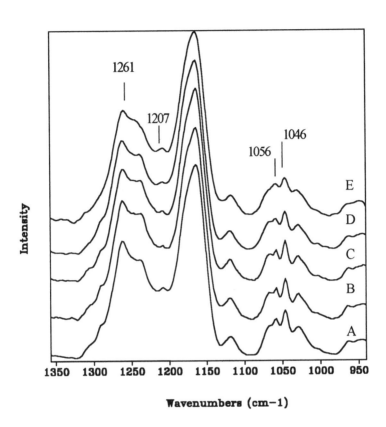

Figure 4.   ATR FT-IR spectra of a 50/50 p-EA/p-Sty blended latex in the 1350-950 cm^{-1} region at the F-A interface with TE polarization at various depths after 72 hours of coalescence. A) 1.3 μm; B) 1.4 μm; C) 1.6 μm; D) 1.9 μm; E) 2.3 μm.

Based on these data, the following scenario can be proposed for 50/50 p-EA/p-Sty blended latex films after 24 hours of coalescence.   Surfactant molecules occupy predominantly the F-S interface, with $SO_3^- Na^+$ hydrophilic groups and hydrophobic tails taking preferentially parallel orientation. The presence of p-Sty is detected at both interfaces, however, higher concentration levels are present at the F-A interface and p-Sty rings of the p-Sty phase are oriented preferentially perpendicular. This is schematically depicted in Figure 6.

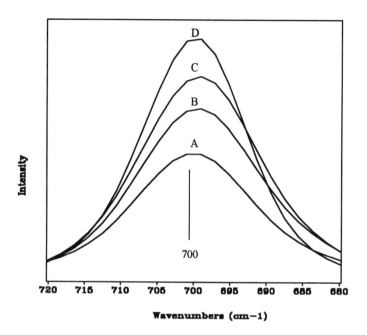

Figure 5. ATR FT-IR spectra of a 50/50 p-EA/p-Sty blended latex in the 720-680 cm^{-1} region after 24 hours of coalescence.   A) F-A, TM polarization; B) F-S, TE polarization; C) F-A, TM polarization; D) F-S, TM polarization.

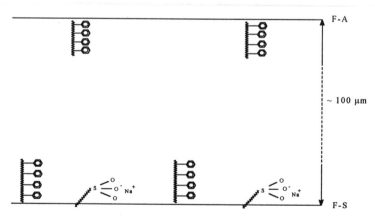

Figure 6.   Schematic representation and location of SDOSS surfactant molecules and styrene in a 50/50 p-EA/p-Sty blended latex after 24 hours coalescence.

With this in mind, let us extend coalescence times to 72 hours. Examination of the 700 cm^{-1} band of the same films after 72 hours of coalescence, which is shown in Figure 7, indicates that the concentration of p-Sty at the F-A interface (traces C and D) is higher as compared to the 24 hour coalescence data (Figure 5), and the orientation of styrene rings changes. Furthermore, the intensity changes of the 700 cm^{-1} band in the TE and TM polarizations shown in traces B and D of Figure 7 indicate that the styrene rings assume a preferentially parallel orientation to the film surfaces. As compared to p-EA phase, p-Sty rings are hydrophobic, and at the early stages of coalescence, will assume perpendicular orientation, in order to facilitate water flux from the film. After water is removed, the styrene rings became parallel.

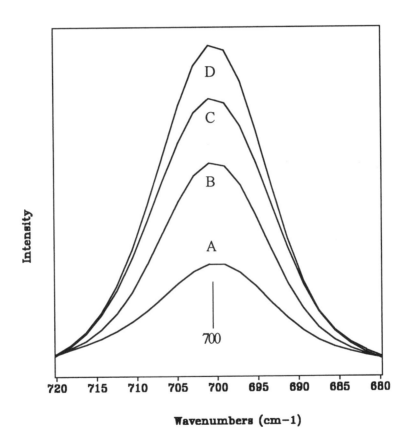

Figure 7. ATR FT-IR spectra of a 50/50 p-EA/p-Sty blended latex in the 720-680 cm^{-1} region after 72 hours of coalescence. A) F-S, TM polarization; B) F-S, TE polarization; C) F-A, TM polarization; D) F-A, TE polarization.

Let us further examine the p-Sty distribution at the F-A interface at various depths. By changing the angle of incidence, thus changing the depth of penetration of light into the specimen, the concentration of p-Sty also changes. Figure 8 shows that the 700 cm^{-1} band reaches its maximum intensity at approximately 1.4 μm from the F-A interface, and decreases at greater depths. Although these results agree with the previous studies for the 50/50 p-BA/p-Sty blended latex films,[14] where p-Sty layers were observed at 1.6 μm, for 50/50 p-EA/p-Sty blended latex films, phase separation of a p-Sty layer is detected at approximately 1.4 μm from the F-A interface. This change in distance from the surface is attributed to the length of the acrylate chains; which exhibit higher compatibility with SDOSS.

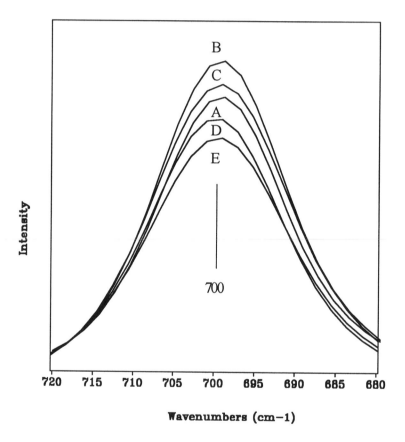

Figure 8. ATR FT-IR spectra of a 50/50 p-EA/p-Sty blended latex in the 720-680 cm^{-1} region at the F-A interface with TE polarization at various depths after 72 hours of coalescence. A) 1.3 μm; B) 1.4 μm; C) 1.6 μm; D) 1.9 μm; E) 2.3 μm.

Extensive research efforts have been made to understand compatibility of components in latex films[1-7] and numerous studies have shown that the compatibility between p-BA latex and SDOSS is higher than that for p-EA latex and SDOSS. This, in turn, reduces the ability of SDOSS to exude in p-BA copolymer latex films. Furthermore, neutralization of the acid groups in p-EA copolymers causes swelling of the latex particles, and extension of the hydrophobic segments into the aqueous phase, which facilitates a greater degree of interactions between copolymer and surfactant hydrophobic ends. Ultimately, this process results in greater overall compatibility. Since p-BA has longer pendant hydrophobic segments (butyl versus ethyl), their presence will enhance hydrophobic interactions between the copolymer and surfactant, and thus will enhance compatibility. For that reason, p-Sty layers in p-EA/p-Sty stratify at 1.4 μm from the F-A interface, whereas for the p-BA/p-Sty blends, stratification was detected at 1.6 μm. Because p-BA has longer hydrophobic segments, it is more compatible with surfactant hydrophobic ends detected at this interface. This reduces exudation of surfactant molecules to the F-A interface, and increases their concentration below the surface.

In summary, these as well as previous studies on p-BA/p-Sty blended latexes demonstrated again that there are significant differences in the latex surface and interfacial properties, depending upon the latex composition. In latex films composed of individual homopolymer particles, hard p-Sty and the soft p-EA particles exhibit a significant degree of phase separation, which occurs near the F-A interface, and stratification of the p-Sty phase is detected. The phase separation during film formation not only influences the mobility and orientation of SDOSS surfactant molecules and p-Sty rings within the latex film, but occurs in the direction normal to the film surface. A schematic representation of stratification processes for 50/50 p-EA/p-Sty blended latex is depicted in Figure 9. After 72 hours of coalescence, the surfactant migrates with the water flux to the F-A interface, and maintains its preferentially parallel orientation. A stratified layer of p-Sty occurs at 1.4 μm from the F-A interface, and the styrene rings change their orientation, and become preferentially parallel to the F-A surface.

Finally, it is appropriate to compare the behavior of the latex systems composed of a 50/50 homopolymer mixture with that of a 50/50 latex copolymer. In both cases, the surfactant presence is initially detected to be the highest at F-S interfaces and its presence has been attributed to the interfacial surface tension between the latex and the substrate.[12,13] Furthermore, hydrophilic $SO_3^-Na^+$ heads of the surfactant are preferentially parallel to the F-S surface. However, as the coalescence process continues, surfactant molecules in both latex systems migrate with the water front toward the F-A interface, and the surfactant distribution is highest near the F-A interface. Once at the F-A interface, the surfactant sulfonate heads maintain their preferentially parallel orientation to the surface. Whereas for the homopolymer mixture, the hydrophobic $(CH_2)_n$ tails exhibit preferentially perpendicular orientation to the surface, in the copolymer latexes, the tails have a random orientation.

Compatibility of latex copolymers and homopolymerized components further affects latex properties. While homopolymerized latex mixtures coalesce into a non-

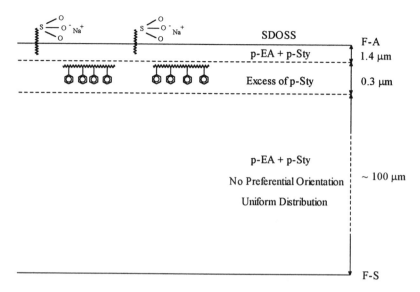

Figure 9. Schematic representation and location of SDOSS surfactant molecules and styrene in a 50/50 p-EA/p-Sty blended latex after 72 hours coalescence.

uniform composite of hard and soft particles near the F-A interface, with two separate T_gs, each representing individual components; the copolymer latex exhibits only a single T_g. Thus, phase separation in the homopolymer mixture is evident. However, it occurs in such a way that the layers with excessive amounts of p-Sty are present at depths of 1.4-1.6 μm from the F-A interface. For the copolymer latexes, the distribution of p-Sty was uniform across the film. As a result, homopolymer mixtures and latex copolymers exhibit different mobility and orientation of surfactants during coalescence.

CONCLUSIONS

In this study, stratification processes that occur during latex film formation and their effect on the distribution of SDOSS surfactant molecules within the latex film were examined. After 24 hours of coalescence, SDOSS surfactant molecules are detected at the F-S interface and $SO_3^-Na^+$ hydrophilic groups are preferentially parallel, whereas the p-Sty phase is present at both interfaces, without stratification at either interface and a preferentially perpendicular orientation to the F-S interface. Upon extending coalescence times to 72 hours, SDOSS surfactant molecules migrate to the F-A interface with the water flux, where the hydrophilic $SO_3^-Na^+$ groups maintain their preferentially parallel orientation to the film surface, and the hydrophobic tails have a preferentially perpendicular orientation. These processes are accompanied by a phase separation of the blended p-EA and p-Sty homopolymers which occurs at the F-A interface. The p-Sty phase separation occurs at a depth of approximately 1.4 μm from the F-A interface and p-Sty rings are oriented parallel to the surface.

These studies also show that ATR FT-IR spectroscopy can be effectively used in the analysis of polymeric surface and interfacial systems when the penetration depths do not exceed 3 - 4 μm. When molecular level information from greater depths is sought, step-scan photoacoustic FT-IR[20] provides a better, although at this stage not a quantitative tool.

ACKNOWLEDGEMENTS

The authors (APC) are thankful to the 1995 North Dakota Governor's School for Science and Mathematics directed by Dr. Allan G. Fischer, Dean of the College of Science and Mathematics, and to numerous industrial sponsors for their financial support (LKT and MWU).

REFERENCES

1. M.W. Urban and K.W. Evanson, *Polym. Comm.*, **31**, 279 (1990).
2. K.W. Evanson and M.W. Urban, in Surface Phenomena and Fine Particles in Water-Based Coatings and Printing Technology, M.K. Sharma and F.J. Micale, Eds., Plenum: New York, 1991, P. 197.
3. K.W. Evanson and M.W. Urban, *J. Appl. Polym. Sci.*, **42**, 2287 (1991).
4. K.W. Evanson, T.A. Thorstenson, and M.W. Urban, *J. Appl. Polym. Sci.*, **42**, 2297 (1991).
5. K.W. Evanson and M.W. Urban, *J. Appl. Polym. Sci.*, **42**, 2309 (1991).
6. T.A. Thorstenson and M.W. Urban, *J. Appl. Polym. Sci.*, **47**, 1381 (1993).
7. T.A. Thorstenson and M.W. Urban, *J. Appl. Polym. Sci.*, **47**, 1387 (1993).
8. T.A. Thorstenson, L.K. Tebelius, and M.W. Urban, *J. Appl. Polym. Sci.*, **49**, 103 (1993).
9. T.A. Thorstenson, L.K. Tebelius, and M.W. Urban, *J. Appl. Polym. Sci.*, **50**, 1207 (1993).
10. J.P. Kunkel and M.W. Urban, *J. Appl. Polym. Sci.*, **50**, 1217, (1993).
11. T.A. Thorstenson, K.W. Evanson, and M.W. Urban, in Advances in Chemistry Series #236, M.W. Urban and C.D. Craver, Eds., American Chemical Society, Washington, DC, 1993.
12. B.-J. Niu and M.W. Urban, *J. Appl. Polym. Sci.*, **56**, 377, (1995).
13. L.K. Tebelius and M.W. Urban, *J. Appl. Polym. Sci.*, **56**, 387 (1995).
14. L.K. Tebelius, E.M. Stetz, and M.W. Urban, *J. Appl. Polym. Sci.*, **62**, 1887, (1996) .
15. L.R. Martin and M.W. Urban, *J. Appl. Polym.Sci.*, **62**, 1893, (1996).
16. M.W. Urban, ATR Spectroscopy of Polymers - Theory and Practice, American Chemical Society, Washington, DC, 1996.
17. M.W. Urban, Vibrational Spectroscopy of Molecules and Macromolecules on Surfaces, Wiley-Interscience, New York, 1993.
18. G. Socrates, Infrared Characteristic Group Frequencies, Wiley-Interscience: New York, 1980, pp. 83-85.
19. B.J.Niu and M.W. Urban, *J. Appl. Polym.Sci.*, **62**, 1893, (1996).
20. B.-J. Niu, L.R. Martin, L.K.Tebelius, and M.W. Urban, in Film Formation in Waterborne Coaitngs, Eds. T.Provder, M.A.Winnik, and M.W.Urban, ACS Symp. Series # 648, American Chemical Society, Washington DC, 1996.

Chapter 13

Development of Porous Structure During Consolidation of Pigmented Paper Coatings

Anna Stanislawska[1] and Pierre LePoutre[2]

[1]Institute of Papermaking and Paper Machines, Technical University of Lodz, Zwirki 36, 90–924 Lodz, Poland
[2]Department of Chemical Engineering, University of Maine, Orono, ME 04469

Recent advances in the study of the consolidation of pigmented coatings, based on clay or ground calcium carbonate and latex or protein, and formulated above the critical pigment volume concentration, are discussed. From SEM examination and measurements of gloss, void fraction, surface area and light-scattering on the freeze-dried coatings the process of development of the porous structure is followed. The overall process is basically the same, independent of pigment shape and binder type or content. Initially the formation of a loose structure with large voids is observed. At a first critical pigment volume concentration (CPVC) the void size decreases to a minimum. Coalescence or adsorption of polymer is accompanied by a decrease in surface area, and shrinkage and an increase in pore size occurs. After a second critical pigment concentration, the pores empty. Additionally, pigment shape and binder type have been shown to affect the consolidation process and the final properties of coating layers.

Pigmented coatings are applied on paper or paperboard to improve final product properties such as brightness, opacity, print quality. Most conventional paper coating formulations consist of a dispersed pigment phase and a dispersed or dissolved binder phase with water as a vehicle. Coating colors also contain small amount of other additives, like dispersant and water retention agents. For most printing papers, the coating formulation is similar to that of a latex-based ceiling paint.

Pigment is the main component of paper coatings and is responsible for properties like brightness, opacity, gloss. The most commonly used pigments are clay, calcium carbonate and titanium dioxide. Binders act as a glue bonding pigment particles together as well as binding the coating to the basesheet to give the coating the required strength. The principal paper coating binders can be divided into two groups:

1. water-soluble polymers (starch, protein, polyvinyl alcohol),
2. water emulsions of synthetic polymers (styrene-butadiene, polyacrylate and polyvinyl acetate).

The amount of various binders and additives usually does not exceed 20 percent of the formulation. The coating is applied on the basesheet at about 55-65% w/w, using a roll or blade coater. It is then dried with hot air and/or infrared radiation. Afterwards, the coated paper is supercalendered. The pigmented coatings provide a surface that is smoother, more uniform, and more receptive to printing inks than an uncoated paper surface.

The performance of a pigmented coating is determined, for the most part, by a group of morphological characteristics, generally referred to as "coating structure". The process of consolidation of this structure is complicated by the influence of many factors, among them: coating composition (pigment type, binder type and concentration and other additives), the basesheet properties (roughness, absorbency), and the drying conditions. Some of the most important properties of the pigments are: shape and size of the particles, chemical composition, and packing characteristics in a dispersion. The amount of binder and its distribution affect both optical and printing properties of the final coated sheet. The quantity of binder should be chosen very carefully, as an excess of binder has an adverse impact on optical and printing properties. The constituents of paper coating formulations are described elsewhere [1, 2] and are out of the scope of this chapter. During drying, the coating structure undergoes shrinkage and migration processes that affect its properties [3, 4], as these depend greatly on the porous structure of the coating.

During the examination of the consolidation of the structure of clay-coatings on an impervious substrate critical concentrations of the clay were defined, corresponding to sudden changes in gloss (first critical concentration, or FCC) and in reflectance (second critical concentration, or SCC) [3, 5, 6]. The FCC corresponds to what Braun [7] defined as surface critical point, and the SCC to the minimum in hiding during the drying of a latex paint [8].

The final structure of the coating, its pore volume, its pore size distribution, the orientation of the pigment particles in the bulk and at the surface, the extent of filming of the binder, etc., are all important parameters that determine the optical and printing properties of coated paper. Understanding the consolidation behavior of pigmented coatings is therefore essential if one hopes to be able to control the structure of the final coating and to optimize it for a particular end-use. While pigment particle size and shape distribution are very important [3, 9-13], the binder may play an equal, if not more important, role [3, 5, 12, 14, 15]. There are basic differences in performance between formulations containing soluble binders and latex binders. Some of these differences have been ascribed to migration processes during drying [4]. There may also be significant differences between latexes, due to differences in their composition that affect colloidal interactions with the pigment, filming temperature etc.

Coating Preparation

This chapter gives emphasis to the development of porous structure during the consolidation of paper coatings. The following pigments and binders were taken into consideration:

Pigments

• Clay (#1) - Ultra White 90, Engelhard, particle size - 90-92% < 2µm, ESD = 0.5µm.

• Calcium Carbonate - Hydrocarb 60, OMYA, Inc., particle size - 60% < 2µm, ESD = 1.4µm.

Binders

• Carboxylated styrene-butadiene latex - DOW 620NA, DOW Chemical Co., particle size - 0.15 µm, $T_g \sim 0^0C$.

• Carboxylated styrene-butadiene latex - RAP 107NA, DOW Chemical Co., particle size - 0.15µm, $T_g \sim 42^0C$.

• Soy protein polymer - SP 2500, Protein Technologies Int., relative molecular mass - 130,000, net negative charge -- 60 moles $COO^{-1}/100$ kg

The dry pigment (clay or $CaCO_3$) is dispersed in water at 75% solids by weight, using 0.03 parts sodium polyacrylate (Dispex N-40, Allied Colloids Inc.) per hundred parts of pigment (pph). The pH is adjusted to 9 using NH_4OH. The binder is then added to the suspension, which is further diluted to 60% w/w for latex coatings, and to 55% w/w for protein coatings. The application solids used in this study for protein coatings is lower than for latex coatings, as higher concentrations resulted in high viscosity, which has created runability problems. The binder content ranged from 0 to 20 pph. The coating suspensions were applied on polyester film with a rod using a laboratory draw-down apparatus. Under the conditions used in this investigation the dewatering takes place by evaporation. As a result, the consolidation time is of the order of minutes as opposed to seconds in commercial coating applications.

In order to preserve the structure existing in the wet stage, the coating was quenched by dipping in liquid nitrogen. Quenching was done immediately after application (called FD-0), at FCC and at SCC. The frozen coating was subsequently freeze-dried by ice sublimation at low temperature (-50^0C) and under vacuum ($5 \cdot 10^{-3}$ *mBar)*. The coatings were also dried at room temperature (RT), under infrared heaters (IR), and in the oven at 105^0C (OV). Selected room temperature dried coatings were also post-treated by exposure to the elevated temperature in the oven (RT+OV).

Since the densities of protein and latex are different, it seems to be more correct to compare coatings of the same volume parts of binder per 100 volume parts of pigment (pph v/v). Table I gives the binder in the usual pph by weight and the corresponding pph by volume for the systems used in this work.

**Table I. Comparison of pph and Volume pph
of Latex and Protein Clay Coatings**

pph w/w	pph v/v latex	protein
1	2.59	
1.38		2.59
5	12.59	9.38
10	25.9	18.77
13.8		25.9
20	51.8	

Critical Concentrations

The two critical concentrations, FCC and SCC, were determined from the changes in gloss and transmittance during the drying of the coatings, using an apparatus based on fiber optics, similar to that described by Young [16], except that the stray light was filtered electronically rather than mechanically. Typical curves of "wet" gloss and transmittance versus time are shown in Figure 1 for clay-protein coatings. The voltage is a measure of gloss or transmittance. The absolute value of the voltage is not of interest, but rather the time of its rapid changes, from which the two critical concentrations, FCC and SCC, are determined.

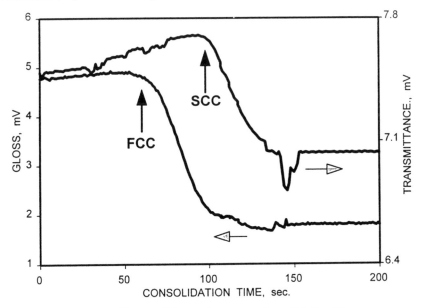

Figure 1. Gloss and Transmittance vs. Consolidation Time.
(Clay Coatings Containing 10 pph of Protein).

Gloss is a measure of the coating surface ability to reflect light specularly. It is expressed as a ratio of the intensity of the light that is reflected specularly from the surface to the incident light intensity. Gloss can be measured at various angles of illumination. For our purpose an angle of 75^0 was chosen, as standard TAPPI gloss measurements are done at this angle. As long as the surface of wet coating is a continuous water film, gloss remains almost constant. The slow increase observed at the beginning is due to the slight increase in the refractive index of the medium, as the solids concentration increases. At FCC the sudden drop in gloss indicates that the water film is no longer continuous, and menisci form at the coating surface.

Light can be scattered, transmitted or absorbed by the coating structure. Usually, for paper coatings based on white pigments, the part of incident light that is absorbed is very small in comparison to the scattered or transmitted fraction, and can be neglected. For our purpose, the light transmittance was recorded as a function of coating consolidation time. At first, the transmittance increases, as a result of increase in solids concentration. At SCC, the transmittance drops sharply, which indicates that less light is transmitted, *i.e.* more light is scattered. At this point of consolidation, air enters the water-filled spaces. This creates a larger difference in refractive index - refractive index of air is 1.0 vs. 1.33 for water and ~ 1.60 for clay - and an increase in light-scattering. Note that the SCC corresponds to the point at which "wet" gloss reaches its lowest, nearly constant value. All coatings examined exhibited the same shape of curves, regardless of pigment type, and binder type and content.

Coating concentrations at FCC and SCC were obtained from porosity measurements on freeze-dried samples. As we can see (Table II), a pure clay coating has the highest solids content at both critical concentrations. The addition of binder, either latex or protein, lowers the solids content at the FCC and SCC. The FCC and SCC occurred at lower solids concentrations with protein-based coatings in comparison to latex-based coatings. The lowering of the immobilization solids (similar to FCC) by the addition of protein has been attributed to adsorption and interactions between protein and clay [17-20]. Although at the pH used in this work (pH $\cong$ 9), the protein net charge is negative, cationic sites are still present and are responsible for attractive interaction with the negatively charged clay plates. As a result, protein molecules adsorb on clay particles. The effect appears to be maximum at 5 pph. When the protein content is increased past 5 pph, the solids content at the critical concentrations increases. If at 5 pph clay surfaces were saturated with adsorbed protein, then additional amount of polymer would create mostly repulsive interactions with polymer covering clay particles. Increasing repulsive interaction would restabilize the system and could be a reason for higher concentration of solids at the FCC and SCC.

Changes in Dimensions During Consolidation

Figure 2 illustrates the changes in volume per unit area of a 10 g/m^2 dry coating during drying. A coating based on clay and 20 pph of latex is used as an example.

Table II. Solids Fraction at Critical Concentrations of Clay Coatings Containing Latex or Protein

Consolidation Stages	Solids Volume Concentration, v/v [%]			
		Latex		
	no binder	5pph	10 pph	20 pph
FCC	72.5	63.9	64.6	64.9
SCC	73.5	62.9	66.4	71.5

Consolidation Stages	Solids Volume Concentration, v/v [%]				
		Protein			
	no binder	1.38 pph	5 pph	10 pph	13.8 pph
FCC	72.5	62.8	49.5	57.4	60.0
SCC	73.5	63.7	50.7	58.2	62.5

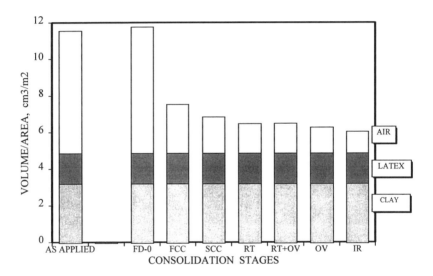

Figure 2. Changes in Volume per Unit Area During Consolidation of Clay Coating. (latex/coating = 20/100, coat weight = 10 g/m2, application solids = 60 %)

The coating structure is composed of clay as a pigment, latex as a binder and air. The amounts of clay and latex are calculated from coat weight and composition of the initial coating and are constant, but the void content decreases during consolidation and its final value depends on drying conditions. The first bar corresponds to the dimensions of the wet coating at the moment of application, so that the top part reflects the water content. The FD-0 coating is the same coating but after freeze-drying. The void fraction in the freeze-dried sample is practically equal to the water fraction in the wet coating. This demonstrates that the quenching and freeze-drying technique preserves the structure existing in the wet coating, and all the water has been replaced by air.

As the consolidation proceeds, the coating thickness decreases as water evaporates. At the first critical concentration (FCC) the amount of air is much lower than at the beginning of the process. This is the point where a loose network has formed. At the second critical concentration (SCC), capillaries start to empty, and water is replaced by air.

With coatings containing 20 pph of latex, the volume per unit area was still decreasing after the SCC. As a result the RT dried coating had lower void volume than the coating freeze-dried at the SCC. With the film - forming latex ($T_g \sim 0^0C$), drying conditions had little effect on coating void volume at 5 and 10 pph (not shown). At 20 pph of latex, on the other hand, intense drying (OV and IR), led to lower void volume (denser structures) - particularly with clay - the densest structures being obtained with IR (Figure 2). Post-treatment did not change void fraction. Once coating was dried, the further exposure to the elevated temperature did not affect final volume of coating.

Generally, a similar behavior was observed with $CaCO_3$ coatings (not shown).

Evolution of Surface Structure During Consolidation

Gloss. The pure clay coatings and coatings containing latex as a binder, freeze-dried at the beginning of the consolidation process (FD-0), had very low gloss (Figure 3). At 60% solids, the pigment particles are expected to quickly recover a random orientation after the application of the low shear field during application with the draw-down rod. At the FCC, the gloss of the freeze-dried structures was the highest, which means that the particles had reached the highest state of order and orientation at the surface and, presumably, in the bulk. If one could prevent further particle rearrangement, one would have a high gloss in the final structure that might make calendering unnecessary. Unfortunately, as drying proceeds, the gloss drops in proportion to the shrinkage [21], and the final value was found to depend on drying conditions and latex content.

With intense drying, as in the case of OV and IR, the gloss dropped much more than for RT drying, the effect being very significant for the coatings with 20 parts of latex. At higher drying temperature, melt flow of the latex polymer is enhanced, and the coating shrinks more, causing more disruption of the surface. After thermal post-treatment of RT samples, the gloss practically did not change, remaining higher than the gloss of coatings dried in the oven or under IR.

The same trends were observed with both pigments, but calcium carbonate coatings had much lower gloss than clay coatings.

With the clay - protein coatings examined here, the evolution of gloss is different. There is no maximum at the FCC, except at the lowest binder content - 1.38 pph (Figure 4). Above 1.38 pph, the structures freeze-dried at the moment of application (FD-0) had the highest gloss and as the coating consolidation proceeded, the gloss decreased continuously. At a protein content of 1.38 pph, the interaction is considered minimal and the clay particles can still rearrange and align, creating the highest gloss at the FCC. At higher protein content, the interaction is strong,

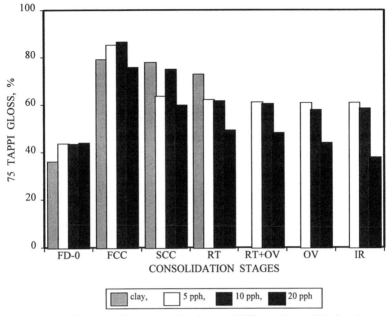

Figure 3. Gloss of Clay-Latex Coatings at Different Consolidation Stages.

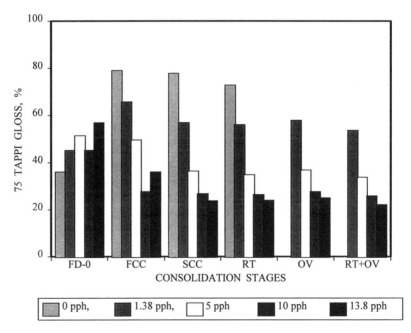

Figure 4. Gloss of Clay-Protein Coatings at Different Consolidation Stages.

preventing the clay plates from aligning. This may also account for the very significant drop in gloss when protein content was increased from 1.38 to 5 pph. The final gloss was greatly influenced by the protein content: Gloss decreased by almost 50% when the protein level was increased form 1.38 to 5 pph and continued to decrease, though at a slower rate.

When compared at the same binder content, at the binder levels examined in this study, clay-protein coatings had lower gloss than clay-latex coatings. Once completely dried, the pure clay coatings had higher gloss than binder-containing coatings.

Coating Shrinkage after FCC

All coatings shrink during drying. The initial stage of the shrinkage, which cannot be prevented and is similar for all coatings, corresponds to the removal of water until the formation of a network or filtercake at FCC. After FCC, shrinkage is a function of a number of parameters [3], including pigment and binder type and amount.

Latex Film Forming Temperature. Table III illustrates the influence of latex level and its T_g on structure consolidation, expressed here in terms of shrinkage. Shrinkage is defined as the percent change in void volume between FCC and the final dry structure, relative to the void volume at the FCC.

Table III. Shrinkage of Clay and CaCO₃ Coatings

	Coating Shrinkage, [%]				
Latex Content pph	Clay RT-NF	Clay RT	Clay OV	CaCO₃ RT	CaCO₃ OV
5	7.1	10.2	17.5	8.9	11.0
10	6.1	10.3	17.9	11.5	16.6
20	8.7	40.4	47.9	35.2	55.4

After the FCC, coatings dried at RT with a latex of T_g around 40°C (RT-NF), have a relatively low shrinkage. It can be assumed that shrinkage is due mostly to the rearrangement of the solid pigment and latex particles under the compressive stresses exerted by the capillary pressure of the receding water front. Shrinkage is independent of latex content. With a filming latex, the clay coatings exhibit a clear difference in behavior at low (<10 pph) and high (20 pph) latex content, as if there were two different consolidation regimes. Shrinkage is same at 5 and 10 pph. At 20 pph, shrinkage is extensive and increases with drying temperature (OV versus RT). The difference in shrinkage between non-filming and filming latexes can be attributed to the deformation and melt flow of the latex polymer under the compressive stresses. Melt flow is enhanced at the higher drying temperature.

The difference in behavior between low and high latex content is important as it confirms a hypothesis put forward earlier [5] of the existence of two different regimes of consolidation, depending on the latex/pigment ratio. Below a critical latex content the consolidation of the structure was largely determined by the packing of the clay. Above a critical binder level, the consolidation was thought to be controlled by the ease of coalescence and melt flow of the latex at the prevailing drying temperature. Indeed, at low binder level (5-10 pph), there would be a continuous pigment matrix with the latex particles distributed within this matrix and the structural arrangement of the pigment particles would control the overall shrinkage during drying. This is illustrated in Figure 5a. At the higher binder level (20 pph), pigment particles are almost completely surrounded by a latex matrix (Figure 5b). The local stresses that develop during coalescence can now be transmitted through the binder matrix. These stresses cause particle reorientation and increase overall structural shrinkage.

Binder Type.   Figure 6 illustrates the influence of binder level and type on shrinkage of the structure that as formed at FCC. Binder content is expressed here in terms of volume concentration to account for the differences in density between latex and protein.

The lowest shrinkage is observed for the binder - free coating. It has been shown that gloss drops in proportion to shrinkage [*21*]. Since the highest gloss was found for pure clay coatings (Figure 3 & 4), this kind of behavior was anticipated. In the absence of binder, the shrinkage is most likely caused by rearrangement of the solid pigment particles. When a binder is added, the extent of shrinkage of the coating increases. With latex coatings, the shrinkage has been attributed to the capillary pressure forces [*6*] as a result of flow of the polymer.

Even though the FCC was reached earlier with protein coatings, for example at 25.9 pph v/v of binder solids, concentrations at the FCC were: 60.0% v/v for protein vs. 64.6% v/v for latex., the void fractions of coatings dried at room temperature were the same (0.33). As a result the shrinkage of the protein-based coatings was higher than that of the latex-based coatings. With protein-based coatings, a sharp increase in shrinkage with binder content was observed up to approximately 10 pph v/v (5 pph). It can be noticed that - for the range of binder level and protein type examined - above 10 pph v/v coating shrinkage remains constant. As it has been hypothesized earlier in this chapter, at 5 pph (~10 pph v/v) the clay surface might be saturated with protein. If the shrinkage is due to reorientation of the clay particles when surfaces are saturated with soy polymer, then further increase in polymer content would not have an effect on coating shrinkage.

The shrinkage increased with increasing drying temperature, but only for latex-based coatings. With protein-based coatings, shrinkage was the same for RT and OV drying. This was anticipated, as protein does not go through such a drastic structural change as latex which goes from a solid to a melt. This latex structural change is enhanced when difference between drying temperature and polymer T_g is increased.

Evolution of Surface Area During Consolidation

Latex coalescence and bridging between pigment particles, as well as adsorption of the protein on clay surfaces are processes expected to decrease the specific surface

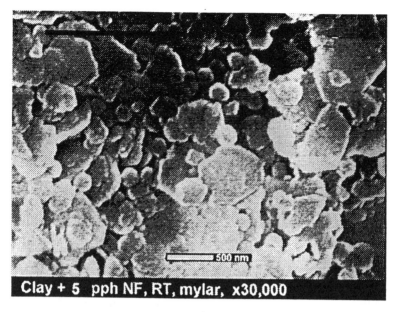

a)

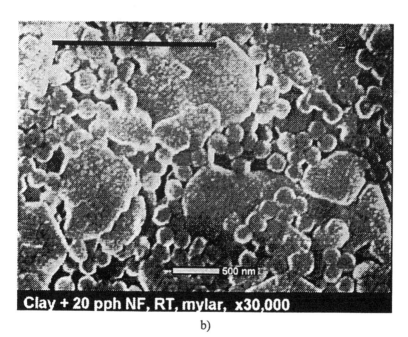

b)

Figure 5. Surface of Uncoalescence Clay Coating Containing 5 (a) and 20 pph of
Latex (b).

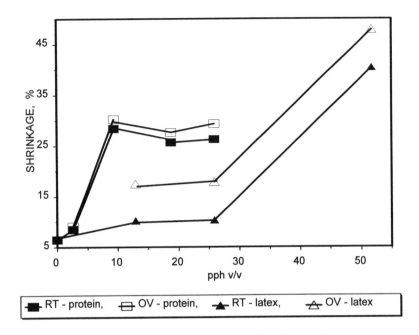

Figure 6. Shrinkage of Clay-Latex and Clay-Protein Coatings.

area of the coating. Therefore changes in surface area during consolidation may shed some light on these processes.   Specific surface area was measured by nitrogen adsorption method.

Pigment Type. The specific surface area of clay and $CaCO_3$ coatings with latex as a binder is shown in Table IV. At the FCC the latex has not coalesced, and the values of surface area are similar to those at the FD-0. As one moves to the SCC, surface area decreases sharply. The earlier work by Watanabe *et al.* [6] had demonstrated that latex coalescence occurred between the FCC and the SCC, that is, in water. The surface area data confirm this nicely.

Table IV. Specific Surface Area of Clay-Latex and $CaCO_3$ - Latex Coatings

	Specific Surface Area, $[m^2/g]$					
Consolidation Stages	Clay 5 pph	Clay 10 pph	Clay 20 pph	$CaCO_3$ 5 pph	$CaCO_3$ 10 pph	$CaCO_3$ 20 pph
FD-0	15.18	15.02	16.25	6.74	6.89	8.52
FCC	15.10	14.95	15.29	6.40	7.63	8.71
SCC	10.36	11.19	9.74	4.84	3.57	3.80
RT	10.04	10.96	7.81	4.34	4.36	2.82
RT + OV	8.06	7.85	5.02	3.84	2.78	2.35
OV	6.80	5.94	4.36	3.31	2.41	1.31

For the coatings containing 5 and 10 pph of latex, there was no further change in surface area after the SCC for coatings dried at room temperature. When the latex level was increased to 20 parts, however, a further Increase in surface area was observed after the SCC, indicating that coalescence was not complete. Again, the differences between low and high binder content lend further support to the two-regimes - pigment-controlled and binder-controlled - consolidation. After thermal post-treatment of the RT dried samples, a further decrease in surface area was observed. However, the post-treated samples still had a much higher surface area than those dried directly in the oven. Even though the coatings were exposed to the same temperature for the same length of time, the driving force for wetting and spreading of the polymer over pigment surfaces is different in the presence or in the absence of water [22].

Binder Type. With latex-based coatings, covering binder level up to 20 pph, a large decrease in surface area occurred between the FCC and the SCC, as a result of latex particles coalescing into a film. With protein-based coatings investigated the most significant drop in surface area is observed after the SCC, (Table V) i.e. after water-filled spaces start to empty, except for the coating containing the largest amount of protein (13.8 pph) where the drop in surface area is observed between the FCC and the SCC.

Table V. Specific Surface Area of Clay Coatings Containing Protein

Consolidation Stages	Specific Surface Area, $[m^2/g]$			
	1.38 pph	5 pph	10 pph	13.8 pph
FD-0	17.0	13.0	11.2	12.0
FCC	17.0	15.5	10.4	14.1
SCC	16.1	13.9	9.0	6.5
RT	13.2	8.8	6.9	5.4
OV	13.3	8.3	6.2	6.1

In the presence of water, there is an equilibrium between molecules associated with water and molecules adsorbed on clay caused a decrease in specific surface area. One can notice that surface area decreases when binder content is increased. A higher level of polymer addition increases the possibility for bridging between pigment surfaces, and this would reduce the final surface area of the coating. The surface area of clay-protein coatings is not affected by drying at 105^0C, as similar values for RT and OV dried coatings were obtained. This is in opposition to what happened in latex coatings, where an increase in drying temperature lowered surface area, as a higher drying temperature resulted in increased polymer melt-flow.

Pore Size and Void Scattering Efficiency

Equivalent Spherical Diameter. Voids present in coating have different shapes and sizes, as well as void size distribution. An equivalent spherical diameter can provide an information about the size of average pore. From surface area and void fraction data one may calculate an equivalent spherical diameter of the pores present in the coating structure. Table VI shows the evolution of the size of the pores as the coating consolidates.

Pore size goes through a minimum at the FCC. The largest ESD values were found for the FD-0 coatings. At the beginning of the process the particles are far apart; and this creates the largest voids. As water leaves, pigment particles are brought closer together, the total volume of the coating decreases. This forces the clay plates to align parallel to each other, what causes the decrease in pore size. This decrease continues till the point at which particles have reached the highest state of order and orientation. Since this occurs at the FCC, the pore size was expected to

Table VI. Equivalent Spherical Diameter (ESD)
of Clay-Latex and CaCO$_3$ - Latex Coatings

Consolidation Stages	*Equivalent Spherical Diameter, [µm]*					
	Clay 5 pph	*Clay 10 pph*	*Clay 20 pph*	*CaCO$_3$ 5 pph*	*CaCO$_3$ 10 pph*	*CaCO$_3$ 20 pph*
FD-0	0.249	0.254	0.254	0.557	0.513	0.430
FCC	0.093	0.097	0.104	0.215	0.156	0.125
SCC	0.142	0.120	0.117	0.259	0.310	0.238
RT	0.126	0.119	0.121	0.289	0.241	0.250
RT + OV	0.172	0.171	0.191	0.319	0.347	0.271
OV	0.171	0.211	0.189	0.375	0.408	0.371

have its minimum at this point. This phenomenon (minimum in pore size at the FCC) was observed with both binder systems used in this study, e.g. particulate latex and soluble protein. As latex filming occurs, latex particles fuse together and void size increases, even though the total void volume decreases.

Thermal post-treatment of the RT coatings increased pore ESD but direct drying in the oven produced - for most of the cases - the largest pores through the lowest void fraction. It would seem that these are conditions leading to the most extensive spreading of the polymer over pigment surfaces. One notes that, overall, the ESD of pores in clay coatings are much smaller than those in CaCO$_3$ coatings.

It can be noticed that - for the formulations examined in this work - pores are larger in protein coatings (Table VII), in comparison to latex coatings. Being amphoteric by nature, protein have the ability to interact electrostatically with clay particles in aqueous suspension. The cationic sites on the protein interact with anionic sites on the clay possibly creating a flocculated structure with large distances between pigment particles, creating large voids.

It can be noticed that there is a large increase (doubling) in the pore size when the protein content is increased from 1.38 to 5 pph. This is in agreement with the hypothesis of a maximum in the interaction with clay discussed in the paragraph on "Critical Concentrations". A Further increase in protein content did not change the void size significantly.

**Table VII. Equivalent Spherical Diameter (ESD)
of the Pores of Clay Coatings Containing Protein**

Consolidation	Equivalent Spherical Diameter, [μm]			
Stages	1.38 pph	5 pph	10 pph	13.8 pph
FD-0	0.291	0.382	0.436	0.410
FCC	0.082	0.159	0.178	0.121
SCC	0.083	0.168	0.199	0.235
RT	0.096	0.201	0.199	0.235
OV	0.094	0.207	0.215	0.199

Void Scattering Efficiency. Light - scattering is a useful technique to probe the void structure of coatings. Measurements at wavelength 457 nm, where brightness is measured, are reported here (Figure 7). In paper coatings based on low refractive index pigments, voids are the scattering medium and light - scattering depends on size of the voids and void fraction. At the beginning (FD-0), when there is a lot of voids in the coating, the light - scattering coefficient (LSC) is very high. A large drop is observed as one reaches the FCC. For coatings containing 5 and 10 parts of latex, the LSC remains almost constant after the SCC, and the value is relatively independent of drying conditions or pigment shape. At 20 pph of latex, the LSC continued to decrease after the SCC and the effect was particularly large with the $CaCO_3$ coatings.

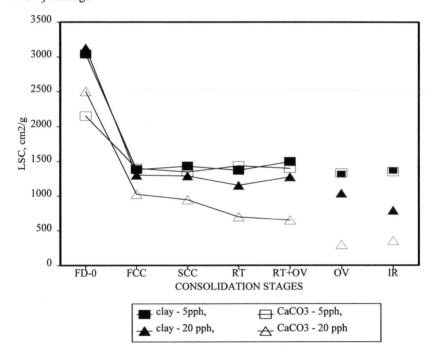

Figure 7. Light-scattering Coefficient at Different Consolidation Stages.

The LSC of the clay coatings increased significantly after thermal post-treatment in the oven at 105^{0}C and the effect increased with the amount of latex. This has been explained by a change in the void size as latex melt flow is more complete [14]. Void scattering efficiency results discussed later confirm this explanation. However, the results presented here show that the phenomenon seems to be dependent on the pigment shape, as there was little change in the LSC of the rhombic $CaCO_3$ coatings after post-treatment. In fact, the LSC of the 20 pph $CaCO_3$ coating *decreased* after post-treatment.

LSC is commonly expressed per unit basis weight. However, in coatings voids are the scattering elements and the light-scattering depends on both the size of the voids and their number per unit basis weight, or specific void fraction. Therefore void scattering efficiency (VSE) seems to be a more useful parameter than the usual specific light-scattering coefficient (LSC). VSE is obtained by normalizing the LSC with the specific void fraction, giving the scattering efficiency per unit volume of voids. The units are reciprocal length.

It can be seen in Figure 8 that, even though the LSC of FD-0 is very high, the scattering efficiency of the voids is rather low. The voids observed on SEM micrographs are very large, probably too large, for efficient scattering of light. Overall, void scattering efficiency increases with consolidation of the structure. A further increase is observed after post-treatment of the clay coating at 20 pph of latex.

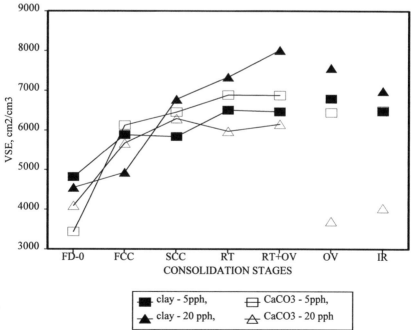

Figure 8. Void Scattering Efficiency (VSE) at Different Consolidation Stages.

Figure 9 compares VSE of latex and protein coatings containing 25.9 pph v/v of binder. It has been shown that there is an optimum pore size for scattering efficiency [23] as a function of light wavelength. Since voids are larger in the protein coatings and the scattering efficiencies are similar in both systems, it is possible that the pore sizes of protein and latex coatings lie on the two opposite sides of the optimum.

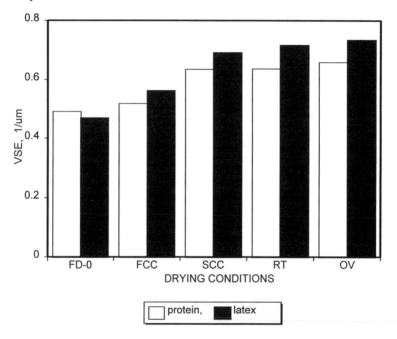

Figure 9. VSE of Clay Coatings Containing 25.9 vol. pph of Binder at Different Consolidation Stages.

Summary

Regardless of pigment shape or binder type and level, pigmented coatings formulated above the critical pigment volume concentration (CPVC) went through the same stages of consolidation, i.e., first and second critical concentrations (FCC, SCC). However, binder type and level influences the final structure properties.

At low latex content the structure formation is determined mostly by the pigment network. When latex level is increased beyond some critical range, a continuous binder matrix is formed, and consolidation, and the final values were found to depend on latex level and drying conditions. The effect of drying conditions was especially significant for coatings containing 20 parts of latex - the more intense the drying the lower the void volume. Void volume did not change after RT coatings were exposed to thermal post-treatment in the oven. Void size went through the minimum at FCC, and its final value depended on drying conditions, pigment shape, and binder type and level. Intense drying and thermal post-treatment yielded the structures with the largest voids, as a result of more extensive melt flow of binder.

With pure clay and latex coatings, gloss of the freeze-dried structures went through the maximum at the FCC and its final value depended on drying conditions and latex level. The more intense the drying and the higher the latex level, the greater the shrinkage and the lower the gloss. The thermal post-treatment did not change gloss, nor shrinkage. With protein coatings this maximum is observed only for low binder formulations. Protein coatings have lower gloss than latex coatings.

Shrinkage was found to be greater when protein was used as a binder, in comparison to latex-based coatings, because the very bulky structure produced at the FCC does not guarantee a bulky dry coating.

A sharp decrease in surface area occurred between the FCC and the SCC, as a direct result of the latex coalescence, which took place between those two critical concentrations. The results are in very good agreement with the other measurements and provide further insight on the extent of melt flow of the latex. The results also indicate that wetting and spreading of latex over pigment is different in the presence or in the absence of water. In clay-protein coatings, the changes in surface area occurred mostly after the SCC. This could be of importance in the selection of a drying "strategy".

Literature Cited

1. Casey, J.P.; *Pulp and Paper. Chemistry and Technology;* 3rd ed.: Willey-Interscience: New York, NY, **1983**; vol. IV, 2023.
2. Hagemeyer, R.W.; *Pigments for Paper;* TAPPI Press: Atlanta, GA; **1984**.
3. Lepoutre, P.; *Prog. in Organic Coatings,* 17, 89 (**1989**).
4. Whalen-Shaw, M.J.; *"Binder Migration in Paper and Paperboard Coatings"*, TAPPI Press, Atlanta, GA, 1993, ch. 4, p. 61.
5. Lepoutre, P., and Rezanowich, A.; *Tappi J.* 60 (11), 86 (**1977**).
6. Watanabe, J., and Lepoutre, P.; *J. Applied Polymer Sci.* 27 (11), 4207 (**1982**).
7. Braun, J.H.; *JCT* 63 (799), 43 (**1991**).
8. Cleveland Society for Coating Technology; *JCT* 62 (786), 43 (**1990**).
9. Hagemeyer, R.W.; *Tappi J.* 43 (3), 277 (**1960**).
10. Hagemeyer, R.W.; *Tappi J.* 47 (2), 74 (**1964**).
11. Hagemeyer, R.W.; *Tappi J.* 47 (10), 595 (**1964**).
12. Reinbold, I., and Ullrich, H.; *Tappi J.* 63 (1), 47 (**1980**).
13. Alince, B., and Lepoutre, P.; *Colloids and Surfaces* 6 (2), 155 (**1983**).
14. Lepoutre, P., and Alince, B.; *J. Appl. Polymer Sci.* 26 (3), 791 (**1981**).
15. Whalen-Shaw, M.; *Proc. 1984 Tappi Coating Conf.,* TAPPI Press, Atlanta, **1984**, p.11.
16. Young, T.S., Weyer, L.G., Pivonka, D.E., and Ching, B.; *Tappi J.* 76 (10), 71 (**1993**).
17. Herbert, A.J., Gautam, N., and Whalen-Shaw, M.; *Proc. 1990 Tappi Coating Conf.,* TAPPI Press, Atlanta, **1990**, p. 431.

18. Coco, C.E.; *1983 TAPPI Coating Conf. Proc.*, TAPPI Press, Atlanta, **1983**, p.109.
19. Coco, C.E.; *1984 TAPPI Coating Conf. Proc.*, TAPPI Press, Atlanta, **1984**, p.131.
20. Whalen-Shaw, M., and Coco, C.E.; *Tappi J.* 68 (5), 63 (**1985**).
21. Lee, D.I.; *Proc. 1974 Tappi Coating Conf.*, TAPPI Press, Atlanta, **1974**, p.97.
22. Lepoutre, P.; *Trends in Polym. Sci.*, 3 (4), 112 (**1995**).
23. Borch, J., and Lepoutre, P.; *Tappi J.* 61 (2), 45 (**1978**).

Chapter 14

The Drying of Waterborne Coatings

Edgar B. Gutoff

194 Clark Road, Brookline, MA 02146

Drying is explained in terms of the constant rate period, where the drying rate is not necessarily constant but is controlled by the conditions in the drying air, and the falling rate period, where it controlled by diffusion to the surface. For aqueous systems most of the drying occurs in the constant rate period. The psychrometric chart is explained, and its use for constant rate drying calculations is demonstrated. The method of modeling of both drying periods is covered. Skinning that can occur during drying is explained, as is the use of moist drying air to avoid it. The causes and cures of drying defects, such as those caused by air motion, stress-related defects, and blisters and pinholes, are covered.

A coating is a relatively thin liquid film containing a binder and perhaps pigments and various additives. Usually water or an organic solvent is present in which the binder and the additives are dissolved or dispersed. During drying the solvent evaporates. The binder, if not dissolved but dispersed (such as a latex), should coalesce into a continuous film. Cross-linking reactions may take place during and after drying. This chapter discusses the drying of water-borne coatings and some of the defects that may occur during the drying process.

In the initial stage of drying the water-borne coating behaves as a pool of water. The vapor pressure of the water at the coating surface is close to that of pure water at the temperature of the coating. The vapor pressure may be slightly less due to the lowering of the vapor pressure by dissolved species, though frequently no significant vapor pressure lowering is seen. Vapor pressure lowering is proportional to the mole fraction of dissolved species. When the binder is a polymer of moderate to high molecular weight, or a dispersed latex, the mole fraction of dissolved material tends to be very low and the vapor pressure lowering is negligible. The initial phase of drying is termed the constant rate period, and when the equilibrium temperature is reached water evaporates as fast as the heat of vaporization can be supplied.

The term *constant rate* in one sense is a misnomer, in that the drying rate is not necessarily constant. It is constant when the system is at equilibrium with the drying air, but in the initial stages when the coating is cooling down or heating up to its equilibrium temperature the drying rate will be changing. The equilibrium constant rate temperature may never be reached when the coating is on a high heat capacity material, such as sheet metal. The term *constant rate period* just means that the surface of the coating is wet with solvent (water) and the solvent can get to the surface (usually by diffusion) at a rate sufficient for the surface to remain wet. Thus the resistance to evaporation is in the vapor phase; in fact, the constant rate period is often defined as the period where conditions in the air (temperature, humidity, air velocity), rather than the diffusion rate in the coating, determine the drying rate.

Once the coating is at its equilibrium constant rate temperature all the heat that is supplied to the coating is consumed in evaporating the water. Then, as long as the air temperature, the coating temperature, and the heat transfer coefficient (a function only of the air velocity for a given geometry) remain constant, the drying rate will remain constant at a value given by

$$Evaporation\ rate = h\ (T_a - T_c)\ /\ \lambda \tag{1}$$

where h is the heat transfer coefficient, W/m^2-K
       T_a is the temperature of the air, $^\circ C$
       T_c is the coating temperature, $^\circ C$
       λ is the heat of vaporization, J/kg

In equation 1 the rate of heat transfer per unit area is just the heat transfer coefficient times the temperature difference between the drying air and the coating. The heat transfer coefficient increases with the air velocity, and therefore using higher velocity air increases the drying rate. So does using hotter air. The coating temperature, as will be shown below, is often equal to the wet bulb temperature of the air. The wet bulb temperature is only a function of the air temperature and water content (humidity). The coating temperature or the wet bulb temperature of the air can be lowered by reducing the moisture content of the air, which lowers the dew point and the absolute and relative humidity of the air. The coating temperature can also be lowered by lowering the air temperature while keeping the dew point constant, but this is counterproductive as the air temperature should be as high as possible for high rates of heat transfer.

In the constant rate period the coating is usually liquid and is easily damaged.

At some point dry spots appear on the wet surface of the coating and then quickly spread over the whole film. The vapor pressure of the water at the surface drops drastically as the surface of the coating is no longer wet; the rate of drying decreases and is limited by the rate at which water can diffuse to the surface. This drying phase is called the falling rate period.

In solvent systems most of the drying occurs in the falling rate period. The higher the solids concentration the shorter the constant rate period. In some cases there is no constant rate period at all; the coating is in the falling rate period from the start. Water, however, has a much higher heat of vaporization than organic solvents; therefore it takes more heat it to evaporate a given amount. Also, water is a very small molecule and can diffuse to the surface more rapidly than the larger solvent molecules. Thus it is relatively easy for water to keep the surface wet even while evaporation takes place. As a result of these two factors most of the drying in water-borne coatings occurs in the constant rate period. Frequently one can get excellent control of the drying process for water-borne coatings by controlling the time or location of the end of constant rate period.

The step growth oligomers crosslinked on metal surfaces by Original Equipment Manufacturers (discussed in chapter 1) could be expected to follow the constant rate model until a high crosslink density is achieved on the metal substrate. Latex coatings, on the other hand, would vary according to the chemical composition of the coating (discussed in chapters 1 & 4) and the porosity of the substrate. For example, water would wick into paper, wood, and wallboard to different degrees. As the volume fraction of the latex increases due to water evaporation, capillary forces bring the latex particles together. Surface tension brings the film down and a thin layer of coalesced particles closes the surface of the drying latex. This may also end the constant rate period. As the composition of the latex is varied from styrene to methyl methacrylate to vinyl acetate the increasing hydrophilicity increases the tendency for water to act as a plasticizer for the latex. Increasing the hydroxyethyl methacrylate comonomer content of the methyl methacrylate latex would also increase the plasticization behavior of water and impede the drying rate (discussed in chapter 4).

Figure 1 is a graph generated by a computer model that illustrates the drying of a water-borne coating on a polyester support in a continuous 6-zone dryer. The air temperatures, the coating temperature, and the residual water are plotted against the location in the dryer. This could represent any type of water-borne coating, such as a latex paint. Any type of drying could be used. With drying in ambient air there would be but one drying zone, the air temperature would be about 20°C, and the time in the single zone would be measured in hours. Usually drying with heated air is used, and sometimes infra-red heating is used in addition. Paper is often dried using drum dryers. Frequently more than one zone is used, in order to adjust the drying conditions to optimize the rate of drying. In Figure 1, in zone 1 the coating temperature rises to approach the equilibrium constant rate temperature with the air but never reaches it. In zones 2, 3, and 4, with their higher air temperatures, the equilibrium temperature seems to be reached. This equilibrium temperature varies with the air temperature and humidity. For single-sided drying with only convection heaters the equilibrium temperature of the coating is the wet bulb temperature of the air. However, hot air can be blown not only at the coating but also at the back side, resulting in double-sided heating. Also, additional heat can be supplied by infra-red heaters. In these cases the equilibrium coating temperature will be higher than the wet bulb temperature of the air.

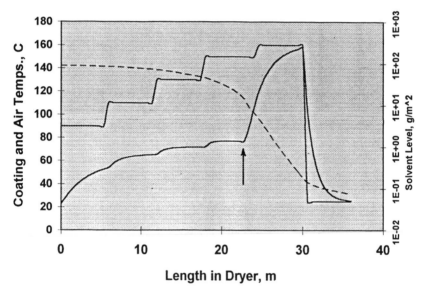

Figure 1. Drying of a water-borne coating in a 6-zone dryer. The end of the constant rate period is indicated by the arrow. These curves were generated by a drying spreadsheet.

In Figure 1 the end of the constant rate period occurs 23 m into the dryer, just before the end of zone 4. It is marked by an arrow, and can be identified by the rapid rise of the coating temperature towards the air temperature. In this particular dryer the last zone, zone 6, is used to cool the coating down to room temperature.

Wet Bulb Temperature

The term *wet bulb temperature* should be explained. If we wrap a wet piece of cloth around the end of a thermometer bulb and use a fan to suck room air past it, the water will evaporate. The fan sucks the air past the thermometer rather than blowing air at it, because we do not want heat from the fan motor to warm the air before it reaches the thermometer. The heat for the evaporation comes from the air and from the wet cloth. For heat to transfer from the air to the wet cloth the cloth has to be at a lower temperature than the air. The temperature of the wet cloth thus drops to allow heat to flow from the bulk air to it. The equilibrium temperature is called the wet bulb temperature of the air. The wet bulb temperature is a function only of the air temperature (termed the dry bulb temperature) and the moisture content of the air - as measured by the dew point of the air (the temperature at which the air is saturated with water) or by the absolute humidity (in units of mass of water per mass of dry air). The relative humidity (the partial pressure of water in the air divided by the vapor pressure of water at the temperature of the air, expressed as a percentage), in conjunction with the air temperature, is also a measure of the moisture content of the air. These relationships can be illustrated on a psychrometric chart, shown in Figure 2.

The abscissa or *x*-axis is the dry bulb temperature of the air. The term *dry bulb* arises because usually a standard thermometer is placed next to the one with the wet cloth around the bulb. The ordinate or *y*-axis on the right side of the chart is the absolute humidity or moisture content of the air in g of water per kg of dry air. The saturation line is the curved line going from the lower left towards the upper center. It gives the absolute humidity of air saturated with moisture at any temperature. Note the temperature numbers on the saturation line, agreeing with the dry bulb temperatures directly below. When the air is saturated it is at its dew point, and any further lowering of the temperature will cause condensation of moisture or dew. Thus these numbers represent the dew point of air at any given absolute humidity. Also note the straight lines sloping downwards to the right from the dew points on the saturation line. These sloping lines are lines of constant wet bulb temperatures. For example, if we find the intersection of an air temperature of 24°C with an absolute humidity of 6 g per kg dry air (or a dew point of 6.6°), and follow it along the line sloping up to the left, we find that the wet bulb temperature of the air is 14°. This will be the equilibrium temperature of the coating when this air is used for single-sided drying, irrespective of the velocity at which the air is blowing at the surface (as long as radiational heat is negligible and the coating is getting almost all of its heat by convection from the air). Conversely, if we know the wet bulb temperature of the air and the dry bulb temperature, we can find the absolute humidity of the air and the dew point. The intersection of the dry bulb temperature

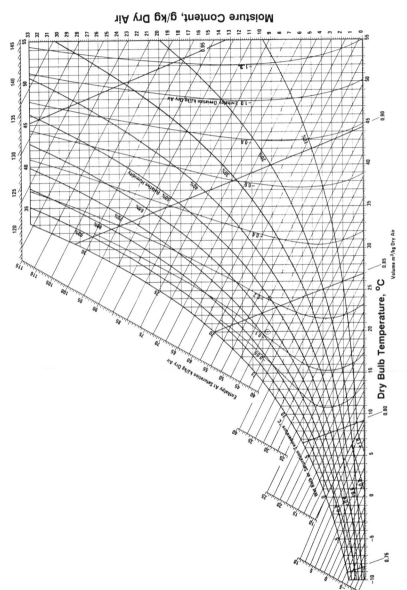

Figure 2a. Psychrometric chart for normal temperatures and for sea level. Reproduced with permission of Carrier Corporation. Copyright 1975 by Carrier Corporation.

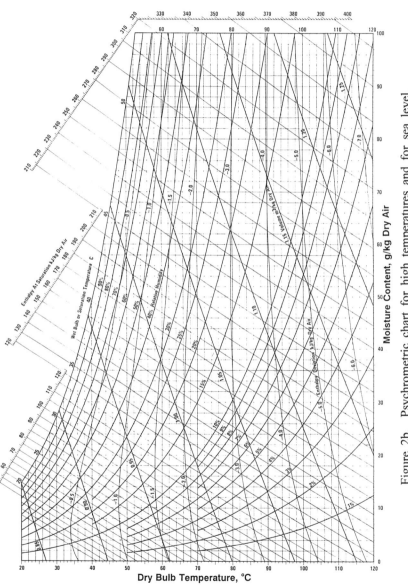

Figure 2b. Psychrometric chart for high temperatures and for sea level. Reproduced with permission of Carrier Corporation. Copyright 1975 by Carrier Corporation.

and the wet bulb temperature (or the dew point or the absolute humidity) gives the condition of the air.

Now note the curved lines roughly parallel to the saturation line. These are lines of constant relative humidity. From the intersection point of the dry and wet bulb temperatures we can find the relative humidity of the air.

Once the constant rate period is over, water cannot diffuse to the surface at the rate it had previously been evaporating, for now the evaporation rate is diffusion limited. Therefore less heat is now consumed per unit time in evaporation, and the excess heat transferred to the coating is used to heat the coating towards the air temperature.

In the falling rate period the rate at which water diffuses through the coating to the surface - the evaporation rate - is equal to the product of the diffusivity of water at the surface of the coating times the concentration gradient at the surface of the coating.

$$Evaporation\ rate = (- D\ \partial C\ /\ \partial X)_{surface} \qquad (2)$$

where c is the concentration of water, g/cm^3

D is the diffusivity of water, cm^2/s

x is the distance from the base of the coating, cm

The diffusivity of the water in the coating increases with the temperature of the coating and increases greatly with water content. The diffusivity tends to be very low when the coating is almost dry. It is very difficult to remove the remaining residual water. Higher air temperatures aid in drying because water diffuses more rapidly at higher temperatures. However, once the coating approaches the air temperature in the falling rate period, higher air velocities do not increase the drying rate as they no longer influence the temperature and thus do not influence the diffusivity. Heat transfer is now of relatively minor importance.

Skinning

Skinning is a frequently observed phenomenon where a dry skin forms on the surface of the coating while the interior of the coating remains un-solidified and very soft. The skin greatly reduces the rate of diffusion of water to the surface - the drying rate. Skinning and its prevention can be explained with the help of Figure 3.

In chemical engineering we normally assume equilibrium conditions at an interface. Thus, in the falling rate period, the water content at surface of the coating will be in equilibrium with the air. If the air is essentially dry (with a very low dew point) then the surface of the coating will be dry. With practically no water at the surface, the diffusivity, being a strong function of water concentration, will be very low. Therefore the rate of diffusion of water to the surface - the rate of drying - will be very low. In the rest of the coating, where the coating is still wet and the diffusivity is relatively high, diffusion will be relatively rapid and therefore the concentration will be fairly uniform, followed by a sharp drop nearly to zero at the surface. This is illustrated in the sketch on the right. We have formed a skin.

In the sketch on the left the drying air contains some moisture. Therefore the moisture concentration at the surface is well above zero. Now the diffusivity will be reasonably high and water can diffuse to the surface at a higher rate than in the previous case. The concentration profile will no longer be flat, but will slope gradually down to the surface concentration. There is no skin.

This qualitatively demonstrates that, in the early stages of the falling rate period, one can dry faster using moist air than by using dry air, and at the same time prevent skin formation. Obviously, if the moisture content is too high no drying will take place; one can even drive moisture from the air into the coating. The moisture content of the air for the maximum drying rate varies with the moisture content of the coating, and decreases as drying proceeds. To reach a low residual water content dry air must be used in the final stages of drying.

One further point should be made concerning skinning. When a skin forms the underlying coating is still very soft and easily damaged. If the air flow is not uniform, such as from many round nozzles, dryer bands can form. This has been observed at the start of the falling rate period.

Modeling Drying

We have discussed the qualitative aspects of the drying of water-borne coatings; now we should discuss some quantitative aspects. In the simplest case the coating will be of moderate to heavy thickness, perhaps up to 100 μm (about 4 mils) or more on a relatively thin web with low heat capacity, such as a plastic film no thicker than several hundred micrometers, perhaps 8 mils. The temperature of the coating will then be at its equilibrium value for almost all of the constant rate period. If drying air is used only on the coating side the coating temperature will be the wet bulb temperature of the air, which can be easily found from a psychrometric chart. The rate of evaporation can then be calculated from equation 1 if we know the heat transfer coefficient. In dryers the heat transfer coefficient is approximately proportional to the air velocity to the 0.78 power or to the pressure drop across the air nozzles (usually equal to the pressure in the air plenum) to the 0.39 power (*1*)

$$h = h_o \, (V / V_o)^{0.78} \tag{3}$$

or

$$h = h_o \, (P / P_o)^{0.39} \tag{4}$$

where h_0 is the heat transfer coefficient at the reference velocity V_0 or at the reference pressure P_0.

When the geometry in top and bottom sections (for two-sided dryers) of all zones are the same then the heat transfer coefficient at reference conditions (arbitrarily chosen, but often 10 m/s or 250 Pa pressure - or 10,000 ft/min or one inch of H_2O) will be same for all sections in all zones. This reference heat transfer coefficient can be estimated, as Martin (2) has done. It also can be determined from measurements of the rate of temperature rise for a thick uncoated web. However, it is frequently found by matching the calculated drying rate to experimental data, as will be discussed after Example 1.

The water content at the end of the constant rate period can be found from laboratory tests. Such a test might consist of blowing warm air at a coating on a top-loading scale and plotting weight versus time, as in Figure 4. The constant rate period is the straight-line portion of the initial curve. Where the straight line portion ends is the end of the constant rate period. This definition, however, is not precise. An unambiguous definition is the intersection of the straight line portions before and after the bend. Obviously the end of the constant rate period, where the rate of diffusion of water to the surface becomes less than the rate of evaporation of a pool of water on the surface, depends on the conditions of the test - the air temperature, humidity, and velocity, and the nature of the coating (the nature of the dissolved solids and the amount of dispersed solids). However, it is often a relatively weak function of these variables, and in many cases changes insignificantly as they change within normal ranges. For a number of coatings we have found that the end of the constant rate period occurs when the coating consists of about 80% solids and 20% water, plus or minus several percent. Although the exact value should be determined by experiment and may differ significantly for widely differing coatings, this value of 20% water can be used as a rough approximation when data are not available.

Latent Heat of Vaporization. Knowing the water content at the end of the constant rate period allows us to calculate the amount of water to be evaporated to reach that condition. Equation 1 can now be used to find the evaporation rate given the latent heat of vaporization of water. The latent heat of evaporation is a function of temperature and is readily available in handbooks and in steam tables. In the range of usual temperatures it was approximated (1) as

$$\lambda \ (in \ J/g) = 2501.7 - 2.38 \ t \ ^\circ C \tag{5}$$

Example 1. A water-borne coating containing 35% solids was coated at a coverage of 50 g/m^2 on a thin plastic base and dried with 90°C air with a dew point of 20°C. The heat transfer coefficient in this dryer is known to be 110 W/m^2-K. The constant rate period ends when the coating contains 20% moisture. How long does it take to reach this condition?

The wet coating contains:
solids   = 0.35 g solids/g wet ctg × 50 g wet ctg/m^2 = 17.5 g/m^2
water   = 0.65 g water/g wet ctg × 50 g wet ctg/m^2 = 32.5 g/m^2

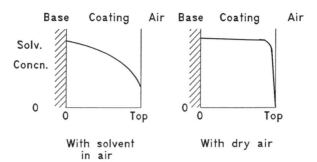

Figure 3. Sketch showing the concentration profile in the early stages of the falling rate period. On the right dry air is used; on the left moist air. See the text for an explanation.

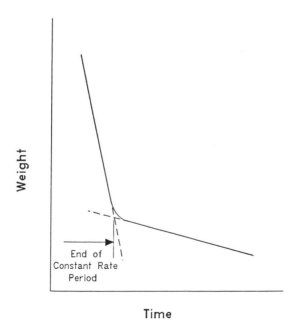

Figure 4. Weight vs. time in a drying test. The end of the constant rate period is taken as the intersection of the two straight lines.

water remaining at end of constant rate =
    17.5 g solids /m^2 × (0.20 g water/0.80 g solids)   = 4.4 g/m^2
water to be removed =
    32.5 g initial water/m^2 - 4.4 g water at end of CR/m^2 = 28.1 g/m^2
From the psychrometric chart in Figure 2b the wet bulb temperature of the
    air, which is the equilibrium coating temperature   = 35.5°C
The heat of vaporization of water is a function of temperature
    and its value at 35.5°C (3) (Equation 5 gives 2417.2) =2417.6 J/g
Assuming temperature equilibrium is rapidly reached, from equation 1, the
    rate of evaporation is

$$Evap.\ rate = 110\ (W/m^2\text{-}K) \times (90 - 35.5)\ (K)\ /\ 2417.6\ (J/g) = 2.48\ g/m^2\text{-}s$$

Note that temperature differences in °C or in °K are the same.
    The time to reach the end of the constant rate period is
    28.1 g water/m^2 ÷ 2.48 g water/m^2-s   = 11.3 s
If the coating were on a continuous sheet moving a 1 m/s, then distance into
    the dryer to the end of the constant rate period   = 11.3 m

If the heat transfer coefficient is not known but the location of the end of the
constant rate period is known, then different values of the heat transfer coefficient
can be tried until the calculated location matches the experimental value. This is
easy to do when using a spreadsheet for the calculations. The end of the constant
rate period is found from a plot of temperature versus location in the dryer, such as in
Figure 1. At the end of the constant rate period the temperature rises sharply. Of
course, the temperature will also rise sharply if the air temperature in a new zone is
higher than in the previous zone. It is relatively rare, however, for the end of the
constant rate period to be exactly at the end of a dryer zone.

Dryers for continuous coatings are frequently divided into a number of zones,
and the air temperature and moisture content controlled independently. The drying
rate in each zone can be found as above, and thus the moisture content of the coating
at the end of each zone if it is in the constant rate period, as well as the location in
the dryer at the end of the constant rate period. When different air velocities or air
pressures are used in the different zones, then the heat transfer coefficients can be
calculated from the heat transfer coefficient at reference conditions using either
equation 3 or 4. If the heat transfer coefficient at reference conditions is not known
it can be found by choosing different values until a match is found with the
experimental location of the end of the constant rate period.

It should be pointed out that in making these calculations an additional
assumption has been introduced: the temperature is uniform from the top to the
bottom of the coating, and down to the bottom of the support as well. This is an
excellent assumption for most coatings on most supports, as has been shown (4).
However, for unusually thick coatings, or for coatings on very thick supports with a
relatively low thermal conductivity, this assumption breaks down. It would hold,

however, for coatings on supports that are good insulators, for then essentially no heat would flow between the coating and the support.

For double-sided or flotation dryers the coating temperature is no longer the wet bulb temperature of the air, and so the calculations become more involved. Double-sided dryers are used in most new dryers for coatings on continuous supports, such as are used in making photographic films and papers, magnetic tapes, adhesive tapes, and specialty coated films and papers. As the support floats on air instead of running over rollers, there is no possibility of it becoming scratched. In addition, because heat is supplied on both sides, the dryers can be considerably shorter. However, more air needs to be supplied and therefore the operating costs are higher. Drum dryers, where a continuous sheet is pressed against heated drums, are commonly used for drying paper.

With both single-sided and double-sided drying the whole process, including the equations representing the psychrometric chart, can be entered into a computer spreadsheet, as has been done by Cary and Gutoff (*1*). We will now modify their equations slightly, and will introduce the possibility of use of infra-red heaters.

Let us make a heat balance on a unit area of coating in the constant rate period. Heat enters the coating by convection on one or both sides, and perhaps also by infra-red radiation. The heat is used to vaporize solvent and, if the system is not at its equilibrium temperature, to heat up the coating and the support. This leads to (*4*)

$$h_{ctg} (T_{a,ctg} - T_{web}) + h_{base} (T_{a, base} - T_{web}) + q_{ir} =$$

$$K_m \lambda (C_{surf} - C_{bulk}) + \Sigma W_i C_{p,i} dT_{web}/d\theta \qquad (6)$$

where c_{surf}, c_{bulk} are the solvent concentrations in the air at the surface and in the bulk

k_m is a mass transfer coefficient

W is the coverage, in wt./area, of the components in the coated web

λ is the latent heat of vaporization

θ is time

First we introduce the Chilton-Colburn relationship (*5*) between the heat and mass transfer coefficients. Next the perfect gas laws are used to find the concentration of water in the air from the vapor pressure of the water. Then this equation is used to find the rate of change of coating temperature in the constant rate period. Cary and Gutoff (*1*) assumed no infra-red heat and the equilibrium constant rate temperature. Thus they dropped out the last terms on the left and right sides of this equation. Inclusion of infra-red heating is relatively simple. However, inclusion of the transient term in the equation changes the method of solution (*4*) - now one has to march forward in time from the initial coating temperature, instead of solving an implicit algebraic equation.

The modeling discussed above can be used to locate the end of the constant rate period, which is more than adequate for most water-borne coatings. However, in some cases it is important to model the complete drying process, including the falling rate period. This has been done by Gutoff (*4*) for all types of solvents, but in that model the temperature variation of the heat of vaporization is not included.

The Falling Rate

In modeling the falling rate period the coating may be considered to consist of a number of slices, with the diffusion of water taking place between the slices. The evaporation rate is the rate of diffusion to the surface (equation 2). The concentration of water at the surface is assumed to be in equilibrium with the water vapor in the drying air at the local temperature. If completely dry air is used then the concentration of water at the surface would be zero.

As the equilibrium relationship between adsorbed water in the coating and water vapor in the air is rarely known, the surface concentration is usually assumed to be zero. When the several parameters for the diffusion model are chosen to match experimental data, the errors introduced by the assumption of zero surface concentration appear to be adequately compensated for.

The diffusivity of water through the coating is a function both of the water content of the coating and the temperature of the coating. Diffusivity increases with temperature, and the temperature effect is expressed as an activation energy of diffusion. This is typically 3 - 5 kcal/mol. To allow for the effect of the water content on the diffusivity in the coating any of a number of models may be used; the model used by Beels and Claes (*6*) is one of the simpler ones.

The drying programs discussed above (*1,4*) are semiempirical, in that the various parameters - the heat transfer coefficient at reference condition, the constants for the diffusivity relationship if falling rate calculations are included, and a vapor pressure factor to correct for vapor pressure lowering - can all be found by matching the spreadsheet to one or two test runs. This makes it relatively easy to use. A number of assumptions are involved, some of which have already been discussed. The temperature is assumed constant across any cross section of coating and web. The coating, which may consist of many layers, is taken as one well-mixed mass. If the water contains other solvents such as ethanol or acetone the solvent properties are taken as unchanging - which while it may be a good approximation it obviously cannot be exact, as the more volatile components will evaporate faster than the water.

There are other drying models which involve fewer approximations but require more data to run and may require more powerful computers. For example, Cairncross et al. (*7*) have a complete finite element model of drying. In the falling rate period each component of the solvent has its own diffusivity equation. The many parameters in the model may be difficult to determine.

It is suggested that the simplest model should be used that still gives the information needed and adequately matches the data.

Drying Defects

During drying a number of defects can occur, whether the coating is dried in a dryer or by standing in ambient air. Skinning has already been thoroughly discussed. In this section we will discuss several other drying defects: those due to air motion, those related to drying stresses, blisters, and pinholes.

Defects Caused by Air Motion. When the coating is soft and easily disturbed it can be easily damaged by non-uniform air motions. Thick layers are easier to disturb than thin layers, and low viscosity layers are easier to disturb than high viscosity layers. At the beginning of the dryer the coated layer is the thickest and has the lowest viscosity, and therefore it is as the start of the dryer that most problems with air motion occur. One possible cure is to coat a more concentrated layer, so that the layer would be thinner and would have a higher viscosity.

Dryer Bands. Non-uniform drying air flows can disturb the coating and cause dryer bands. This occurs most frequently in a dryer when round air nozzles are used. If the air velocity is too high dryer bands with a spacing equal to the diameter of the nozzles will be seen, as indicated in Figure 5a. When the air comes out of slots the width of the coating, dryer bands are rarely seen. Even if the air disturbs the coating a standing wave will form directly under the slot, with a smooth coating reforming as the web moves downstream. This is shown in Figure 5b.

When dryer bands form the air velocity has to be reduced. If the coatings have been gelled by chilling before entering the dryer, as may be the case with gelatin-based layers, the web temperature should be maintained below the melting point of the gel. This can be done by reducing the humidity of the air, which will lower the wet bulb temperature. With single-sided drying, as explained earlier, the equilibrium coating temperature in the constant rate period is the wet bulb air temperature. With double-sided heating, the coating temperature is higher than the web bulb temperature, but the coating temperature is still reduced by lowering the air humidity.

Even with gelled layers dryer bands can still form if the gel is soft and the air velocities are too high.

Mottle (one of many colored spots on a surface). Mottle is another defect caused by air motion, but here the disturbance is random in nature, and occurs over distances on the order of a centimeter or so. It has been known to occur where the wet coating enters the dryer. In one case in a dryer the enclosure had been deformed while the air pressure above the coating had not been properly balanced to be the same as the room pressure. Air was blowing out of the dryer enclosure unevenly, as observed by holding a thread where the coated web entered the dryer.

The cure in this case was to balance the air pressure to be the same as the room pressure, and to repair the dryer enclosure.

Surface Blow-Around. In some cases the air motion can be so severe that large scale movement of the wet coating occurs. Such air motions must be avoided in the early stages of drying.

Stress-Related Defects. When a coating dries there is a tremendous change in volume. If, for example, there were 10% solids by volume, then on drying the volume will decrease by about 90%. The coating will tend to shrink in all dimensions. However, it is restrained in two dimensions by adhesion to the web. As result stresses develop in the coating which can result in curling and cracking, in crazing, in delamination, in windowing (or "starry night" in photographic films), or in reticulation in processed photographic films.

    Curling and Cracking. The shrinking of the film in drying puts the film in tension and the support, which prevents the shrinking in the horizontal dimension, into compression. This is illustrated in Figure 6, which shows why this shrinkage of the coating tends to cause curl of the coating plus support in the direction of the coating. If the support is too stiff to allow appreciable curl, or if the back side has been coated with a similar layer and therefore tends to curl in the opposite direction, then no appreciable curl can occur and the coating remains under tension. If the coating is too weak to support this tension the coating will crack. This is sometimes referred to as mud cracking, as it resembles a field of mud after the sun dries it out. Croll (8), in a simplified analysis, showed that the residual tensile stress in a coating is related to the solvent level at the point of solidification and the final solvent level in the dry coating. He approximated the residual tensile stress by

$$S_t = E\,(\phi_s - \phi_r)\,/\,(\,1 - \upsilon\,) \tag{7}$$

where E is Young's modulus of elasticity
        υ is Poisson's ratio
        ϕ_s is the volume fraction of solvent at the point of solidification
        ϕ_r is the volume fraction of residual solvent at the end of drying

    Young's modulus is the constant of proportionality between stress and strain. Poisson's ratio is another fundamental elastic constant and is a measure of the change in volume on stressing a material. If the volume remains constant Poisson's ratio equals 0.5. If the volume increases on applying tension then the value is less.

    Croll identified the solidification point as the solvent level at which the glass transition temperature rises to the drying temperature, and stress can build up on further solvent removal.

    Note that the stress is independent of the coating thickness and of the initial solution concentration. Thus a thicker coating will have a greater tensile force (equal to the stress times the area) and therefore will more likely to curl or crack.

    It should also be pointed out that during slow drying the molecules have more time to relax before motion effectively ceases; therefore with slower drying there is a reduced tendency to curl or crack. It is generally true that stress-related defects are reduced or eliminated by slower drying. However, for high production rates the highest possible rates of drying are desired.

As the residual solvent level increases the stress is less. Thus, at high humidities, where the equilibrium moisture content is higher, the stress will be less, the curl will be less, and there will be less tendency to crack. In summer when the humidity tends to be high in many parts of the country curl is noticeably less than in winter. Adding humectants or plasticizers increases the mobility of the molecules and reduces curl and cracking. At elevated temperatures the elastic modulus is less and there is greater molecular mobility, both of which tend to reduce curl and cracking. Holding the dried coating at elevated temperatures can, in some cases, reduce the curl and the tendency to form cracks.

There are conflicting reports on the effect of adding dispersed particles. In some cases they reduce curl and cracking, and in some cases they have an opposite effect. Apparently the particle shape, among other factors, plays a role.

Delamination. If the adhesion between a coated layer and the support, or between a coated layer and a lower layer, is poor, than the residual stresses that arise in drying can cause the layer to separate from the support or from the underlying layers. The residual stress tend to concentrate at corners and edges, therefore this is where delamination usually begins. In extreme cases a layer can completely come off.

Poor adhesion is usually due to a mis-match of the surface energies at the interface. It is well-known that the surface tension of a wet coating has to be less than that of the solid support. Surfactants usually aid in promoting adhesion as well as wetting. In water-borne coatings surfactants are often necessary to obtain good wetting. With plastic supports the support itself often requires treatment (flame treatment, or corona treatment, or overcoating) to increase its surface energy to promote wettability. The nature of the dispersed or dissolved solids in the coated layer also has a strong influence on adhesion.

Windowing or "Starry Night". Drying stresses can drive particles deeper into the coated layer. Figure 7 illustrates this. As drying proceeds from the top of the layer down towards the support, the stresses on the top of the particle will be greater than on the bottom. This tends to drive the particle down deeper into the coating. In the paint industry this is called windowing. In the photographic industry, where inert silica particles may be used in a surface layer to lower the reflectivity, the particles can displace silver halide crystals. In a uniformly exposed negative this would show up as white spots on a black background, giving rise to the name "starry night". It is usually cured by drying at a lower rate.

Blisters. Cairncross et al. (7) pointed out that if at any point in the drying process the temperature of the coating exceeds the boiling point of water at its local concentration in the coating then the water will boil and cause blisters. This normally occurs only in the falling rate period because evaporative cooling tends to keep the temperature down in the constant rate period. It may seem strange to speak of the boiling point of water in a relatively dry coating. However, dissolved species lower the vapor pressure of water. It then takes a temperature higher than the normal boiling point for pure water ($100°C$) to have the vapor pressure equal the

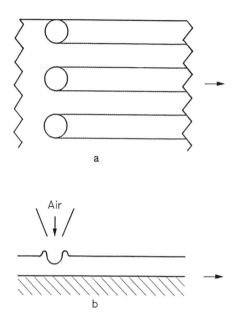

Figure 5. a- Dryer bands forming under round nozzles. b- A standing wave forms under a slot, with a smooth coating downstream.

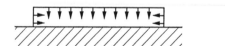

Figure 6. Stress builds up when a coating shrinks during drying but is restrained by adhesion to the support.

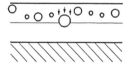

Figure 7. Drying stresses can drive a particle down into the coating. The stresses are greatest near the surface of the coating, where the water content is lowest.

atmospheric pressure, which is the definition of a boiling point.   When the concentration of water in a coating is very low the boiling point elevation is very high.   Thus the boiling point of water in the coating rises greatly as drying proceeds, and it is exceedingly high in an almost dry coating.   Blisters occur when the solvent boils.   Conceivably blisters can also occur by violent bursting of air bubbles, but normally this results in much smaller defects than blisters.   The cure for blisters is to lower the rate of heat transfer to the coating by lowering the air temperature or the air velocity.

Pinholes in the Coating.   The cause of pinholes may or may not be the drying process, but the pinholes appear during drying.   Pinholes can have a number of causes, including air bubbles, craters formed by dirt particles falling onto the coating, and poor surfactancy.   In the coating of a dispersion with an anionic surfactant pinholes were found.   With a non-ionic surfactant giving the same surface tension pinholes also formed.   However, using both the non-ionic and the anionic surfactants, with the same surface tension, the dried coating was pinhole-free.   It is likely that the adsorption of the surfactants onto the dispersed particles may be related to this phenomenon.

Conclusions

The drying of water-borne coatings is now fairly well understood.   The calculation of the location of the end of the constant rate period, were most of the drying occurs in aqueous systems, can be easily done manually for single-sided drying.   Computers speed up the task, and spreadsheets are now available to calculate the complete drying curve, for single- and double-sided drying, with or without the addition of infra-red heating.

The causes and cures for drying defects such as skinning, those caused by air motion, those cause by drying stresses, blisters, and pinholes are explained.

Nomenclature

		Typical SI units
c	concentration	kg/m^3
c_p	heat capacity	J/kg-K
D	diffusivity	m^2/s
E	Young's modulus of elasticity	Pa
h	heat transfer coefficient	W/m^2-K
k_m	mass transfer coefficient	kg/s-m^2-(kg/m^3)
P	air pressure in the plenum	Pa
q_{ir}	rate of radiant energy absorption by the coated web	W/m^2
T	temperature	^0C
V	velocity	m/s
W	coverage per unit area	kg/m^2
x	distance upwards from the bottom of the coating	m

Greek

θ	time	s
λ	latent heat of vaporization	J/kg
ν	Poisson's ratio	-
ϕ_s	the volume fraction of solvent at the point of solidification	-
ϕ_r	the volume fraction of residual solvent at the end of drying	-

Subscripts

a	air
base	back side of web
bulk	bulk air
c, ctg	coating or coating side
i	i^{th} component
o	reference conditions
surf	air next to surface
web	web or coating

Literature Cited

1. Cary, J. C.; Gutoff, E. B., Analyze the Drying of Aqueous Coatings, *Chem. Eng. Prog.* **Feb. 1991**, *87*(2), 73-79.
2. Martin, H., Heat and Mass Transfer Rates Between Impinging Gas Jets and Solid Surfaces, in *Advances in Heat Transfer*; Hartnett, J. P.; Irvine,, T. F., Jr., Eds., Academic Press, New York, NY, 1977, Vol. 13; pp. 193-195.
3. Smith, J. M.; Van Ness, H. C., *Introduction to Chemical Engineering Thermodynamics*, 4th ed., McGraw-Hill, NY, 1987, Appendix C.
4. Gutoff, E. B., Modeling Solvent Drying of Coated Webs Including the Initial Transient, *Drying Technology*, **1996**, *14*, 1673-1693.
5. Chilton, T. H.; Colburn, A. P., Mass Transfer (Adsorption) Coefficients, *Ind. Eng. Chem.* **1934**, *26*, 1183; see also, Perry, R. H.; Chilton, C. H., Eds., *Chemical Engineers' Handbook*, 5th ed., McGraw-Hill, New York, NY, 1973, pp. 12:2.
6. Beels, R; Claes, F. H., Diffusion Phenomena in Gelatin Sheets, *Photogr. Sci. Eng.* **1977**, *21*, 336-342.
7. Cairncross, R. A.; Jeyadev, S.; Dunham, R. F.; Evans, K.; Francis, L. F.; Scriven, L. E., Modeling and Design of an Industrial Dryer with Convective and Radiant Heating, *J. Appl. Polymer Sci.* **1995**, *58*, 1279-1290.
8. Croll, S. G., The Origin of Residual Internal Stress in Solvent-Cast Thermoplastic Coatings, *J. Appl. Poly. Sci.* **1979**, *23*, 847-858.

Chapter 15

Spray Application of Waterborne Coatings

Lin-Lin Xing[1], J. Edward Glass[1,3], and Raymond H. Fernando[2]

[1]Department of Polymers and Coatings, North Dakota State University,
Fargo, ND 58105
[2]Research Development Center, Armstrong World Industries, Inc.,
Lancaster, PA 17604

In this final chapter, the application of coatings by spray is addressed. This method provides a rapid means for covering a substrate, but the efficiency of coverage and substrate appearance are influenced by the distribution of drop sizes. The types of spray guns and nozzles used, and the parameters defined as important in droplet generation are reviewed. Rayleigh's linear analysis cannot account for the wide distribution of droplet sizes produced in a practical spray process. This can be accounted for by the growth of non-sinusoidal surface waves as a non-linear effect. Given the complexity and interaction of the many variables involved, a universal concept for the spray application behavior of Newtonian fluids, mostly hydrocarbons and glycerine/water mixtures, has not been realized. In the latter part of this chapter, the behavior of non-Newtonian water-borne coatings applied by conventional-air and airless sprays are examined and evidence for the importance of dynamic uniaxial extensional viscosities in spray behavior is presented.

Spraying is a process in which a quantity of fluid emerges from a nozzle as a sheet which rapidly disintegrates into ligaments, and then into a large number of small droplets. It is utilized in many processes that include combustion (liquid fuel injection), agriculture (pesticide applications), and the chemical industries (i.e., spray drying, spray painting, etc.). Because it is rapid, spraying is widely used in large volume coating applications such as paper products, wood, wallboard and stucco, automobiles and appliances.

This chapter on the spray application of coatings is organized in three parts: **I**, discussion of the methods and equipment used; **II**, discussion of previous studies that approach the process as an engineering endeavor using dimensionless numbers,

[3]Corresponding author

and; **III**, discussion of the application of water-borne coatings from a component influence on the rheological properties and spray behavior of the fluid.

TYPES OF SPRAYING APPLICATIONS, NOZZLE DESIGN AND THEIR PROCESS VARIABLES

Types of Spraying Applications and Their Process Variables (*1-3*).   A wide variety of spray equipment is available. The most commonly used types are air (conventional), airless, air-assisted airless, electrostatic and combinations of these approaches.

Conventional Air Spray. Spray applications of coatings using air pressure as the only power source is the most popular method. In the application of a fluid by air pressure the viscosity of the fluid, fluid and air pressure, air to fluid pressure ratio, nozzle shape and size, and gun-to-surface distance are the important variables influencing the spray pattern and its behavior. To keep the overspray to a minimum, a fluid pressure of 5-25 psi is usually required; the air pressure is set, typically at 30-85 psi(*1*). The lowest air pressure which will atomize the coatings should be used. Nozzle openings in an air spray application range from 1-3.5 mm in diameter for various coating viscosities (generally defined by a flow time through a No.4 Ford cup of 25-30 sec.). The gun should be kept perpendicular to the surface during its entire movement, and between 6-10 in. from the surface.

A primary problem is the overspray created, and a 20 - 40 % loss in transfer efficiency. The primary advantage is the versatility associated with the choice in air pressure, coating pressure, and spray pattern, which is limited with most of the other devices.

Airless Pressure Spray. In an airless pressure spray application, the coating is forced by a high fluid pump pressure (1,000-6,000 psi) through a very small orifice (0.18-1.2 mm in diameter) causing it to atomize into very fine droplets. The same variables influencing the spray pattern in air spray are important in airless spray patterns: the viscosity of the fluid, fluid pressure, gun-to-surface distance, nozzle shape and size. The high fluid pressure allows application of high viscosity fluids such as high-solid coatings. The gun is held 12-14 in. from the surface. The shape and opening of the nozzle determines the width of the spray pattern and film thickness. The nozzle is frequently constructed with a tungsten carbide insert to minimize abrasive wear. Different nozzle types may be used, such as hollow-cone, solid-cone, and fan-spray nozzles, discussed in the next subsection.

The advantages of an airless spray relative to an air spray is its speed of application with less overspray; however, it is less versatile in that the spray pattern is not easily adjusted, the extremely small nozzle is easily clogged by foreign matter and the technique is limited to large areas. It is also dangerous to clean the airless spray gun with high pressure.

Air-assisted Airless Spray. This is different from airless spray in that two hose connections are required, one is the air-hose which supplies air to the gun through a small air compressor, another is the fluid hose that supplies fluid to the gun through hydraulic pressure. The fluid pressure is less than that in an airless spray (e.g., below 1000 psi.). The advantages of air-assisted airless spray is rapid application but with superior atomization and less overspray and drift.

Electrostatic Spray. When the coating leaves the spray gun, an electrode at the tip charges the atomized paint droplets. The article to be coated must be conductive and grounded and of opposite electrical charge to attract the charged spray droplets. Spray nozzles or guns are specially constructed for electrostatic spraying with an electrode extension in the center of the fluid nozzle to create a high voltage (it may be as high as 60,000 volts) that delivers a charge to the atomized coating. From a charged rotational disc and bell, the coating may also be sprayed centrifugally. Alternatively, a suitable electric field may be generated within a high voltage wire frame or grid through which sprayed droplets travel to acquire a charge. Atomization of the coating may be by air, airless (Figure 1) or rotational techniques.

Electrostatic spray has a number of distinct advantages: complete coverage of odd shapes with minimal overspray. A very uniform film on a moving substrate is obtained because the deposited coating acts as an insulator and will not accept further material when it reaches a definite film thickness. These features are countered by the limitations that the substrate needs to be conductive and the coating needs to be specially formulated to accept the electrostatic charge. Maintenance of the equipment is expensive, and only one coat may be applied, since the applied film insulates the substrate.

Nozzle Types (*3-5*). The major factors affecting droplet size are nozzle type and capacity, spray angle and pressure, and the fluid's properties. Nozzles are designated by a flow number which is a convenient way of comparing their output:

$$\text{Flow number} = \text{flow rate} / (\text{fluid pressure differential})^{1/2}$$

Different energy forms (pressure, centrifugal, kinetic and sonic energy) are applied to break up the bulk liquid during the spraying process. Accordingly the spray nozzles may be classified into the following types: pressure, rotary atomizers, pneumatic and sonic nozzles (*5*). In the spray application of a coating, pressure nozzles (fan, solid- and hollow-cone nozzles) and rotary atomizers (spinning discs) are often used (*4*).

Fan Nozzle. The internal shape of a fan nozzle is designed to cause the liquid to move in a single direction and to curve inwards so that two streams of liquid forming a fan meet at a lenticular or elliptical orifice. The shape of the orifice is particularly important in determining not only the amount of liquid emitted but also the shape of the sheet emerging from it, particularly the spray angle. The type used in this study and the elliptical pattern generated are illustrated in (Figure 2).

Cone Nozzle. For a given output and pressure, a cone nozzle produces a finer spray than the equivalent fan nozzle because of the swirling motion of fluid

AIRSPRAY ELECTROSTATIC

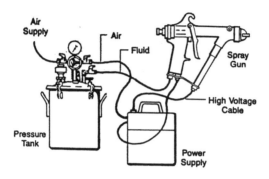

AIRLESS ELECTROSTATIC

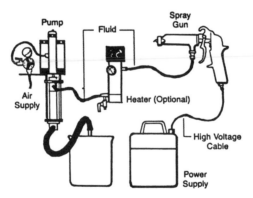

Figure 1. Diagram of Electrostatic Spray
(reprinted with permission from the Federation of Societies for Coatings Technology).

through the nozzle. For cone-type nozzles (hollow or solid cone), a wide range of through puts, spray angles and droplet sizes can be obtained with various combinations of orifice size, number of slots or holes in the swirl plate, depth of the swirl chamber and the pressure on the liquid.

Liquid is forced through a swirl plate, having one or more tangential or helical slots or holes, into a swirl chamber. An air core is formed as the liquid passes with a high rotational velocity from the swirl chamber through a circular orifice. In a hollow-cone nozzle (Figure 3), the thin sheet of liquid emerging from the orifice forms a hollow cone because of the air core present as it moves away from the orifice. The spray angle ranges from 30-160°, depending on design.

A solid-cone nozzle is similar in design to the hollow-cone nozzle except that the spray pattern is achieved by passing fluids centrally through the nozzle (Figure 4) to fill the air core. The resulting full-volumetric coverage enhances the rates of mass and heat transfer between the spray liquid and gas passing through the cone. The solid cone gives a narrower angle of spray and larger droplets than does the hollow cone. The spray angle is 30-120°. Example of the nozzles and spray pattern to be expected with a solid and hollow cone nozzles are illustrated in (Figures 3 and 4).

Centrifugal-Energy Nozzle (e.g., spinning discs). Centrifugal-energy nozzles (Figure 5) are rotated by a separate power source and are generally used in electrostatic spray applications. Liquid is fed near the center of a rotating surface so that centrifugal force spreads the liquid to the edge where the droplets are formed. The main types of centrifugal energy nozzles are disc, cups and cylindrical sleeves or wire mesh cages. These nozzles are capable of handling slurries and other materials which may clog the narrow passages of other nozzles. Single droplet, ligament and sheet formation from a spinning disc is illustrated in Figure 6.

THE PHYSICAL ASPECTS OF THE SPRAYING PROCESS

A wide variety of studies have been conducted on the spray application of fluids. They can be classified into the following groups: mechanisms of sheet disintegration; theories of ligament breakup and dimensional analysis, to relate droplet size with operating parameter such as air and fluid pressures, nozzle shape and size, and the physical properties of fluids (such as surface tension, viscosity and density).

Mechanisms of Sheet Disintegration. The high-speed photographic studies of Dombrowski and Fraser (6,7) identified three distinct modes of sheet disintegration: wavy-sheet (Figure 7), perforated, and rim disintegration (Figure 8). The predominate mechanism depends on fluid properties, nozzle design features and process operating conditions. Fraser and coworkers have suggested (7) that the wavy-sheet disintegration mechanism is the more usual, and gives smaller drops than the perforated one. The perforated disintegration, however, is still promoted as a dominant mechanism by other researchers (8,9).

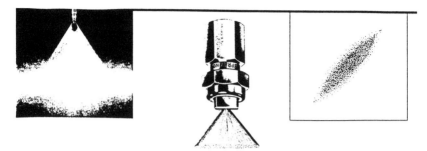

Figure 2. Fan-Spray Nozzle
(Figures 2-4 were provided courtesy of the Industrial Spray Products Catalog,
8447 Lake Street, Omaha, Nebraska 68134).

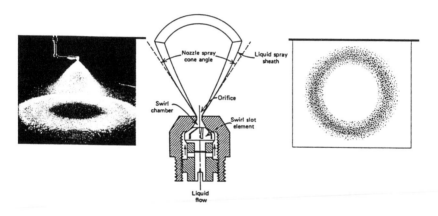

Figure 3. Hollow-Cone Nozzle

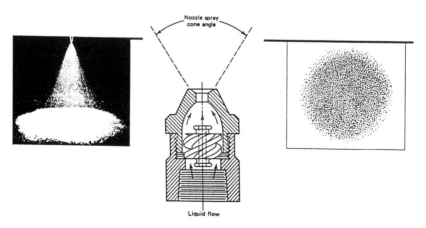

Figure 4. Solid-Cone Nozzle

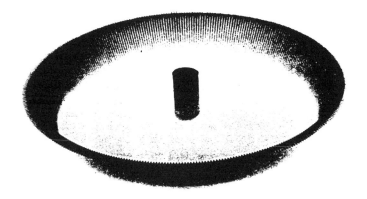

Figure 5. Grooved Toothed Spinning Disc
(Reproduced with permission from reference 5. Copyright 1976 Longman Scientific & Technical.)

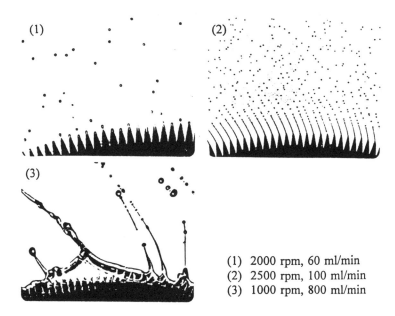

(1)   2000 rpm, 60 ml/min
(2)   2500 rpm, 100 ml/min
(3)   1000 rpm, 800 ml/min

Figure 6. The Formation of Single droplet, Ligament and Sheet from Spinning Disc
(Reproduced with permission from reference 5. Copyright 1976 Longman Scientific & Technical.)

Wavy-sheet disintegration. According to the general model (7) for the disintegration process, a wavy perturbation builds up in the sheet, giving it the typical form of a "waving flag" (Figure 7), that disintegrates when the unstable wave perturbation grows, at right angles to the direction of liquid flow, to a critical value. A wavy-sheet disintegration is produced under flow conditions of low turbulence in the orifice and with fluids of low viscosity and low surface tension.

The sheet dissipates into roughly parallel ligaments (Figure 7). Driven by surface tension, the ligaments contract into cylindrical segments and then into drops. The mechanism defined by Rayleigh for a filament breakup to drops (discussed below) is likely in play for the wavy-sheet disintegration process in total.

Perforated sheet disintegration. A perforated sheet is produced in the orifice under highly turbulent flow conditions (i.e., high Reynolds number, defined in the next subsection, > 20, 000) or fluids having high surface tension, high density and low viscosity or unwettable particles existing in the suspension fluids. As holes develop in a sheet they are driven to expand by surface tension forces, their boundaries form unstable network of ligaments which eventually break into chains of droplets.

In these first two mechanisms of sheet disintegration, surface tension, viscous and inertia and aerodynamic forces are involved in the process of producing spray droplets. These variables are encompassed in dimensionless parameters referred to as Reynolds and Weber numbers, defined in a subsection to follow.

Rim disintegration. In rim disintegration, surface tension contracts the edge of the liquid sheet and forms rims (Figure 9) which produce large droplets at low pressure. At higher pressures threads of liquid are thrown from the edge of the sheet.

Theories of Ligament Breakup

Linear theory (first-order). Rayleigh (10) undertook the first theoretical treatment of capillary jet   stability. He examined the stability of a stationary, infinitely long, inviscid jet with a circular cross section. Neglecting the effects of surrounding air and assuming that an initially sinusoidal perturbation remained sinusoidal (i.e. a first-order perturbation), Rayleigh, found that only axisymmetrical surface disturbances with a wavelength (λ) to jet diameter (d) ratio (λ/d) > π would grow (i.e., wavelengths must be longer than the circumference of the jet), while disturbances with $\lambda/d < \pi$ would be stable. The unstable waves were shown to grow in an exponent manner with time and the frequency of vibration.

Weber (11) later incorporated liquid viscosity and liquid-air interactions into this linear analysis and more accurately predicted the disturbance growth rates of a jet. Sterling and Sleicher (12) modified this theory by including the effects of the gas phase viscosity and found improved agreement with experimental results. Although the addition of viscous and aerodynamic forces modified the growth rate of the surface wave, the linear analysis (uniform drop model) can not account for the results observed. A wide distribution of droplet sizes is produced in a practical spray process. Commonly, one observes that large drops are interspersed with smaller drops (satellites).

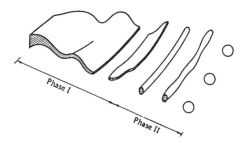

Figure 7. The Model for Wavy-sheet Disintegration

(Figures 7–9 adapted from reference 7.)

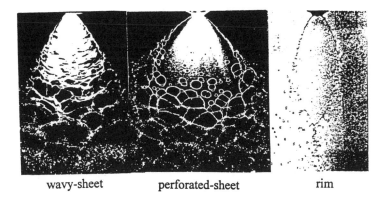

wavy-sheet   perforated-sheet   rim

Figure 8. Three Mechanisms of Disintegration of Spray Sheet.

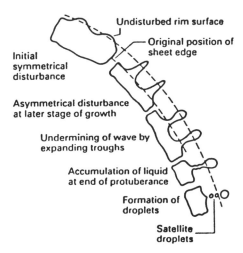

Figure 9. Rim Disintegration

Nonlinear theory (second, third-order). It is a common observation that the waveform of an initially sinusoidal perturbation becomes nonsinusoidal close to the point of drop formation (*13*). In an experimental and analytical study of the instability, Emmons, Chang and Watson (*14*) found that the growth of the non-sinusoidal surface wave is a non-linear effect. Nonlinear capillary instability has been studied by numerous researchers (*15-20*) since this epoch study.

Dimensional Analysis. Dimensionless analysis is a technique used to normalized variables to assess the relative contribution of components in a complex system. For example, Krieger (*21*) used such analyses to define the relative contributions of median particle size, adsorbed layer thickness, and electroviscous effects on the rheology of latex dispersions. Various dimensional analyses (*11, 22-28*) have been made in spray studies to relate droplet size with operating parameters such as nozzle shape and size, and the physical properties of fluids.

For a non-viscous liquid, Rayleigh (*10*) first predicted that the filament would break up into essentially spherical drops with a uniform diameter D_{AV} for a given orifice diameter, d_n

$$D_{AV} = 1.89 \, d_n \tag{1}$$

For more viscous liquid, Weber (*11*) used a modified diameter ratio:

$$D_{AV} / d_n = 1.89 \, [\, 1 + 3 \, We_L^{1/2} / Re_L \,]^{1/6} \tag{2}$$

The liquid Reynolds (Re) and the liquid Weber (We) numbers (two dimensionless groups), are defined in equations 3 and 4:

$$Re_L = \rho_L \, V_j \, d_n \, / \, \eta_L \tag{3}$$

$$We_L = \rho_L \, V_j^2 \, d_n / \, \sigma \tag{4}$$

where: V_j = liquid jet velocity, σ = surface tension,
ρ_L = liquid density, η_L = liquid viscosity.
d_n = orifice diameter,

The Re number represents the ratio of inertia forces to the viscous force, and the We number represents the ratio of the disruptive aerodynamic forces to the fluid's surface tension.

Ohnesorge (*5*) also observed that jet stability is a function of Reynolds number (i.e., jet dissipation is a function of liquid viscosity, density, surface tension and nozzle size). He observed that the mechanism of liquid break-up could be expressed in three stages, each stage characterized by the magnitude of the ratio of the (Weber)$^{0.5}$/Reynolds dimensionless numbers, reflecting the ratio of viscous to surface tension forces operating on the fluids (equation 5).

$$We^{1/2} / Re = \eta_L / (\, \rho_L \sigma d_n \,)^{1/2} \tag{5}$$

When this analysis is plotted against the Reynolds number (Figure 10), the graph defines three zones. In the first zone at low Re numbers, the break-up of liquid jets exhibits the Rayleigh mathematical prediction. In zone 2, at intermediate Re numbers, the break-up of liquid is by oscillations with respect to the jet axis. The magnitude of these oscillations increases with air resistance until complete disintegration of the jet takes place. In zone 3, at high Re numbers, complete atomization occurs at the orifice from which the jet emerges.

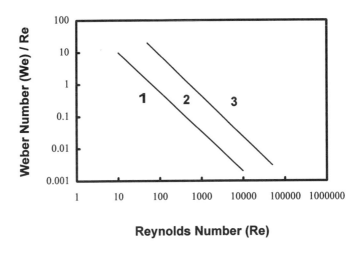

Figure 10. Ohnesorge Chart showing Liquid Jet Disintegration as a Function of Re Number (Adapted from reference 5).

From an experimental viewpoint there has been only one systematic study of fluid properties across the spectrum of nozzles and application processes. Atomizer types, atomizer geometries, liquid physical properties (surface tension and viscosity) and operational settings were examined(26-31), that included air, airless(28), rotary atomizers(29), and electrostatic spray (31) equipment. In the first of these studies, Wang and Lefebvre (22) examined the spray process in a hollow-cone nozzle. Diesel oil and water were chosen to provide the variation in surface tension and mixtures of diesel oil and liquid polybutene provided the variations in viscosity. Their observations were described in the following equations:

$$SMD = SMD_1 + SMD_2 \qquad (6)$$

where:

SMD = Sauter mean diameter of droplets
(diameter of a droplet whose ratio of volume
to surface area is equal to that of the entire
spray)

SMD_1 represents the first stage of the spray process in which the instability of a liquid sheet is generated due to the combined effects of hydrodynamic and aerodynamic forces.

$$SMD_1 / t_s \propto (Re * We^{1/2})^{-0.5} \tag{7}$$

where:

t_s = the initial sheet thickness at nozzle exit, $t_s = t*\cos\theta$
(t^* = film thickness in the orifice; θ= half the spray-cone angle; the Reynolds and Weber numbers are defined in equations 3 and 4).

SMD_2 represents the second stage of the spray process in which the unstable liquid breaks into ligaments and then droplets.

$$SMD_2 / t_s \propto We^{-0.25} \tag{8}$$

Lefebvre suggested several modifications in his study: the Re number, which relates to the bulk liquid, should be based on V_j -- the absolute velocity generating the turbulence and instabilities within the bulk liquid; the Weber number, which is associated with events occurring through the action of the surrounding gas on the liquid surface, should be based on V_R --- the relative velocity which is promoting the atomization mechanisms that occur on the liquid surface and in the adjacent ambient gas. The proportionality in the term $SMD_1 / t_s \propto (Re * We^{1/2})^{-0.5}$ is intended to represent the manner in which surface tension and viscous forces act together in opposing the disruptive actions of the hydrodynamic and aerodynamic momentum forces. This model is superior to Ohnesorge's model in the pressure atomization process, where velocity is of paramount importance.

The Purdue group studied (26) the effect of operating conditions and liquid properties (viscosity and surface tension) on the drop sizes and distributions produced by a fan-airless spray atomizer. The fluids employed were water, water/glycerine mixtures, silicone oils, and an unspecified commercial enamel coating. The following equation described their observations:

$$SMD/d_h = 2.83 \ (\sigma \ \mu_L^2 / \rho_A \ d_h^3 \ \Delta P_L^2)^{0.25} + 0.26 \ (\sigma \ \rho_L / \rho_A \ d_h \ \Delta P_L)^{0.25} \tag{9}$$

where:

SMD---- Sauter mean diameter, m
d_h----- hydraulic mean diameter of nozzle orifice, m

σ----- surface tension, N/m

μ_L---- absolute viscosity of liquid, Ns/m^2

ρ_A---- density of air, kg/m^3

ρ_L---- density of liquid, kg/m^3

ΔP_L----- nozzle injection pressure differential, Pa

This equation is unsuitable for liquids which combine high surface tension (>73 mN/m) with very high viscosity ($100*10^{-6}$ m^2/s, or 100 cst). The author explained that when high surface tension is accompanied by an abnormally high viscosity, a change in the mode of sheet disintegration into drops occurs.

In a high-speed, rotary-bell atomization study, using a laser-based diffraction technique and high-speed photography to quantify nozzle exit behavior (a technique used in all of the Purdue studies), this group examined (29) several Newtonian fluids: water, glycerine/water, corn syrup/water and hydrocarbon oil that differ in surface tension (by a factor of 2.5) and viscosity (by 100). High-viscosity fluids film the bell very evenly and produce long regular ligaments, whereas low-viscosity fluids film the bell incompletely and produce very irregular ligaments that disintegrate near the bell edge. The mean drop size was fairly insensitive to large changes in flow rate and viscosity at bell speeds higher than 20,000 rpm. Increasing the flow rate or bell speed at lower values while the other was held constant, lead to broader distributions of drop sizes in the spray pattern, similar to that observed in air and airless applications. The latter were referenced to theses, but not reported in literature publications.

An examination of the influence of electrostatic forces (30) on a commercial enamel paint (Newtonian between 80-8,000 s^{-1}) on the size of droplets produced from fan-electrostatic airless spray revealed that at low injection pressures, the average size of droplets is decreased and the droplet size distribution is narrowed with an increase in the applied voltage. The possible reasons are due to internal electrostatic forces countering surface tension, or external repulsive forces which reduce droplet coalescence. As the injection pressure is increased, the electrostatic influence on drop size and distribution decreases.

Other studies of possible peripheral interest to coating applications are listed in Table I.

PREVIOUS and CURRENT STUDIES OF SPRAY APPLIED COATINGS

Prior Studies.   Although the application of coatings by spray has been practiced for over 70 years, the process is still not well understood because of the complexity of the atomization process, differences in the design, size and operating conditions of the nozzles tested, and variations in the fluid's properties. Studies of industrial coatings (e.g., high-solid, anti-corrosive vinyl coatings, NVV= 30 %, PVC= 15-34 %), applied (32) by airless spray, have focused on maintaining the sag behavior of the coating on a substrate after spray application. It was concluded that the best compromise between sprayability and sagging is obtained with a fluid having a shear thinning index between of -0.50 and -0.60. This study did not address the fluid's

Table I. Spray Droplet Size Relationships to Process and Fluid Variations

Material & Nozzle	Dimensional Analysis	Reference
black liquor, hollow-cone spray nozzle	$D_m = 3.47 \, \eta^{0.14} \, V_n^{-0.47}$ D_m=median droplet size η=viscosity of black liquor V_n=nozzle velocity	Samuels (23)
water & kerosine (low viscosity liquid), fan spray nozzle	$SMD \propto (FN\sigma/\theta\Delta P_L)^{1/3}$ SMD=Sauter mean diameter, m FN=nozzle flow number,m^2 σ= surface tension, N/m θ=spray angle, radians ΔP_L= nozzle injection pressure differential, Pa	Dorman[a] (24)
glycerine / water (finite viscosity liquid), fan-spray nozzle	$SMD = 0.071 \, [\, t_s \chi \sigma \, \mu_L^{0.5} / \rho_L^{0.5} \, U_L^2 \,]^{1/3}$ SMD= Sauter mean diameter, m t_s=the sheet thickness at breakup χ=the distance downstream from the nozzle at which the sheet breaks apart, c.g.s. σ = surface tension, N/m μ_L= absolute viscosity of liquid, Ns/m^2 ρ_L= density of liquid, kg/m^3 U_L=velocity of liquid, m/s	Dombrow-ski & Johns[b] (25)
syrup aqueous solution (Newtonian liquids), fan-spray nozzle	$MMD \propto (FN\sigma^2/r \, \theta \, \Delta P_L^2)^{1/3}$ MMD=Mass median diameter, m FN=nozzle flow number,m^2 σ= surface tension, N/m r=length of liquid sheet to point of breakup,m θ= spray angle, radians ΔP_L= nozzle injection pressure differential, Pa	Ford & Furmidge[c] (26)

Comments:

[a] The low viscosity liquid used is not realistic one.

[b] Unfortunately, t_s (the sheet thickness at breakup) is unknown.

[c] The equation has a stronger mean drop size dependence on surface tension and injection pressure. The drawback of this expression is its inclusion of a unknown term r (length of liquid sheet to point of breakup). It is also confined to relatively low injection pressures. [< 0.38 Mpa (40 psi)]

behavior under high deformations, and therefore its behavior when exiting the spray nozzle, and it was limited to a specific formulation. In another study (33), the atomization of a high-solid vinyl coatings (NVV=25-40 %, PVC=13.8 %) by airless spray (the type of nozzle was not reported) measured the rheological behavior in the very high shear rate region (60,000 sec^{-1}) through the use of a Severs Extrusion Rheometer. It was reported that the atomization appears to be largely dependent on their rheological properties at ultra-high shear rates and to only a minor extent on the surface tension characteristics. The need for thixotropy in the high build coating and the importance of thermal effects also were mentioned.

The studies described in the previous section were on Newtonian fluids. In spray applications it is common to relate the sprayability of a coating to its viscosity at high shear rates. This was also true in early roll applied coatings that related viscosities at high shear rates to spatter(*34*); however, in a more detailed examination of the misting behavior of roll applied coatings, spatter was clearly related (*35, 36*) to dynamic uniaxial extensional viscosities (DUEVs). As Strivens (*37*) has noted, studies of this nature have been ignored, and most who review these areas pretend that coatings do not exhibit viscoelastic behavior. NonNewtonian behavior is noted in the industrial coating spray studies (*32, 33*) cited above, and most water-borne coatings exhibit NonNewtonian flow. Viscoelastic behavior should be expected in "real world" formulations.

Studies in Progress. Most water-borne coatings are non-Newtonian fluids, and they exhibit viscoelastic behaviors. Our studies in this area (discussed below) support a proposal that it is the dynamic uniaxial extensional viscosity of a fluid that is significant in spray behavior. The formulation contains an excess of surfactant which imparts a constant surfaces tension to all formulations, and the thickener does not influence the surface tension of the formulation. The formulations are pigmented with TiO$_2$ which equalizes the formulation's density. Thus in the first phase of our study the viscosity (most are nonNewtonian) of the coatings is the only variant. The formulations are examined in air spray studies with a pressure of 55 psi (see Table II for flow velocities for water). The formulation containing 6*10^6 molecular weight (M.W.) polyoxyethylene (POE) also is examined in high pressure (2,000 psi.) airless spray. The formulations are visualized as they are exited from fan, solid-cone and hollow-cone nozzles.

The formulations (Table III) contain an acrylic latex and TiO$_2$. This study is done within the practices used in formulating an architectural coating. The formulation is prepared at a 32 % NVV level of disperse phase, with the ratio of

Table II. The Spray Nozzles Used in Air-Spray Application

Nozzle Type	Spray Angle (°)	gpm[a]/40 psi	gpm/60 psi	Equivalent Orifice Diameter (in)
Fan	73	0.154	0.19	0.034
Solid Cone	53	0.19	0.23	0.025
Hollow Cone	70	0.0417	0.057[b]	0.031

[a] gpm: gallons per minute, [b] at 75 psi

Table III. The Formulation of Latex Coatings

Material	Function	wt.%
TiO$_2$-R900	pigment	17.8
Tamol 731(25 wt.%)	dispersant	0.5
Tergitol 15-S-9	surfactant	0.25
Ethylene Glycol	freeze-thaw stabilizer	1.25
Texanol	coalescing aid	1.25
NDW	antifoaming agent	0.29
PhHgAc	anti-fungus agent	0.0375
Latex E-1698(46 wt.%)	acrylic copolymer(binder)	38.5
Water & thickener		40.2
Total		100.0

TiO$_2$ to latex at 0.21 (pigment volume concentration). A water-soluble polymer (W-SPs) is used to thicken the formulation to a 90 Kreb Unit (KU) viscosity. The amount of W-SP required is inversely related to its molecular weight. Thus, to achieve a 90 KU formulation requires ca. 5X the amount of 68,000 M.W. hydroxyethyl cellulose (HEC) than of the 950,000 M.W. HEC. All formulations had to be diluted to obtain a sprayable formulation. The amounts of the thickeners used in the final formulation are given in Table IV and in the Figures. The HEC and POE W-SPs thicken primarily through chain entanglements. Hydrophobically-modified, Ethoxylated urethane (HEUR) thickeners (38) also were used to thicken the formulation. With these "associative thickeners" the amount required is determined by the size of the terminal hydrophobe, that determines the amount of association among the thickener. With a C$_{18}$H$_{37}$- or C$_{14}$H$_{29}$- size hydrophobe only low amounts of thickener are required to achieve a 90 KU formulation viscosity, even though their molecular weights are low; however, with a C$_6$H$_{13}$- terminal hydrophobe size, the amounts required are high, near those required for the low M.W. HEC. With this understanding, the rheology of the formulations are examined.

The viscosity dependence on shear rate is illustrated for the HEC and POE thickened formulations (28% NVV, 20% PVC to obtain sprayability) in Figure 11 and for HEUR thickened formulations in Figure 12. The viscosity of the traditional HEC thickened formulations are very shear thinning, a phenomenon related to depletion flocculation (discussed in chapter 6). The HEUR thickened formulations with large hydrophobes also are shear thinning. This is related to a critical shear rate at which the hydrophobic interactions are disrupted. The small hydrophobe HEUR

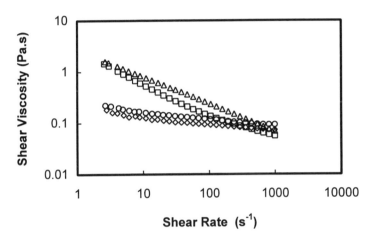

Figure 11. Viscosity vs. Shear Rate of Latex Coating
Thickener: ◊, no thickener; O, POE (Mv = 6.0 * 10^6);
□, HEC (Mv = 6.8 * 10^4); Δ, HEC (Mv = 9.5 * 10^5).

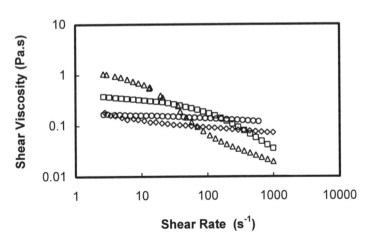

Figure 12. Viscosity vs. Shear Rate of Latex Coating
HEUR Thickener: ◊, no thickener; O, HEUR-45; □, HEUR-40; Δ, HEUR-60.

Table IV. Characterization of the Thickeners Used in the Latex Coatings

Thickener	Mv.	wt.%
POE-309	$6.0*10^6$	0.015
HEC-100M	$9.5*10^5$	0.25
HEC-QP09	$6.8*10^4$	1.20
HEUR-60[a]	$2.9*10^4$	0.25
HEUR-40[b]	$2.7*10^4$	0.24
HEUR-45[c]	$2.5*10^4$	1.00

[a-c] The synthesis and characterization of model HEUR thickeners is a laborious task and spray studies demand significant amount of materials. We choose to use these commercial materials from King Industries, and our assessment of $C_{18}H_{37}$-, $C_{14}H_{29}$- and C_6H_{13}- terminal sizes for these respective materials is an approximation from our studies of model HEURs, not a disclosure of King Industries.

thickened formulation is nearly Newtonian, despite the large amount of thickener added. This latter formulation will not be discussed in detail as high viscosities at high shear rates are undesirable in spray applications; they promote air entrapment and surface defects in coatings. As cited earlier by Striven (37) and suggested by the shear thinning behaviors of most of the formulations in Figures 11 and 12, these coatings should have an elastic characteristic. Oscillatory flow in shear deformation studies, permit the determination of an elastic (G', the storage modulus) and a viscous (G", the loss modulus) contribution to the flow behavior (Chapter 6). There is no elastic behavior detectable in the oscillatory study of the formulation without thickener (Figure 13A). Among the HEC and POE thickeners, the high M.W. HEC thickened formulation exhibits the largest elasticity, i.e., storage modulus, G', Figure 13D) despite the small amount of thickener used. With the larger hydrophobe sizes in the HEUR thickeners, the elastic character also increases (Figure 14C-D), despite the low amounts of thickener used. The unusual response in Figure 14D needs additional study. The HEUR with the C_6H_{13}- terminal hydrophobe size is the least elastic (Figure 14B). Even at high (10 Hz) frequency, the storage modulus, G', does not approach the loss modulus (G") curve. Both the low M.W. HEC and the small hydrophobe HEUR require a large amount of thickener, yet both formulations are significantly lower in elastic character.

   The jet stability behavior of an aqueous solution of very high molecular weight acrylamide/acrylic acid copolymer (HPAM) has been studied at low concentrations (39). This type of fluid has a very low shear viscosity, below that of the glycerol/water or other type of Newtonian fluid discussed in the previous section, yet the jet stability of the HPAM fluid is notably greater than those of the Newtonian fluids studied. The HPAM fluid, however, has a notable DUEV. Interestingly, the dimensionless breakup length of the HPAM fluid was observed to fit the equation cited by Weber (equation 2), but the application of dimensionless numbers allows for

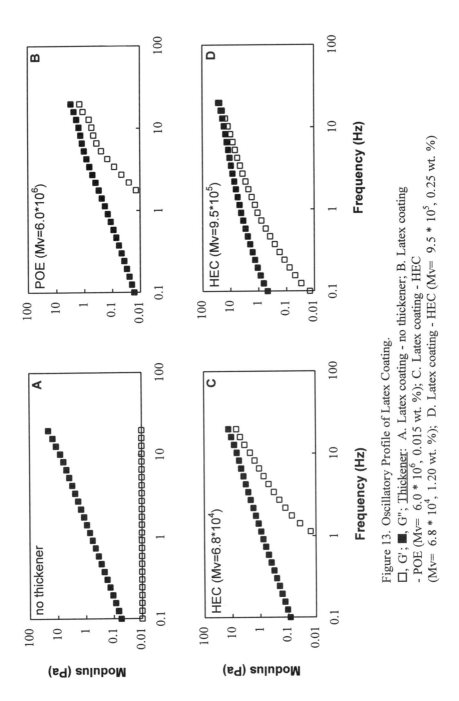

Figure 13. Oscillatory Profile of Latex Coating.

□, G'; ■, G", Thickener: A. Latex coating - no thickener; B. Latex coating
- POE (Mv= 6.0 * 10⁶, 0.015 wt. %); C. Latex coating - HEC
(Mv= 6.8 * 10⁴, 1.20 wt. %); D. Latex coating - HEC (Mv= 9.5 * 10⁵, 0.25 wt. %)

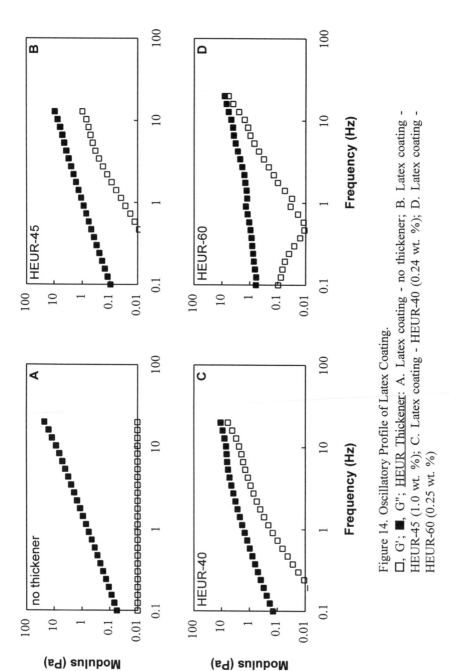

Figure 14. Oscillatory Profile of Latex Coating.
□, G'; ■, G"; HEUR Thickener: A. Latex coating - no thickener; B. Latex coating -
HEUR-45 (1.0 wt. %); C. Latex coating - HEUR-40 (0.24 wt. %); D. Latex coating -
HEUR-60 (0.25 wt. %)

a number of adjustable parameters. Matta and Tytus (*40*) have observed that the fluid's shear rate viscosity in the nozzle has no effect on the resultant drop size. They discuss First Normal Stress Difference and Die Swell behavior in terms of the mass mean diameter of the droplets, and suggest that the latter may be related to the extensional viscosity of the polymer solutions.

The spray pattern of the pigmented latex dispersion without thickener (Figure 15) is illustrated for air sprayed formulations using a fan nozzle (the A Figures through the following discussions), a solid cone nozzle (B Figures) and a hollow cone nozzle (C Figures). The sheet of fluid is evident exiting the fan nozzle; however, the sheet is smaller (the photo is not definitive) and may have dissipated into drops on exiting the solid and hollow cone nozzles. When 1.2 wt.% of the low M.W. HEC is added, the sheet may remain in tact for a slightly longer distance from the nozzle tip (Figure 16), then the fluid dissipates into filaments and then into drops with distance from the nozzle. As noted in Figures 4 and 5, the film applied through the solid cone covers the entire surface of the circle defined by the spray profile; with the hollow cone only the circumference of the circle is covered. In the fan nozzle an elliptical area is covered (Figure 3). This pattern is, in general, followed with the other thickened latex coatings. The formulations with the more efficient HEUR thickeners are illustrated in Figures 17 and 18. They are notably more sprayable than the formulation containing the large amount of low M.W. HEC, even though the latter has a lower storage modulus characteristic.

When the high M.W. nonhydrophobe-modified, water-soluble polymers, HEC and POE, thickened formulations are sprayed the patterns exhibit a remarkable change (Figures 19 and 20). With the POE formulation the wavy flag pattern is distinctly visible from the fan nozzle. The filaments do not dissipate into droplets in the POE formulation; they dissipate into a distribution of large and small drops in the HEC formulation, when 5X the distance from the nozzle head (Figure 19). With the cone nozzles there is insufficient deformation to promote spray patterns (Figure 20). It is likely that the flow pattern is biaxial and the higher viscosities under this mode of deformation retards sheet formation. The asymmetric wave is evident in the HEC formulation with the solid cone. The wavelength in the sprayed fluid is reduced in the hollow cone and irregular structures, similar to shark skin behavior observed in polyolefin extrusion, are observed. This behavior is muted in the cone studies of the POE formulations. The steady state shear viscosity and elastic character (G', storage modulus determined by the shear deformation oscillatory studies) of the respective formulations can not account for the observed differences in spray behavior. They are understandable in terms of dynamic extensional viscosities (DEV). The reversal in uniaxial DEV and steady state shear viscosities for POE and high M.W. HEC aqueous solutions are illustrated in Figure 21 (*41*).

Airless Spray (2,000 psi)

Higher deformation rates during the spray process are present in the higher pressure (2,000 psi, with a 0.017 inch diameter fan nozzle) airless spray, and the formulation thickened by POE of $6*10^6$ M.W. (with the highest DUEV) was examined. At the low POE concentration (0.015 wt %) used in the air spray study

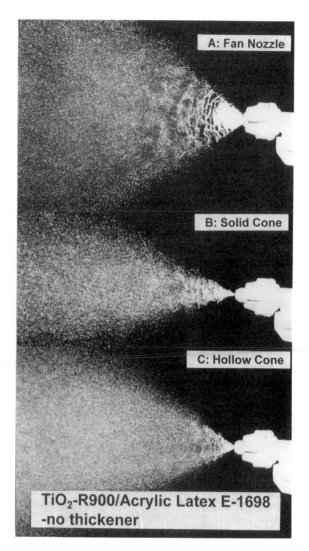

Figure 15. The Spray Patterns of Latex Coating. No Thickener
(TiO$_2$ -R900 / Acrylic Latex E-1698 , NVV=28 %, PVC = 20 %)
A. Fan Nozzle; B. Solid Cone Nozzle; C. Hollow Cone Nozzle

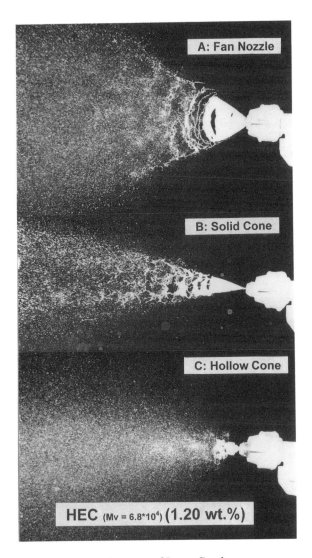

Figure 16. The Spray Patterns of Latex Coating
Thickener: HEC QP-09 (Mv= 6.8 * 10^4) (1.20 wt. %)
A. Fan Nozzle; B. Solid Cone Nozzle; C. Hollow Cone Nozzle

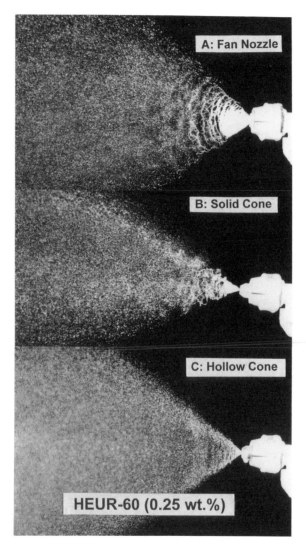

Figure 17. The Spray Patterns of Latex Coating
HEUR-60 (Mv= 2.9 * 10^4) (0.24 wt. %)
A. Fan Nozzle; B. Solid Cone Nozzle; C. Hollow Cone Nozzle

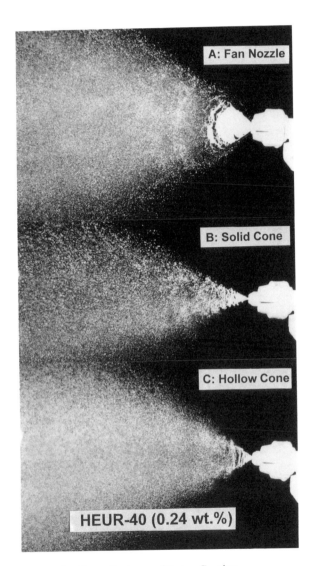

Figure 18. The Spray Patterns of Latex Coating
HEUR-40 (Mv= 2.7 * 10^4) (0.25 wt. %)
A. Fan Nozzle; B. Solid Cone Nozzle; C. Hollow Cone Nozzle

A

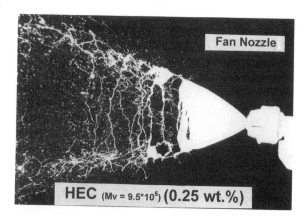

Figure 19. The Spray Patterns of Latex Coating
High M. W. Thickeners Through a <u>Fan Nozzle</u>
A. Thickener: HEC (Mv= 9.5 * 10^5) (0.25 wt. %);
B. Thickener: POE (Mv= 6.0 * 10^6) (0.015 wt. %)

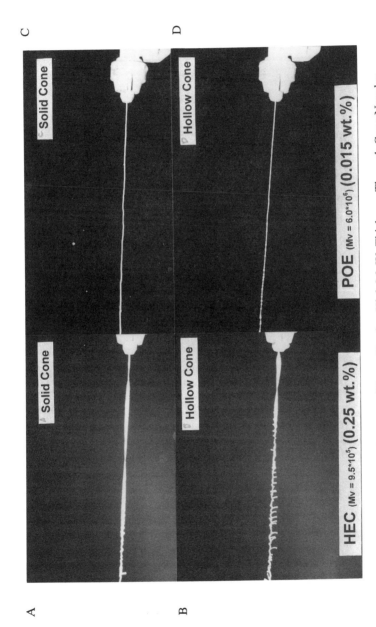

Figure 20. The Spray Patterns of Latex Coating High M. W. Thickener Through Cone Nozzles

A. Thickener: HEC (Mv= $9.5 * 10^5$) (0.25 wt. %), Solid Cone. B. Thickener: HEC (Mv= $9.5 * 10^5$) (0.25 wt. %), Hollow Cone.
C. Thickener: POE (Mv= $6.0 * 10^6$) (0.015 wt.%), Solid Cone. D. Thickener: POE (Mv= $6.0 * 10^6$) (0.015 wt.%), Hollow Cone.

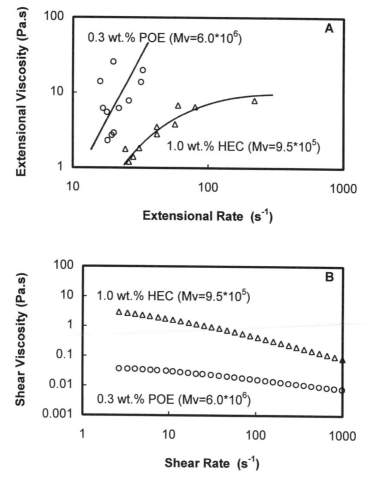

Figure 21. The Dynamic Uniaxial Extensional Viscosity & Shear Viscosity
Dependence on Deformation Rate of Aqueous Thickener Solutions
O, 0.3 wt.% POE (Mv = 6.0 * 10⁶); Δ, 1.0 wt.% HEC (Mv = 9.5 * 10⁵).
A. Dynamic Uniaxial Extensional Viscosity (41); B. the Shear Viscosity

(Figure 19B) the wavy flag gives way to a sprayable paint (Figure 22B); however, at a slightly higher concentration (0.2 wt.%) of POE, the paint does not give an acceptable spray pattern (Figure 22C).

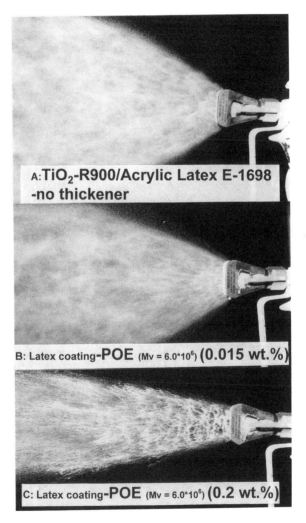

Figure 22. The Airless Spray Patterns of Latex Coatings through a Fan nozzle (0.017 in. orifice diameter, 2000 psi.)
A. Latex Coating (TiO$_2$ -R900 / Acrylic Latex E-1698 , NVV=28 %, PVC = 20 %) with no Thickener ;
B. Latex Coating. Thickener: POE (Mv= 6.0 * 10^6) at 0.015 wt. %;
C. Latex Coating. Thickener POE (Mv= 6.0 * 10^6) at 0.2 wt. %.

These studies mark only the beginning of our studies into water-soluble polymer thickened disperse phase interactions and their influence on spray application behavior.

Acknowledgment

The authors gratefully acknowledge the financial support of the National Science Foundation Industry/University Cooperative Research Center in Coatings at North Dakota State University, Michigan Molecular Institute, and Eastern Michigan University.

Literature Cited

1. Levinson, S.B., *Application of Paints and Coatings,* Federation Series on Coatings Technology, **1988**, Published by Federation of Societies for Coatings Technology, Philadelphia.
2. McBane, B.N., *Automotive Coatings,* Federation Series on Coatings Technology, **1987**, Published by Federation of Societies for Coatings Technology, Philadelphia.
3. Zink, S.C., in *"Encyclopedia of Chemical Technology"* , 3rd Ed., **1983**, vol.6, 414, 466.
4. Matthews, G.A., *Pesticide Application Methods,* 2nd. Ed., **1992**, John Wiley & Sons, Inc., New York.
5. Masters, K., *Spray Drying, An Introduction to Principles, Operational Practice and Application,* 2nd. Ed., **1976**, John Wiley & Sons Inc., New York.
6. Dombrowski, N., and Fraser, R.P., *Phil. Trans. Roy. Soc. London A.,* **1954**, 247(924), 101.
7. Fraser, R.P., Eisenklam, P., Dombrowski and Hasson, D., *A.I.Ch.E. Journal,* **1962**, 8(5), 672.
8. Spielbauer, T.M. and Aidun, C. K., *Tappi Journal,* **1992**, 75(3), 195.
9. Spielbauer, T.M. and Aidun, C. K., *Tappi Journal,* **1994**, 77(9), 95.
10. Rayleigh, L., *The theory of sound,* vol.2, Dover Publications, New York, **1945** (republication of 1896 edn.), pp.343-75
11. Weber, C., *Z. Angew. Math. Mech.,* **1931**, 11(2), 136.
12. Sterling, A.M. and Sleicher, C.A., *Journal of Fluid Mech.,* **1975**, 68(3), 477.
13. Taub, H.H., *The Physics of Fluids,* **1976**, 19(8), 1124.
14. Emmons, H.W., Chang, C.T. and Watson, B.C., *Journal of Fluid Mech.,* **1960**, 7, 177.
15. Pimbley, W.T. and Lee, H.C., *IBM Journal Res. Develop.,* **1977**, 21, 21.
16. Weihs, D. and Frankel, I., *Journal of Fluid Mech.,* **1982**, 116, 393.
17. Yuen, M., *Journal of Fluid Mech.,* **1968**, 33(1), 151.
18. Lafrance, P., *The Physics of Fluids,* **1975**, 18(4), 428.
19. Chaudhary, K.C. and Redekopp, L.G., *Journal of Fluid Mech.,* **1980**, 96(2), 257.
20. Torpey, P.A., *Phys. Fluids A,* **1989**, 1(4),661.
21. Kreiger, I.M., *Adv. Colloid Interf. Sci.,* **1972**, 3, 111
22. Wang, X.F., and Lefebvre, A.H., *Journal of Propulsion,* **1987**, 3(1), 11.
23. Empie, H.J., Lien, S.J., Yang, W.and Samuels, D.B., *Journal of Pulp and Paper Science,* **1995**, 21(2), 63.

24. Dorman, R.G., *British Journal of Applied Physics*, **1952**, 3, 189.

25. Dombrowski, N. and Johns, W.R., *Chem. Eng. Sci.*, **1963**, 18, 203.

26. Ford, R.E. and Furmidge, C.G.L., *British Journal of Applied Physics*, **1967**, 18, 335.

27. Janna, W.S. and John, J.E.A., *ASME Journal of Engineering for Industry*, **1979**, 101, 171.

28. Snyder, H.E.; Senser, D.W.; Lefebvre, A.H., *ASME Journal of Fluids Engineering*, **1989**, 111, 342.

29. Corbeels, P.L.; Senser, D.W. and Lefebvre, A.H., *Atomization and Sprays*, **1992**, 2, 87.

30. Snyder, H.E.; Senser, D.W.; Lefebvre, A.H., *IEE Transactions On Industry Applications*, **1989**, 25(4), 720.

31. Lefebvre, A.H.; Senser, D.W., *Industrial Finishing*, **1990**, June, 16.

32. Plla, S., *J. Oil Col. Chem. Assoc.*, **1973**, 56, 195.

33. Ginsberg, T., *Fatipek*, 497-502, 1974.

34. Dorman, J.D. and Stewart, D.M.D., *J. Oil Col. Chem. Assoc.*, **1976**, 59, 115.

35. Glass, J.E., *Journal of Coating Technology*, **1978**, 50(641), 56.

36. Fernando, R.H.; Glass, J.E., *J. Rheology*, **1988**, 32(2), 199.

37. Strivens, T.A., in *"Paint and Surface Coatings, Theory and Practice"*, Lambourne, R., Ed., Ellis Horwood Limited Publishers, Chichester, England, **1988** Chapter 15.

38. Tarng, Ming-Ren; Ma, Zeying; Alahapperuma, Karu; Glass, J.E., in *"Hydrophilic Polymers: Performance with Environmental Acceptance"*, Glass, J.E., Ed., Advances in Chemistry 248, American Chemical Society: Washington, DC, **1996**, Chapter 24.

39. Gordon, M.; Yerushalmi, J.; Shinnar, R., *Trans. Soc. Rheol. (N.Y.)*, **1973**, 17, 303.

40. Matta, J.E. and Tytus, R.P., *J. of Applied Polymer Science*, **1982**, 27, 397.

41. Soules, David Andrew, *Ph.D. thesis*, **1989**, North Dakota State University.

Author Index

Affiliation Index

Subject Index